Memoirs of the American Mathematical Society
Number 390



Donna M. Testerman

# Irreducible subgroups of exceptional algebraic groups



**Published by the**
**AMERICAN MATHEMATICAL SOCIETY**
**Providence, Rhode Island, USA**

**September 1988 • Volume 75 • Number 390 (second of 3 numbers)**

1980 *Mathematics Subject Classification* (1985 *Revision*).
Primary 20E28, 20G05, 20G15.

---

**Library of Congress Cataloging-in-Publication Data**

Testerman, Donna M., 1960–
Irreducible subgroups of exceptional algebraic groups/Donna M. Testerman.
p. cm. – (Memoirs of the American Mathematical Society, ISSN 0065-9266; no. 390)
"September 1988. Volume 75. Number 390 (end of volume)."
Bibliography: p.
ISBN 0-8218-2453-8
1. Linear algebraic groups. 2. Representations of groups. 3. Embeddings (Mathematics)
I. Title. II. Series
QA3.A57 no. 390
[QA171]
510 s–dc19 88-15554
[512'.22] CIP



---

**Subscriptions and orders** for publications of the American Mathematical Society should be addressed to American Mathematical Society, Box 1571, Annex Station, Providence, RI 02901-9930. *All orders must be accompanied by payment.* Other correspondence should be addressed to Box 6248, Providence, RI 02940.

**SUBSCRIPTION INFORMATION.** The 1988 subscription begins with Number 379 and consists of six mailings, each containing one or more numbers. Subscription prices for 1988 are $239 list, $191 institutional member. A late charge of 10% of the subscription price will be imposed on orders received from nonmembers after January 1 of the subscription year. Subscribers outside the United States and India must pay a postage surcharge of $25; subscribers in India must pay a postage surcharge of $43. Each number may be ordered separately; *please specify number* when ordering an individual number. For prices and titles of recently released numbers, see the New Publications sections of the NOTICES of the American Mathematical Society.

**BACK NUMBER INFORMATION.** For back issues see the AMS Catalogue of Publications.

**MEMOIRS** of the American Mathematical Society (ISSN 0065-9266) is published bimonthly (each volume consisting usually of more than one number) by the American Mathematical Society at 201 Charles Street, Providence, Rhode Island 02904. Second Class postage paid at Providence, Rhode Island 02940. Postmaster: Send address changes to Memoirs of the American Mathematical Society, American Mathematical Society, Box 6248, Providence, RI 02940.

CONTENTS

## ABSTRACT

Let Y be a simply connected, simple algebraic group of exceptional type, defined over an algebraically closed field k of characteristic $p > 0$. The main result describes all semisimple, closed connected subgroups of Y which act irreducibly on some rational kY module V. This extends work of Dynkin who obtained a similar classification for algebraically closed fields of characteristic 0. The main result has been combined with work of G. Seitz to obtain a classification of the maximal closed connected subgroups of the classical algebraic groups defined over k.

AMS subject classification (1980).

20E28, 20G05, 20G15.

Key words and phrases.

exceptional algebraic groups, representation theory, embeddings of algebraic groups

## INTRODUCTION

Our purpose here is to study triples $(A,Y,V)$, where $Y$ is a simply connected, simple algebraic group of exceptional type, defined over an algebraically closed field $k$ of characteristic $p > 0$, $V$ is an irreducible rational $kY$ module, and $A$ is a semisimple, closed connected subgroup of $Y$ such that $V|A$ is irreducible. (We refer to the above set of hypotheses as the "main problem.") In our main result, we obtain a precise description of the triples $(A,Y,V)$.

Before stating our result, we introduce the following notation. Let $T_A$ be a maximal torus of $A$, $T_Y$ a maximal torus of $Y$, with $T_A \leq T_Y$. Let $\Pi(A) = \{\alpha_1,\alpha_2,\ldots\}$ and $\Pi(Y) = \{\beta_1,\beta_2,\ldots\}$ be bases of the root systems $\Sigma(A)$ and $\Sigma(Y)$, respectively, with $\mu_i$ the fundamental dominant weight corresponding to $\alpha_i$ and $\lambda_i$ the fundamental dominant weight corresponding to $\beta_i$. Let $\lambda$ be the high weight of $V$. (Our labelling of Dynkin diagrams is described on page 8.) Finally, we write $A = G_2$, for example, to mean that $\Sigma(A)$ has type $G_2$.

<u>Main Theorem.</u> If $V|Y$ is tensor indecomposable, one of the following holds:

(i) $A = A_1$, $Y = G_2$, $\lambda|T_A = 6\mu_1$, $\lambda|T_Y = \lambda_1$ and $p \geq 7$.

(ii) $Y = G_2$, $p=3$, $\Sigma(A)$ is a subsystem of $\Sigma(Y)$ containing all long (respectively, short) roots of $\Sigma(Y)$, and $\lambda|T_Y$ has long (short) support.

---

Received by the editors February 8, 1987 and, in revised form September 4, 1987.

(iii) $A = G_2$, $Y = F_4$, $p=7$, and $\lambda|T_A = 2\mu_1$ and $\lambda|T_Y = \lambda_4$.

(iv) $Y = F_4$, $p=2$, $\Sigma(A)$ is a subsystem of $\Sigma(Y)$ containing all long (respectively, short) roots, and $\lambda|T_Y$ has long (short) support.

(v) $A = A_2$, $Y = E_6$, $\lambda|T_A = 2\mu_1+2\mu_2$, $\lambda|T_Y = \lambda_1$ or $\lambda_6$, and $p\neq 2,5$.

(vi) $A = G_2$, $Y = E_6$, $\lambda|T_A = 2\mu_1$, $\lambda|T_Y = \lambda_1$ or $\lambda_6$, and $p\neq 2,7$.

(vii) $A = C_4$, $Y = E_6$, $\lambda|T_A = \mu_2$, $\lambda|T_Y = \lambda_1$ or $\lambda_6$, and $p\neq 2$.

(viii) $Y = E_6$, $A = F_4$ is the fixed point subgroup of the graph automorphism of Y and

(a) $\lambda|T_Y = \lambda_1 + (p-2)\lambda_3$ or $(p-2)\lambda_5 + \lambda_6$, for $p>2$, or

(b) $\lambda|T_Y = (p-3)\lambda_1$ or $(p-3)\lambda_6$, for $p>3$.

Moreover, if the pair (A,Y) is as in (ii), (iv) or (viii) $V|A$ is irreducible. As well, if $p\geq 7$ (respectively, $p = 7$, $p \neq 2,5$, $p \neq 2,7$, $p \neq 2$) and Y has type $G_2$ (respectively, $F_4$, $E_6$, $E_6$, $E_6$), there exists a subgroup $B \leq Y$, of type $A_1$ (respectively, $G_2$, $A_2$, $G_2$, $C_4$) such that B acts irreducibly on $V(\lambda_1)$ (respectively, $V(\lambda_4)$, $V(\lambda_1)$, $V(\lambda_1)$, $V(\lambda_1)$) with the high weight described in (i) (respectively, (iii), (v), (vi), (vii)).□

The results of (i), (ii) and (iv) are proven in [12], where G. Seitz considered the main problem in case Y is a classical group. We establish (iii), (v), (vi) (vii) and (viii) and the existence of an irreducible $A_1$ in $G_2$ in this paper. The proof of the existence of an irreducible $C_4$ in $E_6$ was communicated to the author by Seitz and is also included here. The remaining existence proofs ($A_2 < E_6$, $G_2 < E_6$ and $G_2 < F_4$) are given in [16], where the conjugacy classes of the irreducible subgroups are also determined.

For an arbitrary irreducible rational kY module V, Steinberg's tensor product theorem ([15]) implies $V|Y = V_1{}^{q_1}\otimes\cdots\otimes V_k{}^{q_k}$, where each $V_i$ is a nontrivial irreducible kY module with restricted high weight and $\{q_1,\ldots,q_k\}$ are distinct p-powers. (We refer to $V_i{}^{q_i}$ as a conjugate of $V_i$.) If $V|A$ is irreducible, for some subgroup A, then $V_i|A$ is irreducible for each i and the triple $(A,Y,V_i)$ is described in the above theorem. Hence, there is no loss of generality in assuming throughout that $V|Y$ is tensor indecomposable.

The consideration of triples (A,Y,V) in the case where char(k) = 0 was undertaken by E.B. Dynkin in [7]. Given A, a semisimple algebraic group and $\varphi: A \longrightarrow SL(V)$ an irreducible rational representation, Dynkin determined all overgroups of A in SL(V), Sp(V) or SO(V). In a straightforward way, this information yielded a classification of all maximal, proper, closed connected subgroups of the classical algebraic groups. In our situation, where char(k) = p, the Main Theorem has been combined with the results obtained by Seitz in [12] to obtain a similar classification of the maximal proper closed connected subgroups of the classical algebraic groups over k. (This is perhaps the most striking application of the results to date.)

Theorem (A). Let A be a simple algebraic group and $\varphi: A \longrightarrow SL(V)$ an irreducible, rational representaion which is tensor indecomposable. Then with specified exceptions, the image of A is maximal among proper, closed connected subgroups in one of SL(V), Sp(V) or SO(V). Moreover, any other maximal, proper closed connected subgroup of the isometry group of V arises naturally as the stabilizer of a subspace of V or the stabilizer of a tensor product decomposition of V.□

For a more precise statement and the proof, see Theorem (3) in [12]. By far, the major portion of the proof of Theorem (A) lies in describing the "specified exceptions." These fall into two categories, as follows:

Theorem (B). Let Y be a simple algebraic group and $\varphi: Y \rightarrow SL(V)$ an irreducible rational representation which is tensor indecomposable. If A is a proper, closed connected subgroup of Y with $V|\varphi(A)$ irreducible, then one of the following holds:

(i) $\varphi(Y) = SL(V)$, $Sp(V)$ or $SO(V)$.

(ii) $(\varphi(A),\varphi(Y),V)$ appears in Table 1.□

Table 1 contains the combined results of this paper and [12], and lists all embeddings $A < Y < SL(V)$ where A and Y are irreducible, $V|Y$ is tensor indecomposable and $Y \neq Sp(V)$ or $SO(V)$. For a complete explanation of the notation in Table 1, see the end of the introduction; we make a few remarks here. To describe the modules $V|A$ and $V|Y$, we give the high weights. To describe the embedding of A in Y, we indicate the action of a covering group of A on the irreducible kY module W, where W is the natural, classical module for Y, if Y is classical, and W is an irreducible, restricted kY module of minimal dimension, if Y is exceptional. Finally, we note that there are examples for arbitrarily large primes for which there are no counterparts in the characteristic zero result; e.g. $I_1'$ and $T_1$ in Table 3. Hence, interestingly enough, the philosophy that the answer to the sort of question studied here should be the same for large primes p as the answer to the analogous zero characteristic question fails to be justified.

The methods in [12] and this paper differ greatly from those of Dynkin, by necessity. Since in characteristic p, rational modules for simple groups need not be completely reducible nor tensor indecomposable, as in zero characteristic, some of Dynkin's key reductions do not carry over. Though we may assume that in the triple $(A,Y,V)$, $V|Y$ is tensor indecomposable, it happens that $V|A$ can be tensor decomposable. One may notice that throughout the paper, case-by-case analysis is required whenever this possibility persists. If we desired only to prove Theorem (A) or to give a new proof of Dynkin's result, we could assume $V|A$ to be tensor indecomposable and shorten much of our work. As well, for a new proof of the zero characteristic result the small prime analysis of Chapter 9 and the difficulty created by the absence of formulae for the dimensions of and multiplicities of weights in irreducible modules could be avoided.

We now give a survey of the methods used in this paper. Let $(A,Y,V)$ statisfy the hypotheses of the main problem. We obtain preliminary information about the triple $(A,Y,V)$ via induction. Choose a maximal parabolic $P_A$ of A, with unipotent radical $Q_A$ and Levi factor $L_A$. By the Borel-Tits theorem [2], there exists a parabolic subgroup $P_Y$ of Y with $P_A \leq P_Y$ and $Q_A \leq Q_Y = R_u(P_Y)$. If $L_Y$ is a Levi factor of $P_Y$, a result of Smith ([13]) implies that $L_A'$ and $L_Y'$ act irreducibly on the fixed point space $V_{Q_A}$. Hence, considering the projection of $L_A'$ into the quasisimple components of $L_Y'$ which act nontrivially on $V_{Q_A}$, we obtain a smaller rank version of the original problem. Theorem (7.1) of [12] is a complete solution of the main problem in the case where rankA = 1. Working inductively, we may describe $V_{Q_A}$ (so partially describe V) and partially describe the embedding of $L_A'$ in $L_Y'$. Though we are inducting on the rank of A, we handle the case where rank(A) = 2 and $Y = E_n$ in Chapters 6 - 9. Hence, in Chapters 4 and 5, we assume the results of the later chapters.

In Chapter 2, we establish machinery for studying general parabolic embeddings. As well, we prove results applicable only in the context of irreducibility on some module. (Several of the results are proven in [12].) Through this work, we can see the influence of the inductive information on (1) the projections of $L_A'$ in the components of $L_Y'$ which act trivially on $V_{Q_A}$ and (2) the embedding of $Q_A$ in $Q_Y$. Our considerations are as follows. With $L_Y$ acting on $Q_Y$ via conjugation, certain quotients of $Q_Y$ may be regarded as modules for $L_Y'$, and hence as modules for $L_A'$. We consider the image of the $L_A'$ module $Q_A/[Q_A,Q_A]$ in these quotients. Of course, in an arbitrary parabolic embedding, $Q_A$ may appear in few $L_Y'$

composition factors of $Q_Y$. But another consequence of Smith's result is the equality of the commutator subspaces $[V,Q_A]$ and $[V,Q_Y]$. The existence of particular weight spaces in $[V,Q_Y]$ often forces $Q_A/[Q_A,Q_A]$ to appear in particular quotients of $Q_Y$. Moreover, in most cases $Q_A/[Q_A,Q_A]$ must appear as an $L_A'$ submodule. This will place restrictions on the projection of $L_A'$ in the quasisimple components of $L_Y'$ which act nontrivially on particular composition factors of $Q_Y$. We compare this with the inductively given information and perhaps produce a contradiction, or at least broaden our knowledge of the embedding $P_A \le P_Y$.

Throughout the paper, various numerical methods are employed as well. Since $[[V,Q_A],Q_A] \le [V,Q_Y],Q_Y]$ and $[V,Q_A] = [V,Q_Y]$, $\dim([V,Q_Y]/[[V,Q_Y],Q_Y]) \le \dim([V,Q_A]/[[V,Q_A],Q_A])$. Moreover, if $Z(L_A)° \le Z(L_Y)°$ (which is usually implied by a suitable choice of $P_Y$), then the dimension of a $Z(L_Y)°$ weight space of $[V,Q_Y]/[[V,Q_Y],Q_Y]$ is bounded by the dimensions of $Z(L_A)°$ weight spaces of $[V,Q_A]/[[V,Q_A],Q_A]$. Seitz gives an explicit upper bound on the dimensions of the latter. This yields further restrictions on the high weight of $V|Y$. When the high weights of $V|A$ and $V|Y$ are almost explicitly determined, we attempt to show that $\dim V|Y$ exceeds the upper bound on $\dim V|A$ given by the Weyl degree formula. For this purpose, various methods for obtaining lower bounds on dimensions of $kY$ modules are discussed in Chapter 1.

The absence of a "natural" module for the exceptional group Y gives rise to (expected) differences between Seitz's work in [12] and our work here. If Y is a classical group with natural (classical) module W, Seitz proves that in most cases, $W|A$ is irreducible and tensor indecomposable. This provides information about the restriction of elements of $\Sigma(Y)$ to a maximal torus of A and, coupled with an inductive hypothesis, usually implies that $V|A$ is a conjugate of a restricted module. As mentioned before, the tensor indecomposable situation is much easier to handle.

In our situation, where Y is an exceptional group, we may think of the "natural" module W as a restricted rational kY module of minimal dimension. However, there is no complete theory relating the subgroup structure of Y to its action on W. We do however use the module W whenever possible. We consider the action of $L_A'$ on W, in particular the $L_A'$ composition series of W. (This can be determined only when we have a fairly complete knowledge about the image of $L_A'$ in $L_Y'$) If dim(W) is relatively small (e.g., 26, 27 or 56) we can list all rational kA modules of this dimension, determine their $L_A'$ composition series and compare with the given $L_A'$ composition series of $W|L_A'$. Though fruitful in specific situations, this analysis does not serve the purpose that the natural module does for the classical groups. Rather, the bounded rank of the exceptional groups and our extensions of Seitz's results on parabolic embeddings enable us to restrict to the few possibilities of the Main Theorem.

For the convenience of the reader, many of the preliminary results from [12] are listed in this paper. It is useful to see that some of the results in Chapter 2 are natural extensions of the results on parabolic embeddings in [12]. A few essential theorems from [12], which we do not state, are often referenced. Theorem (7.1), mentioned already, is the solution of the main problem in case rank(A) = 1. Theorem (4.1) is a solution for the case where rank A = rank Y. (See Chapter 3 for a partial statement.) Theorem (8.1) gives the solution of the main problem for certain natural embeddings of classical groups. And finally, we refer to the list of all triples (A,Y,V), where Y has classical type, as the Main Theorem of [12].

Throughout the paper, we use the following labelling of Dynkin diagrams.

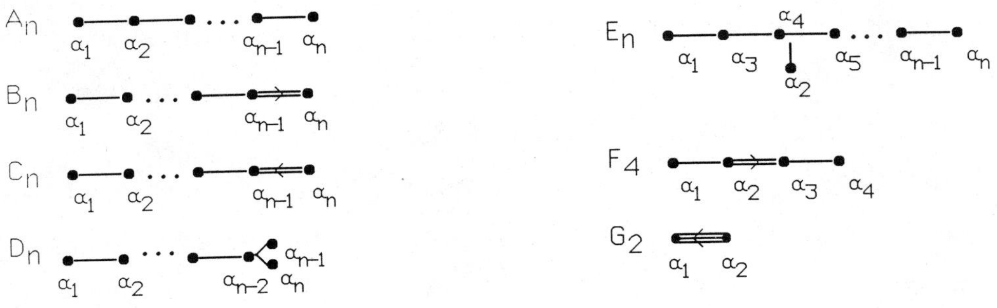


Let us make a few remarks about the notation in Table 1. The second column indicates the types of the groups A and Y, respectively. When the symbol " $\longrightarrow$' " occurs, $A \leq B < Y$, for a closed, connected subgroup B, which is a commuting product of quasisimple groups as indicated. The notation means that either A projects surjectively to each of the simple factors of Y or some factor is of type $B_2$ and the projection is an $A_1$ acting irreducibly on the spin module for $B_2$. Moreover, in order to make sure $V|A$ is irreducible, it may be necessary for the projections to involve distinct field twists.

The third column describes the action of a covering group of A on a particular irreducible kY module, W. If Y is classical, W is the natural module for Y; if Y has type $G_2$, $F_4$, or $E_6$ ($E_7$ and $E_8$ do not arise), W is a restricted module of dimension 7 (6 if p=2), 26 (25 if p=3) or 27, respectively.

In the fourth and fifth columns the actions of A and Y on the module V are described, and in the last column any prime restrictions are indicated. Column 1 associates with each example a number. In the cases where there is an analogous zero characteristic example, Dynkin's numbering has been used. So $I_1-I_{12}$, $II_1-II_9$, $III_1$, $IV_1-IV_{10}$, $V_1$ and $VI_1-VI_3$ appear in [7]. Notation such as $VI_1'$ refers to a variant of

Dynkin's $VI_1$. Examples $MR_i$ are those where rank$A$ = rank$Y$ and examples labelled $S_1$–$S_9$ are special examples occuring only when $p = 2$ or $3$; these were found by Seitz in [12]. Examples $T_1$ and $T_2$ are found in this paper.

In conclusion, the author would like to express thanks to Gary Seitz, who suggested the problem, read an earlier version of this paper and offered useful advice throughout. As well, special thanks are given to Mark Reeder for numerous mathematical insights.

## CHAPTER 1: PRELIMINARY LEMMAS

Let V be a finite dimensional vector space over an algebraically closed field k of characteristic $p > 0$, and let X be a semisimple, closed, connected subgroup of SL(V) with fixed maximal torus T. Let $\{\alpha_1, \ldots, \alpha_n\}$ be a base for the root system $\Sigma(X)$ and let $e_{\alpha_i}$ and $f_{\alpha_i}$ denote the corresponding elements of the Lie algebra L(X). Labelling Dynkin diagrams as in Table 1, let $\lambda_i$ be the fundamental dominant weight corresponding to $\alpha_i$. Assume $V|X$ is irreducible and let $\lambda$ be the high weight of V. Then $\langle\lambda,\alpha_i\rangle \geq 0$, for each i and V is said to be *restricted* if $\langle\lambda,\alpha_i\rangle < p$, for $1 \leq i \leq n$. For a subgroup $N < X$, let $V_N$ denote the space of fixed points of N on V and $[V,N]$ the commutator subspace $\langle v-nv \mid v \in V, n \in N\rangle$.

<u>(1.1).</u> (i) $V = V_1{}^{q_1} \otimes \cdots \otimes V_k{}^{q_k}$, where each $V_i$ is an irreducible restricted module for X and $q_1, \ldots, q_k$ are distinct powers of p.

(ii) If V is restricted, then V is also irreducible when viewed as a module for L(X).

<u>Proof</u>: (i) is Steinberg's tensor product theorem (see [15]). For (ii) see Section A of [1].□

<u>(1.2).</u> ([13]) Let P be a proper parabolic subgroup of X with unipotent radical Q and Levi factor L. Then $L \cong P/Q$ acts irreducibly on $V_Q$.□

<u>(1.3).</u> ((1.7) of [12]) Let P be a proper parabolic subgroup of X with unipotent radical Q and Levi factor L. Then $V/[V,Q]$ is irreducible for L. In fact, this quotient is L-isomorphic to $((V^*)_Q)^*$.□

<u>(1.4).</u> ([2]) Let $X \leq Y$, where Y is a closed, connected subgroup of SL(V) and let P be a parabolic subgroup of X with unipotent radical Q. There is a parabolic subgroup $P_Y$ of Y with unipotent radical $Q_Y$ such that

$P \leq P_Y$ and $Q \leq Q_Y$.□

(1.5). Let $X$, $Y$, and $P$ be as in (1.4), and choose $P_Y$ as in (1.4) minimal such that $P \leq P_Y$ and $Q \leq Q_Y = R_u(P_Y)$. Suppose $L_Y' = L_1 \cdots L_r$, where $L_i$ is a simple normal subgroup of $L_Y$, with root system of classical type for $1 \leq i \leq r$. Then, $Z(L)^\circ \leq Z(L_Y)^\circ$.

Proof: This follows from the proof of (2.8) in [12].□

(1.6). ((1.4) in [12]) Let $X \leq Y$, $P \leq P_Y$, $Q \leq Q_Y$ be as in (1.4). Then, $V_Q = V_{Q_Y}$. So $L$ and $L_Y$ are reductive groups both acting irreducibly on $M = V_Q$ and the image of $L$ in $SL(M)$ is contained in the image of $L_Y$.□

(1.7). ((1.6) of [12]) Suppose $X$ is simple. Then $V$ can be expressed as the tensor product, $V = V_1 \otimes V_2$, of two nontrivial restricted $kX$-modules if and only if $V$ is restricted and the following conditions hold:

(i) $X$ has type $B_n$, $C_n$, $F_4$, or $G_2$, with $p = 2,2,2,3$, respectively.

(ii) $V_1$, $V_2$ may be arranged such that each $V_i$ has high weight $\lambda_i$, $\lambda = \lambda_1 + \lambda_2$, and $\lambda_1$ (respectively, $\lambda_2$) has support on those fundamental dominant weights corresponding to short (long) fundamental roots.□

(1.8) Definition: Suppose $X$ is simple.

(A) We say $V$ is *basic* (respectively, *p-basic*) if the following conditions hold.

(i) $V$ is restricted.

(ii) If $X$ has type $B_n$, $C_n$, $F_4$, or $G_2$ with $p = 2,2,2,3$, respectively, then $\lambda$ has short (respectively, long) support.

(B) If $X$ and $p$ are as in (ii), we say the pair $(X,p)$ is *special*.

(1.9). Let $X$, $P$, $Y$, and $V$ be as in (1.4) and suppose $V|X$ is basic. Also, if $(X,p)$ is special (respectively, $(G_2,2)$), assume $\Pi(X) - \Pi(L)$ is $\{\alpha\}$ (respectively, long). Then there exists a parabolic subgroup $P_Y$, of $Y$, such that the following hold:

(i) $P \leq P_Y$ and $Q \leq Q_Y = R_u(P_Y)$.

(ii) $L \leq C_Y(Z(L)^\circ) \leq L_Y$, a Levi complement to $Q_Y$ in $P_Y$.

(iii) If $T_Y$ is any maximal torus of $Y$ containing $T$, then $T_Y \le L_Y$.

Proof: This follows from the first two paragraphs of the proof of (2.8) in [12].□

(1.10). ((2.16) of [12]) Let $Y$ be a simple algebraic group and $\varphi: Y \longrightarrow SL(M)$ a basic representation. Suppose $X$ is a simple, closed subgroup of $Y$ and $\varphi|X$ is an algebraic conjugate of a restricted representation of $X$. Then $\varphi|X$ is restricted.□

(1.11). (i) $V^*$ is irreducible with high weight $-w_0\lambda$, where $w_0$ is the long word in the fundamental reflections generating the Weyl group of $X$.

(ii) $X$ leaves invariant a nondegenerate bilinear from on $V$ if and only if $\lambda = -w_0\lambda$.

(iii) If $X$ has type $B_n$, $C_n$, $D_n$ for n even, $E_7$, $E_8$, $F_4$, or $G_2$, then $X$ necessarily stabilizes a nondegenerate bilinear from on $V$.

(iv) If $X$ has type $A_n$, $D_n$ for n odd, or $E_6$, then $X$ stabilizes a nondegenerate bilinear form on $V$ if and only if $\lambda = \tau\lambda$, where $\tau$ is the graph automorphism of the Dynkin diagram of $\Sigma(X)$.

Proof: See Section 31 of [10].□

(1.12). ((1.14) of [12]) Let $X = SL_{n+1}$. For any integer $p>c>0$, the irreducible module $V$ having high weight $c\lambda_1$ or $c\lambda_n$ is isomorphic to the space of homogeneous poylnomials of degree c in a basis of the usual module for $X$, or its dual. Thus, $\dim V = (1/n!)(c+1)(c+2)\cdots(c+n)$.□

(1.13). ((1.13) of [12]) Suppose $X = SL_2$. Then the weight spaces of $T$ on $V$ are of dimension 1.□

(1.14). Suppose $X = SL_2$. Let $\Pi(X) = \{\alpha\}$ and let $\lambda_\alpha$ be the fundamental dominant weight corresponding to $\alpha$. Then $X$ fixes a symplectic (respectively, orthogonal) form on the restricted, irreducible $kX$-module with high weight $n\lambda_\alpha$, where n is odd (even).

Proof: This follows immediately from Lemma 79 of [14].□

(1.15). Suppose $X = SL_2$. Let $\Pi(X) = \{\alpha\}$ and $\lambda_\alpha$ the fundamental dominant weight corresponding to $\alpha$. Let $W$ be the rational $kX$-module

$V(a_1\lambda_\alpha) \otimes \cdots \otimes V(a_m\lambda_\alpha)$ where $V(a_i\lambda_\alpha)$ is an irreducible, rational $kX$-module with high weight $a_i\lambda_\alpha$, for $a_i = \Sigma c_{ij}p^j$, $0 \le j \le m_j$, $c_{ij} \in Z^+$ and $0 \le c_{ij} < p$. If $p>2$ and $\Sigma\Sigma c_{ij}$ is even (respectively, odd), then there is no submodule of W isomorphic to a conjugate of $V(\lambda_\alpha)$ or $V(3\lambda_\alpha)$ (respectively, $V(2\lambda_\alpha)$).

<u>Proof</u>: Suppose $\Sigma\Sigma c_{ij}$ is even. Then $W = \Sigma W_{2k\lambda_\alpha}$, $-r \le k \le r$, a sum of T weight spaces. Hence, since p is odd there is no weight vector with weight $q\lambda_\alpha$ or $3q\lambda_\alpha$, for any p-power q. Similarly, if $\Sigma\Sigma c_{ij}$ is odd, $W = \Sigma W_{(2k+1)\lambda_\alpha}$, $-r \le k \le r-1$, and there is no weight vector with weight $2q\lambda_\alpha$, for any p-power q.□

<u>(1.16).</u> Let X be simple and $\lambda = q_1\gamma_1 + \cdots + q_\ell\gamma_\ell$, where $\gamma_1, \ldots, \gamma_\ell$ are restricted dominant weights and $q_1, \ldots, q_\ell$ are distinct powers of p. Then X leaves invariant a nondegenerate bilinear form on $V = V(\lambda)$ if and only if X leaves such a form invariant on each of $V(\gamma_1), \ldots, V(\gamma_\ell)$.

<u>Proof</u>: This is immediate from (1.11).□

<u>(1.17).</u> ((1.12) of [12]) Let X be simple and $\lambda = \gamma_1 + \cdots + \gamma_\ell$, where $\gamma_1, \ldots, \gamma_\ell$ are arbitrary dominant weights. Suppose that X leaves invariant a nondegenerate symplectic form on $V(\gamma_1), \ldots, V(\gamma_k)$ and a nondegenerate orthogonal form on $V(\gamma_{k+1}), \ldots, V(\gamma_\ell)$. Then

(i) X leaves invariant a nondegenerate bilinear form on $V(\gamma_1) \otimes \cdots \otimes V(\gamma_\ell) = D$.

(ii) There is a singular subspace S of D such that V is X-isomorphic to a nondegenerate subspace of $S^\perp/S$.

(iii) X leaves invariant a symplectic form on V if k is odd and an orthogonal form if k is even.□

<u>(1.18).</u> ((1.9) of [12]) Let X be simple and $L = L(X)$ and suppose that $0 < I < L$ is an ideal of L not containing each $e_\alpha$, $f_\alpha$ for $\alpha \in \Sigma^+(X)$. Then one of the following holds:

(i) $I \le Z(L) \le L(T)$, the Lie algebra of T.

(ii) L has Dynkin diagram of type $B_n$, $C_n$, $F_4$, or $G_2$, $p = 2,2,2,3$

respectively, and I contains all $e_\beta$ for $\beta$ a short root in $\Sigma(X)$.□

Fix a parabolic subgroup $P = QL$ of $X$, where $Q = R_u(P)$ and $L$ is a Levi complement containing $T$. Choose $P$ such that $Q$ is the product of those $T$ root subgroups corresponding to the roots $\Sigma^-(X) - \Sigma(L)$. Let $U$ be the product of all $T$-root subgroups for roots in $\Sigma^+(X)$. Let $\Pi(L) = \Pi(X) \cap \Sigma(L)$ and $Z = Z(L)°$. We now establish notation and list certain useful results regarding the series $V > [V,Q] > [[V,Q],Q] > \cdots > 0$, where $[V,Q^0] = V$ and $[V,Q^i] = [[V,Q^{i-1}],Q]$. Note that $[V,Q^j]$ is $T$ invariant for all $j$ and hence has a decomposition into a sum of $T$ weight spaces. For $d \geq 1$, set $V^d(Q) = [V,Q^{d-1}]/[V,Q^d]$.

<u>(1.19).</u> ((2.1) of [12]) $V/[V,Q]$ is irreducible as a module for $L'$ of high weight $\lambda|T \cap L'$.□

<u>Definition</u>: Let $\mu$ be a weight of $V$, say $\mu = \lambda - \Sigma c_\ell \alpha_\ell$, with each $c_\ell \geq 0$. Then, the *Q-level* of $\mu$ is $\Sigma c_j$, where the sum ranges over those $j$ for which $\alpha_j \in \Pi(X) - \Pi(L)$. Let $V_T(\mu)$ denote the subspace of $V$ consisting of $T$-weight vectors of weight $\mu$.

<u>(1.20).</u> Suppose $X$ is simple and $V$ is basic, and let $d \geq 0$. If $(X,p)$ is special, assume $\Pi(X) - \Pi(L) = \{\alpha\}$ for some $\alpha \in \Pi(X)$.

(i) If when $(X,p) = (G_2,2)$, $\Pi(L)$ is short, $[V,Q^d] = \oplus V_T(\mu)$, the sum ranging over those weights $\mu$ having $Q$-level at least $d$. Consequently, $V^{d+1}(Q)$ is isomorphic to the direct sum of those weights having $Q$-level $d$.

(ii) If when $(X,p) = (G_2,2)$, $\Pi(L)$ is short, $\dim([V,Q^d]/[V,Q^{d+1}]) \leq s \cdot \dim([V,Q^{d-1}]/[V,Q^d])$, where $s$ is the number of positive roots $\beta$ such that $U_{-\beta} \leq Q$ and $\beta = \alpha_j + \beta'$, for some $\alpha_j \in \Pi(X) - \Pi(L)$ and $\beta'$ is 0 or a sum of roots in $\Pi(L)$.

(iii) If when $(X,p) = (G_2,2)$, $\Pi(L) = \{\alpha_2\}$ is long, $V^2(Q) = V^2(Q)_{\lambda-\alpha_1} \oplus V^2(Q)_{\lambda-2\alpha_1}$, a direct sum of $Z$ weight spaces. Moreover, $\dim V^2(Q)_{\lambda-\alpha_1} \leq 2 \cdot \dim V^1(Q)$ and $\dim V^2(Q)_{\lambda-2\alpha_1} \leq \dim V^1(Q)$.

Proof: Statements (i) and (ii) follow from (2.3) of [12]. So consider the case where $(X,p) = (G_2,2)$ and $\Pi(L) = \{\alpha_2\}$. Since $V$ is restricted, $V$ is irreducible as a module for $L(X)$ and so is spanned by the weight vectors $f_{\gamma_1}\cdots f_{\gamma_m}v^+$, for $\gamma_i \in \Sigma^+(X)$. Order $\Sigma^+(X) = \{r_1,\ldots,r_N\}$ so that the roots $\beta \in \Sigma^+(X) - \Pi^+(L)$ occur first. Then, we may write $f_{\gamma_1}\cdots f_{\gamma_m}v^+$ as $f_{\gamma_1}\cdots f_{\gamma_j}w$, where $w \notin [V,Q]$ and $\gamma_i \in \Sigma^+(X) - \Sigma^+(L)$ for $1 \le i \le j$. Say $w \in V_T(\mu)$.

Claim: With $f_{\gamma_1}\cdots f_{\gamma_m}v^+$, $w$ and $j$ as above, $f_{\gamma_1}\cdots f_{\gamma_j}w \in [V,Q^j]$.

Reason: We use induction on $j$. If $j = 0$, there is nothing to show. So suppose $j > 0$ and $0 \neq f_{\gamma_1}\cdots f_{\gamma_j}w$. By induction, $f_{\gamma_2}\cdots f_{\gamma_j}w \in [V,Q^{j-1}]$. But $x_{-\gamma_1}(1)(f_{\gamma_2}\cdots f_{\gamma_j}w) = f_{\gamma_2}\cdots f_{\gamma_j}w + f_{\gamma_1}\cdots f_{\gamma_j}w + \Sigma w_\ell$, where the sum ranges over $\ell \ge 2$ and $w_\ell \in V_T(\mu-\gamma_2-\cdots-\gamma_j-\ell\gamma_1)$. In particular, $\{f_{\gamma_1}\cdots f_{\gamma_j}w,\ w_\ell \mid \ell \ge 2\}$ is a set of weight vectors for distinct $T$ weights and so is a set of linearly independent vectors. Since $[x_{-\gamma_1}(1), f_{\gamma_2}\cdots f_{\gamma_j}w] \in [V,Q^j]$ and $[V,Q^j]$ is a sum of $T$ weight spaces, we have $f_{\gamma_1}\cdots f_{\gamma_j}w \in [V,Q^j]$. Thus the claim holds.

In the situation where $(X,p) = (G_2,2)$ and $\Pi(L)$ is long, the above claim implies that $V^2(Q)$ is spanned by the images of $f_\gamma w$, for $w \notin [V,Q]$ and $\gamma \in \Sigma^+(Y) - \Sigma^+(L)$. But in fact, since $f_{3\alpha_1+2\alpha_2} = \pm(1/3)[f_{2\alpha_1+\alpha_2}, f_{\alpha_1+\alpha_2}]$ and $f_{3\alpha_1+\alpha_2} = \pm(1/3)[f_{2\alpha_1+\alpha_2}, f_{\alpha_1}]$, we obtain a spanning set for $V^2(Q)$ from the images of $f_\gamma w$ for $\gamma \in \{\alpha_1, \alpha_1+\alpha_2, 2\alpha_1+\alpha_2\}$ and $w \notin [V,Q]$. Hence, $V^2(Q)$ is spanned by weight vectors of $Q$-levels 1 and 2. So we may decompose $V^2(Q)$ as described in (iii). Moreover, the above remarks imply that the $Z$ weight space $V^2(Q)_{\lambda-\alpha_1}$ (respectively, $V^2(Q)_{\lambda-2\alpha_1}$) is spanned by vectors of the form $f_\gamma w$ where $w \notin [V,Q]$ and $\gamma \in \{\alpha_1, \alpha_1+\alpha_2\}$ (respectively, $\gamma = 2\alpha_1+\alpha_2$). Thus, we have the given bounds on the dimensions of the $Z$ weight spaces of $V^2(Q)$.□

(1.21). Suppose $X$ is simple and $V = (V_1)^{q_1} \otimes \cdots \otimes (V_k)^{q_k}$, where each $V_i$ is restricted and $q_1,\ldots,q_k$ are distinct powers of $p$. Then for each

$d \geq 0$, $[V,Q^d] = \Sigma[V_1,Q^{d_1}]^{q_1} \otimes \cdots \otimes [V_k,Q^{d_k}]^{q_k}$, the sum ranging over sets of nonnegative integers $d_1,\ldots,d_k$ with $\Sigma d_i = d$.

Proof: This follows from (2.5) of [12].

(1.22). Let $V$ and $X$ be as in (1.21), with $\Pi(X) - \Pi(L) = \{\alpha\}$. Set $W_i = V_i^{q_i}$, for $i = 1,\ldots,k$.

(i) As modules for $L$, $V^2(Q) \cong \oplus(W_1{}^1(Q) \otimes \cdots \otimes W_{i-1}{}^1(Q) \otimes W_i{}^2(Q) \otimes W_{i+1}{}^1(Q) \otimes \cdots \otimes W_k{}^1(Q))$, the sum over $i = 1,\ldots,k$.

(ii) Assume $V_i$ is basic for each $i$ and that $\alpha$ is long when $(X,p) = (G_2,2)$. The above summands of $V^2(Q)$ are the $Z$ weight spaces for the respective weights $(\lambda-q_1\alpha)|Z,\ldots,(\lambda-q_k\alpha)|Z$. Each such weight space has dimension at most $s\cdot\dim V^1(Q)$, where $s$ is as in (1.20).

(iii) Any $Z$ weight space of $V^2(Q)$ has dimension at most $d\cdot\dim V^1(Q)$, where

$d = s$ as in (1.20), if one of the following holds: $V_i$ is basic for all $i$; or $(X,p)$ is special and $\Pi(X) - \Pi(L)$ is long; or $(X,p) = (G_2,2)$ and $\Pi(X) - \Pi(L)$ is long.

$d = \frac{1}{2}n(n+1)$, $2n-1$, $14$ or $4$, if $(X,p) = (B_n,2)$, $(C_n,2)$, $(F_4,2)$ or $(G_2,3)$, respectively, with $\Pi(X) - \Pi(L)$ short.

$d = 3$, if $(X,p) = (G_2,2)$ with $\Pi(X) - \Pi(L)$ short.

Proof: The proof of this result is found in the proofs of (2.12), (2.13) and (15.3) of [12], except for statement (iii) when $(X,p) = (G_2,2)$ and $\Pi(X) - \Pi(L) = \{\alpha_1\}$ is short. So we will consider this case. Let $N_j$ be the $j^{th}$ summand of the decomposition of $V^2(Q)$ given in (i). Then (1.20)(iii) implies that $N_j = N_j{}^1 \oplus N_j{}^2$, where $N_j{}^\ell$ lies in the $Z$ weight space $V^2(Q)_{\lambda-\ell q_j\alpha_1}$, for $\ell = 1,2$. Hence, the $Z$ weights in $V^2(Q)$ have the form $(\lambda-q_i\alpha_1)|Z$ or $(\lambda-2q_i\alpha_1)|Z$, for $1 \leq i \leq k$. Suppose a $Z$ weight space of $V^2(Q)$ intersects $N_m$ and $N_\ell$ for some $m \neq \ell$. Then $q_i = 2q_j$ for some $1 \leq i \neq j \leq k$. Then $q_i \neq 2q_a$ for $a \neq j$ and $q_i \neq q_b$ for $b \neq i$. Thus, the $Z$ weight space $V^2(Q)_{\lambda-q_i\alpha_1} = N_i{}^1 \oplus N_j{}^2$, and by (1.20)(iii), $\dim(N_i{}^1) \leq 2\cdot\dim V^1(Q)$ and $\dim(N_j{}^2) \leq \dim V^1(Q)$. So the result of (iii) follows.□

Let $X \leq Y$, for $Y$ a closed subgroup of $SL(V)$ and let $P_Y = Q_Y L_Y$ be a parabolic subgroup of $Y$ such that $P \leq P_Y$, $Q \leq Q_Y = R_u(P_Y)$, and $T \leq T_Y$, for $T_Y$ a maximal torus of $L_Y$. Set $Z_Y = Z(L_Y)^\circ$. Let $\Sigma(Y)$ be the root system of $Y$ and $\Pi(Y)$ a fundamental system of $\Sigma(Y)$. We choose $\Pi(Y)$ of $\Sigma(Y)$ such that $U \cap L \leq Q_Y(U_Y \cap L_Y)$, where $U_Y$ is the product of all $T_Y$ root subgroups for roots in $\Sigma^+(Y)$ and $Q_Y$ is the product of $T_Y$ root subgroups for roots in $\Sigma^-(Y) - \Sigma(L_Y)$. Set $\Pi(L_Y) = \Pi(Y) \cap \Sigma(L_Y)$.

<u>(1.23).</u> (i) $[V,Q] = [V,Q_Y]$.

(ii) $V^1(Q) = V^1(Q_Y)$ is an irreducible module for $L$ and $L_Y$.

(iii) $V^2(Q_Y)$ is an $L$-invariant quotient of $V^2(Q)$.

(iv) If $w_0$ (respectively, $s_0$) is the long word in the Weyl group of $Y$ (respectively, $X$), then $-w_0(\lambda)|T \cap L' = -s_0(\lambda)|T \cap L'$.

<u>Proof</u>: (i) – (iii) follow from (2.10) in [12].

Let $W = V^*$. Then, by (1.11), $W$ has high weight $-w_0(\lambda)$ (respectively, $-s_0(\lambda)$) as a $Y$ (respectively, $X$) module. By (ii) and (1.19), $W^1(Q) = W^1(Q_Y)$ is an irreducible $L'$ (respectively, $L_Y'$) module with high weight $-s_0(\lambda)|T \cap L'$ ($-w_0(\lambda)|T_Y \cap L_Y'$). Since, $U \cap L \leq U_Y \cap L_Y$, if $\langle w^+ + [W,Q_Y]\rangle$ is the unique 1-space of $W^1(Q_Y)$ invariant under $U_Y \cap L_Y$, then $\langle w^+ + [W,Q_Y]\rangle$ is also the unique 1-space of $W^1(Q)$ invariant under $U \cap L$. Recalling that $T \leq T_Y$, the result follows.□

<u>Definition</u>: For $\gamma \in \Pi(Y) - \Pi(L_Y)$, set $V^\gamma(T_Y) = \Sigma V_{T_Y}(\mu)$, the sum ranging over those $\mu$ for which $\lambda - \mu - \gamma$ is a sum of roots in $\Pi(L_Y)$. Since the $T_Y$ weights in $V^1(Q_Y)$ all differ from $\lambda$ by a sum of roots in $\Pi(L_Y)$, it follows that $V^\gamma(T_Y) \leq [V,Q_Y]$ and we let

$$V_\gamma(Q_Y) = (V^\gamma(T_Y) + [V,Q_Y{}^2])/[V,Q_Y{}^2].$$

<u>(1.24).</u> ((2.15) in [12]) Assume $V|Y$ is basic and $\gamma \in \Pi(Y) - \Pi(L_Y)$.

(i) If $\langle \lambda,\gamma\rangle \neq 0$, then some $L_Y'$ composition factor of $V_\gamma(Q_Y)$ has high weight $\lambda - \gamma$.

(ii) Suppose $\langle \lambda,\gamma\rangle = 0$, but $(\Sigma L_0,\gamma) \neq 0$ for some simple factor $L_0$ of $L_Y'$ satisfying $\langle \lambda,\Sigma L_0\rangle \neq 0$. Then there exist distinct roots

$\beta_1,\ldots,\beta_\ell \in \Pi(L_Y)\cap\Sigma L_0$ such that some $L_Y'$ composition factor of $V_\gamma(Q_Y)$ has high weight $\lambda-\beta_1-\cdots-\beta_k-\gamma$.□

(1.25). Assume $V|Y$ is basic, $X$ is simple, and that $V|X = V_1{}^{q_1}\otimes\cdots\otimes V_k{}^{q_k}$ with each $V_i$ restricted and $q_1,\ldots,q_k$ distinct powers of p. Then

(i) $V^2(Q_Y) = \oplus V_\gamma(Q_Y)$, the sum ranging over $\gamma\in\Pi(Y)-\Pi(L_Y)$.

(ii) For each $\gamma\in\Pi(Y)-\Pi(L_Y)$, $V_\gamma(Q_Y)\cong V^\gamma(T_Y)$ (as vector spaces).

(iii) For each $\gamma\in\Pi(Y)-\Pi(L_Y)$, $V_\gamma(Q_Y)$ is a weight space for $Z_Y$ of weight $(\lambda-\gamma)|Z_Y$. The decomposition in (i) is the decomposition of $V^2(Q_Y)$ into distinct weight spaces for $Z_Y$.

(iv) If, in addition, $\Pi(X)-\Pi(L)=\{\alpha\}$ and $Z\le Z_Y$, then for each $\gamma\in\Pi(Y)-\Pi(L_Y)$, $\dim V_\gamma(Q_Y)\le \dim V^1(Q_A)\cdot d$, where d is as in (1.22)(iii).

Proof: This follows from (2.11) and (2.14) in [12] and (1.22) (iii) above.□

In the remainder of this section, we list and establish certain results which will be used in obtaining upper and lower bounds on the dimensions of modules. Beyond the notation established in the next paragraph, this material will be explicitly referenced when necessary and is not required for the reading of the next chapter.

Assume, for the remainder of this chapter, that X is simple. Let $W(\lambda)$ be the Weyl module corresponding to the dominant weight $\lambda$. Let $\delta$ be the half sum of the positive roots in $\Sigma(X)$. Write $\mathbb{Z}\Sigma(X)$ for the $\mathbb{Z}$-span of $\Sigma(X)$ and normalize the inner product on $\mathbb{Z}\Sigma(X)$ so that long roots have length 1. Write $\mu\preceq\lambda$ if $\mu=\lambda-\Sigma c_i\alpha_i$, for $c_i\in\mathbb{Z}^+$. Let $P(\lambda-\mu)$ denote the number of distinct ways of writing $\lambda-\mu$ as an integral linear combination of elements of $\Sigma^+(X)$ with nonnegative coefficients. Also, let $m(\mu)$ denote the multiplicity of the weight $\mu$ in $W(\lambda)$. Finally, let $W(X)$ denote the Weyl group of X.

(1.26). (Weyl degree formula): $\dim W(\lambda) = (\prod(\lambda+\delta,\alpha))/(\prod(\delta,\alpha))$, where each product is taken over $\alpha \in \Sigma^+(X)$.

Proof: See Section 24 of [9].□

(1.27). Suppose rank $X = 2$ and $\lambda = m_1\lambda_1 + m_2\lambda_2$.

(a) If $X$ has type $A_2$, $\dim W(\lambda) = (1/2)(m_1+1)(m_2+1)(m_1+m_2+2)$.

(b) If $X$ has type $B_2$, $\dim W(\lambda) =$ $(1/6)(m_1+1)(m_2+1)(m_1+m_2+2)(2m_1+m_2+3)$.

(c) If $X$ has type $G_2$, $\dim W(\lambda) =$ $(1/5!)(m_1+1)(m_2+1)(m_1+m_2+2)(m_1+2m_2+3)(m_1+3m_2+4)(2m_1+3m_2+5)$.

Proof: This follows directly from (1.26).□

(1.28). (Freudenthal): $m(\mu) = (2\Sigma m(\mu+i\alpha)(\mu+i\alpha,\alpha))/((\lambda+\delta,\lambda+\delta) - (\mu+\delta,\mu+\delta))$, where the sum is taken over $\alpha \in \Sigma^+(X)$ and $i \geq 1$.

Proof: See Section 22 of [9].□

(1.29). Let $\langle v^+\rangle$ be the unique 1-space of $V$ invariant under $U = \langle U_r \mid r \in \Sigma^+(X)\rangle$. Assume $\lambda$ is restricted and $\mu \preceq \lambda$. Let $N = |\Sigma^+(X)|$ and let $\{s_1, s_2, \ldots, s_N\}$ denote any sequence of nonnegative integers. Given a fixed ordering in $\Sigma^+(X) = \{\beta_1, \beta_2, \ldots, \beta_N\}$, $V_T(\mu) = \langle (f_{\beta_1}{}^{s_1} \cdots f_{\beta_N}{}^{s_N})v^+ \mid \lambda - \mu = \Sigma s_i\beta_i\rangle$.

Proof: By (1.1), $V$ is irreducible as a module for $L(X)$. As $L(U)$ leaves $\langle v^+\rangle$ invariant, $V_T(\mu) = \langle f_{\gamma_1} \cdots f_{\gamma_t} v^+ \mid \lambda - \mu = \Sigma \gamma_i\rangle$. The result then follows from the Poincare-Birkhoff-Witt Theorem. (See Section 17 of [9].)□

(1.30). Suppose $V$ is a restricted $kX$-module, $\alpha \in \Sigma^+(X)$ and $\mu \preceq \lambda$, such that $V_T(\mu) \neq 0$. Assume $0 < \langle \mu, \alpha\rangle < p$.

(i) $V_T(\mu - d\alpha) \neq 0$, for each $0 \leq d \leq \langle \mu, \alpha\rangle$.

(ii) If $\alpha \in \Pi(X)$, $\dim V_T(\lambda - d\alpha) = 1$ for $0 \leq d \leq \langle \lambda, \alpha\rangle$.

Proof: View $V$ as an irreducible module for $L(X)$. (See (1.1).) Let $\mathcal{H}$ be a Cartan subalgebra of $L(X)$ and let $\mathfrak{N} = \langle e_\alpha, f_\alpha, \mathcal{H}\rangle$. Then $\mathfrak{N} = \mathfrak{N}_0 \oplus \mathcal{H}_0$ (direct sum of Lie subalgebras), where $\mathfrak{N}_0 = \langle e_\alpha, f_\alpha\rangle$ and $\mathcal{H}_0 = C_{\mathcal{H}}(\mathfrak{N}_0)$. Also, $\mathfrak{N}_0 \cong sl_2$.

Now, let $0 \neq v \in V_T(\mu)$ and take a composition series of V under the action of $\mathfrak{N}$, such that $v \in V_{i+1} - V_i$. Then, $\mathcal{H}_0$ induces scalars on $V_{i+1}/V_i$, so $\mathfrak{N}_0$ acts irreducibly on $V_{i+1}/V_i$. The irreducible modules for $\mathfrak{N}_0$ are all restricted (see Section A of [1]); in particular, the weights form a chain. Since $\langle \mu,\alpha \rangle < p$, (i) holds.

The result of (ii) follows from (i) and (1.29).□

(1.31). For $\alpha \in \Pi(X)$, $\dim V_T(\lambda - k\alpha) \leq 1$, for any $k \in \mathbb{Z}^+$.

Proof: Write $V = V_1^{q_1} \otimes \cdots \otimes V_k^{q_k}$, where $V_i$ is a restricted irreducible kX-module with high weight $\nu_i$, and $q_1, \ldots, q_k$ are distinct powers of p. Suppose $\lambda - k\alpha = \Sigma(q_i\nu_i - n_iq_i\alpha) = \Sigma(q_i\nu_i - m_iq_i\alpha)$, for some integers $0 \leq m_i, n_i < p$. Then $k = \Sigma n_iq_i = \Sigma m_iq_i$ is the p-adic expansion of the integer k. So $m_i = n_i$ for all i and $V_T(\lambda - k\alpha) = (V_1^{q_1})_T(q_1(\nu_1 - n_1\alpha)) \otimes \cdots \otimes (V_k^{q_k})_T(q_k(\nu_k - n_k\alpha))$. So $\dim V_T(\lambda - k\alpha) = \prod \dim((V_i^{q_i})_T(\nu_i - n_i\alpha)) \leq 1$ by (1.29) and (1.30).□

(1.32). ((1.10) in [12]) Let $\mu$ be a dominant weight of T and $W_0 \leq W(X)$ be generated by those fundamental reflections corresponding to simple roots $\alpha \in \Pi(X)$ with $\langle \mu,\alpha \rangle = 0$. Then $W_0$ is the stabilizer of $\mu$ in $W(X)$; so there are $|W(X):W_0|$ distinct conjugates of $V_T(\mu)$ in V.□

(1.33). (i) (Linkage principle) Assume X is simply connected and let X(T) be the group of rational characters of T. If $\mu$ and $\nu$ are high weights of composition factors of an indecomposable kX module, then

(a) $w(\mu+\delta) - (\nu+\delta) \in pX(T)$, for some $w \in W(X)$, and

(b) $\mu$ and $\nu$ lie in the same coset of $X(T)/X_r(T)$, where $X_r(T)$ is the sublattice generated by $\Sigma(X)$.

(ii) Suppose that $\mu$ is a dominant weight and that $W(\lambda)$ contains an X-composition factor of high weight $\mu$. Assume that $p > 2$, and that $p > 3$ when A has type $G_2$. Write $\mu = \lambda - \Sigma c_i\alpha_i$, where each $c_i \geq 0$. Then

(i) $2(\lambda+\delta, \Sigma c_i\alpha_i) - (\Sigma c_i\alpha_i, \Sigma c_i\alpha_i) \in (p/2)Z$, if $X \neq G_2$.

(ii) $2(\lambda+\delta, \Sigma c_i\alpha_i) - (\Sigma c_i\alpha_i, \Sigma c_i\alpha_i) \in (p/6)Z$ if $X = G_2$.

Proof: The statement of (i)(a) is Theorem 3 of [11]. Statement (ii) is (6.2) of [12]. Part (b) of (i) follows from the fact that $U_r(V_T(\eta)) \subset \Sigma V_T(\eta+ir)$, for $i \geq 0$, $\eta$ a weight in a rational kX module V, and $r \in \Sigma(X)$. See Lemma 72 of [14].□

(1.34). ((8.6) of [12]) Let $X = SL_{n+1}$ and let V be a restricted irreducible kX-module with high weight $\lambda$. Suppose $1 \leq r \leq i < j \leq s \leq n$ and $\lambda = a\lambda_i + b\lambda_j$ for $a \neq 0 \neq b$. Then $V_T(\lambda - (\alpha_r+\cdots+\alpha_s))$ is spanned by vectors $v_t = f_{\alpha_r+\cdots+\alpha_{t-1}} f_{\alpha_t+\cdots+\alpha_s} v^+$, for $i \leq t \leq j$. Moreover, $v_i, \ldots, v_j$ are linearly independent unless $a+b+j-i \equiv 0 \pmod p$, in which case they span a j-i space and $bv_i + v_{i+1} + \cdots + v_j = 0$.□

(1.35). Let V be a restricted kX-module. Let $\alpha, \beta \in \Pi(X)$, with $(\alpha,\beta) < 0$, such that $\langle \lambda,\alpha \rangle = c$, $\langle \lambda,\beta \rangle = d$, for $0 < c,d$. Then $2 \geq \dim V_T(\lambda-\alpha-\beta) > 0$ and

(i) if $(\alpha,\alpha) = (\beta,\beta)$, $\dim V_T(\lambda-\alpha-\beta) = 1$ if and only if $c+d = p-1$;

(ii) if $(\alpha,\alpha) = 2(\beta,\beta)$, $\dim V_T(\lambda-\alpha-\beta) = 1$ if and only if $2c+d+2 \equiv 0 \pmod p$; and

(iii) if $(\alpha,\alpha) = 3(\beta,\beta)$, $\dim V_T(\lambda-\alpha-\beta) = 1$ if and only if $3c+d+3 \equiv 0 \pmod p$.

Proof: This follows from (1.28) and the final proposition of [4].□

(1.36). Let Y be a simple closed subgroup of SL(V). Let $P_Y = Q_Y L_Y$, $T_Y$, $U_Y$ be as in results (1.23) - (1.25). Let $\gamma \in \Pi(Y) - \Pi(L_Y)$ and $\beta \in \Pi(L_Y)$ be such that $(\gamma,\gamma) = (\beta,\beta)$, $(\gamma,\beta) < 0$, $\langle U_{\pm\beta} \rangle$ is a simple normal subgroup of $L_Y'$, $\langle \lambda,\gamma \rangle \neq 0$ and $\langle \lambda,\beta \rangle = p-1$. Then, there exists $0 \neq w \in V_{T_Y}(\lambda-\gamma-\beta)$ such that $f_\gamma v^+$ and w afford distinct $L_Y'$ composition factors of $V_\gamma(Q_Y)$.

Proof: Note first that for $0 \neq v \in V_{T_Y}(\lambda-\gamma-\beta)$, either v is a maximal vector for $L_Y' \cap U_Y$ or v lies in an $L_Y$ composition factor of $V_\gamma(Q_Y)$ with high weight $\lambda-\gamma$. Clearly, $f_\gamma v^+$ is a maximal vector for $L_Y' \cap U_Y$, and so affords an $L_Y'$ composition factor of $V_\gamma(Q_Y)$. Moreover, by (1.31), there exists a unique $L_Y$ composition factor of $V_\gamma(Q_Y)$ with high weight $\lambda-\gamma$. But $\langle \lambda,\beta \rangle = p-1$ and (1.35) imply that $\dim V_{T_Y}(\lambda-\gamma-\beta) = 2$.

Hence, there exists $0 \neq w \in V_{T_Y}(\lambda-\gamma-\beta)$ as claimed.□

(1.37). Let rank $X = 2$, with $\Pi(X) = \{\alpha,\beta\}$. Let P, Q, and L be as in (1.19) with $L' = \langle U_{\pm\beta}\rangle$. Suppose $V = V_1{}^{q_1} \otimes \cdots \otimes V_k{}^{q_k}$, for each $V_i$ is a basic module for X and $q_1,\ldots,q_k$ are distinct powers of p. Then $\dim((V_T(\lambda - q_i\beta - q_i\alpha) + [V,Q^2])/[V,Q^2]) \leq 2$, for each i.

Proof: By (1.22), $0 \neq w \in V_T(\lambda - q_i\beta - q_i\alpha)$ corresponds to a nonzero vector in $W_1{}^1(Q) \otimes \cdots \otimes W_i{}^2(Q) \otimes \cdots \otimes W_k{}^1(Q)$, where $W_i = V_i{}^{q_i}$. Let $\mu_\ell$ be the high weight of $W_\ell$, for $1 \leq \ell \leq k$. Suppose $w \notin (W_1)_T(\mu_1) \otimes \cdots \otimes (W_i)_T(\mu_i - q_i\beta - q_i\alpha) \otimes \cdots \otimes (W_k)_T(\mu_k)$. Then w projects nontrivially into some weight space of the form $(W_1)_T(\mu_1 - n_1q_1\beta) \otimes \cdots \otimes (W_i)_T(\mu_i - q_i\alpha) \otimes \cdots \otimes (W_k)_T(\mu_k - n_kq_k\beta)$, for $0 \leq n_\ell < p$. Hence, $\lambda - (\Sigma n_\ell q_\ell \beta) - q_i\alpha = \lambda - q_i\beta - q_i\alpha$. So $\Sigma n_\ell q_\ell = q_i$. Dividing this equation by the highest power of p which occurs, and taking congruences modulo p, we obtain $n \equiv 0 \pmod p$ for some $0<n<p$. Contradiction. Thus, $w \in (W_1)_T(\mu_1) \otimes \cdots \otimes (W_i)_T(\mu_i - q_i\beta - q_i\alpha) \otimes \cdots \otimes (W_k)_T(\mu_k)$, which has dimension equal to $\dim((W_i)_T(\mu_i - q_i\beta - q_i\alpha))$. But this is at most 2, by (1.29). Thus, the result holds.□

Using the methods of (1.26), (1.30), (1.32), (1.34) and referring to [8], we obtain the following lower bounds on dimensions of irreducible kX modules.

(1.38). For V a restricted kX-module with high weight $\lambda$, let $d(\lambda,p) = \dim V$, where p is the characteristic of k. Then

(i) $X = F_4$, $d(2\lambda_2,p) \geq 2^6\cdot 3\cdot 5$

(ii) $X = E_6$, 1. $d(\lambda_1+\lambda_3+\lambda_5+\lambda_6,2) \geq 2^6\cdot 3^3\cdot 5$

2. $d(\lambda_1+\lambda_3+\lambda_6,2) \geq 2^4\cdot 3^3\cdot 19$ 3. $d(\lambda_1+\lambda_2+\lambda_6,2) \geq 2^4\cdot 3^3\cdot 13$

4. $d(\lambda_3+\lambda_4+\lambda_5,3) \geq 2^4\cdot 3^3\cdot 5\cdot 13$ 5. $d(2\lambda_1+2\lambda_6,p) \geq 2\cdot 3^3\cdot 5\cdot 11$

(iii) $X = E_7$, 1. $p \neq 3$, $d(\lambda_6+\lambda_7,p) \geq 2^3 \cdot 3^2 \cdot 7 \cdot 19$

2. $d(2\lambda_1,p) \geq 2 \cdot 3^2 \cdot 7 \cdot 17$

3. $x \neq 0$, $d(\lambda_4+x\lambda_7,3) \geq 2^6 \cdot 3^2 \cdot 7 \cdot 19$

4. $d(3\lambda_7,p) \geq 2^5 \cdot 5^2 \cdot 7$

5. $d(2\lambda_2,p) \geq 2^5 \cdot 3^2 \cdot 37$

6. $d(\lambda_1+\lambda_5,2) \geq 2^6 \cdot 3^2 \cdot 7 \cdot 17$

7. $d(\lambda_2+\lambda_5,2) \geq 2^8 \cdot 3^2 \cdot 5 \cdot 7$

8. $d(\lambda_2+\lambda_7,2) \geq 2^4 \cdot 3 \cdot 7 \cdot 37$

9. $d(\lambda_1+\lambda_2,2) \geq 2^6 \cdot 3^2 \cdot 7 \cdot 11$

10. $d(2\lambda_2+\lambda_7,3) \geq 2^6 \cdot 3^2 \cdot 7 \cdot 13$

11. $d(\lambda_4,3) \geq 2^5 \cdot 3^2 \cdot 5 \cdot 7$

12. $d(2\lambda_7,p) \geq 2^2 \cdot 7 \cdot 29$

13. $d(\lambda_1+2\lambda_7,3) \geq 2^4 \cdot 3^4 \cdot 7$

(iv) $X = E_8$, 1. $\{a,b\} = \{1,2\}$, $d(a\lambda_1+b\lambda_8,3) \geq 2^5 \cdot 3^3 \cdot 5 \cdot 241$

2. $d(\lambda_2+\lambda_3+\lambda_8,2) \geq 2^{10} \cdot 3^3 \cdot 5 \cdot 7 \cdot 13$

3. $d(2\lambda_8,p) \geq 2^4 \cdot 3 \cdot 5 \cdot 37$

4. $d(\lambda_1+\lambda_2+\lambda_7,2) \geq 2^{10} \cdot 3^5 \cdot 5 \cdot 7$

5. $d(4\lambda_8,p) \geq 2^7 \cdot 3^2 \cdot 5 \cdot 7 \cdot 11$

6. $d(\lambda_1+\lambda_2+\lambda_8,2) \geq 2^{10} \cdot 3^3 \cdot 5^2 \cdot 7$

7. $p>2$, $d(\lambda_2,p) \geq 2^6 \cdot 3 \cdot 5^3$

8. $d(2\lambda_1,p) \geq 2^4 \cdot 3^4 \cdot 5 \cdot 11$

9. $d(\lambda_2+\lambda_6+\lambda_8,2) \geq 2^9 \cdot 3^3 \cdot 5 \cdot 7 \cdot 11$

10. $d(3\lambda_8,p) \geq 2^5 \cdot 3^3 \cdot 5 \cdot 17$

11. $d(\lambda_5+\lambda_8,2) \geq 2^{10} \cdot 3^3 \cdot 5 \cdot 7$

12. $d(\lambda_2,2) \geq 2^8 \cdot 3^2 \cdot 5^2$

13. $d(\lambda_2+\lambda_8,2) \geq 2^8 \cdot 3^2 \cdot 5^2 \cdot 11$

14. $d(\lambda_2+\lambda_8,3) \geq 2^{12} \cdot 3^3 \cdot 5$

15. $d(\lambda_1+\lambda_8,3) \geq 2^5 \cdot 3^3 \cdot 5^2 \cdot 7$

16. $d(2\lambda_2+\lambda_8,3) \geq 2^9 \cdot 5 \cdot 37$

17. $d(\lambda_1+\lambda_6+\lambda_8,2) \geq 2^7 \cdot 3^3 \cdot 5^2 \cdot 7^2$

## CHAPTER 2: PARABOLIC EMBEDDINGS

In this chapter, we establish certain results concerning the embeddings of parabolic subgroups, and in particular, embeddings of unipotent radicals. We will adopt the following notation for this entire chapter.

Notation and Hypothesis (2.0). Let Y be a simply connected, simple algebraic group over an algebraically closed field k of characteristic $p > 0$. Let $\theta: Y \longrightarrow SL(V)$ be a nontrivial, finite dimensional, irreducible rational representation. Suppose $A = A^\circ$ is a simple closed subgroup of Y such that $V|A$ is irreducible.

Let $\Sigma(A)$, $\Sigma(Y)$ denote the root systems of A, Y respectively, and take $\Pi(A) = \{\alpha_1,\alpha_2,\ldots\}$ to be a fundamental system of $\Sigma(A)$, with $\mu_i$ the fundamental dominant weight corresponding to $\alpha_i$. Let $B_A = U_A T_A$ be a Borel subgroup of A with maximal torus $T_A$ and unipotent radical $U_A$, chosen so that $U_A$ is the product of $T_A$-root subgroups corresponding to roots in $\Sigma^+(A)$. Write $B_A^-$ for $(U_A^-)T_A$, where $U_A^-$ is the opposite unipotent radical. If $\Sigma(A)$ or $\Sigma(Y)$ has only one root length, we will refer to all roots as being "long." Assume $\Sigma(Y)$ has type $G_2$, $F_4$ or $E_n$.

Fix a maximal parabolic subgroup $P_A = Q_A L_A$ of A, where $Q_A = R_u(P_A)$ and $L_A$ is a Levi factor containing $T_A$. Set $\Pi(L_A) = \Pi(A) \cap \Sigma(L_A)$ and $\Pi(A) - \Pi(L_A) = \{\alpha\}$. We will choose $P_A$ such that $\alpha$ corresponds to an end node of the Dynkin diagram and $Q_A$ is the product of $T_A$-root subgroups corresponding to the roots in $\Sigma^-(A) - \Sigma(L_A)$. Let $T(L_A') = T_A \cap L_A'$ and set $Z_A = Z(L_A)^\circ$. We will abuse notation and write $\mu_i$ for $\mu_i|T(L_A')$.

Let $P_Y = Q_Y L_Y$ be a parabolic subgroup of Y such that $P_A \le P_Y$,

$Q_A \le Q_Y = R_u(P_Y)$. (The existence of such a parabolic $P_Y$ is given by the Borel–Tits theorem.) Choose $P_Y$ minimal with these properties. Let $T_A \le T_Y$ for $T_Y$ a maximal torus of $L_Y$. We choose an ordering of $\Sigma(Y)$ and a corresponding base $\Pi(Y) = \{\beta_1, \ldots \beta_n\}$, such that $U_A \cap L_A \le Q_Y(U_Y \cap L_Y)$, where $U_Y$ is the product of those $T_Y$ root subgroups corresponding to the roots in $\Sigma^+(Y)$ and $Q_Y$ is the product of $T_Y$-root subgroups for roots in $\Sigma^-(Y) - \Sigma(L_Y)$. Write $\Pi(L_Y)$ for $\Pi(Y) \cap \Sigma(L_Y)$ and set $Z_Y = Z(L_Y)^\circ$.

We will write $U_r$ for the $T_A$ (respectively, $T_Y$) root subgroup corresponding to the root $r \in \Sigma(A)$ (respectively, $\Sigma(Y)$). Also, let $x_r(t)$ denote elements of $U_r$, for $t \in k$ and $h_\gamma(c)$ denote the element of $T_A$, or $T_Y$, corresponding to the root $\gamma \in \Pi(A)$, or $\Pi(Y)$, for $c \in k^*$. As well, let $e_r$ and $f_r$ denote the corresponding elements of the Lie algebra $L(Y)$ or $L(A)$. For $Y$ of type $E_n$, we will sometimes abbreviate the above notation in the following manner: For $r \in \Sigma^+(Y)$ such that $r = \beta_{i_1} + \cdots + \beta_{i_t}$, $\{\beta_{i_1}, \ldots, \beta_{i_t}\} \subset \Pi(Y)$ with $i_1 < i_2 < \cdots < i_t$, we will write $U_{\pm i_1 i_2 \cdots i_t}$ for $U_{\pm r}$, $e_{i_1 i_2 \cdots i_t}$ for $e_r$ and similarly for $f_r$, $x_{\pm r}(t)$ and $h_r(c)$. For $r \in \Sigma^+(Y)$, $r = \Sigma a_i \beta_i$, $a_i \in \mathbb{Z}^+$ with some $a_i > 1$, we will write $U_{\pm(a_1, a_2, \ldots, a_n)}$ for $U_{\pm r}$, etc.

Write $L_Y' = L_1 \times \cdots \times L_r$, a direct product of simple algebraic groups. We will refer to $L_i$ as a *component* of $L_Y'$. By (1.23), $L_A'$ and $L_Y'$ are irreducible on $V^1(Q_Y)$. Then $V^1(Q_Y) = M_1 \otimes \cdots \otimes M_r$, where each $M_i$ is an irreducible $L_i$ module. The embedding $\rho: L_A \to P_Y/Q_Y \cong L_Y$ gives an embedding of $L_A'$ in $L_1 \times \cdots \times L_r$ and we let $\rho_i: L_A' \to L_i$ be the corresponding projection. Then any module for $L_i$, in particular $M_i$, can be regarded as a module for $L_A'$.

Remark: If $L_i$ is of classical type, with natural module $W_i$, the proper parabolic subgroups of $L_i$ correspond to stabilizers of flags of totally singular subspaces of $W_i$. Thus, $P_Y$ minimal implies $W_i|\rho_i(L_A')$ is either irreducible or $\rho_i(L_A')$ stabilizes a nonsingular subspace of $W_i$. Hereafter, we will use this fact without reference to this remark.

Write $V = V(\lambda)$, where $\lambda$ is a dominant weight of $T_Y$. Let $\lambda_i$ denote the fundamental dominant weight corresponding to the root $\beta_i$. Let $\langle v^+\rangle$ be the unique 1-space of $V|Y$ invariant under $U_Y T_Y = B_Y$. We may assume, as discussed in the introduction,

(i) $V|Y$ is a restricted module, and

(ii) $V|A = V_1^{q_1} \otimes \cdots \otimes V_k^{q_k}$, where $V_1, \ldots, V_k$ are nontrivial restricted $kA$-modules and $q_1, \ldots, q_k$ are distinct powers of p.

As well, note that

(iii) $V|Y \not\cong L(Y)$, the Lie algebra of Y, as L(X), the Lie algebra of X, is always a proper invariant kX-submodule of L(Y). We will use this fact frequently without reference.

For each $\gamma \in \Pi(Y) - \Pi(L_Y)$, we define a certain normal subgroup $K_\gamma$ of $P_Y$, which in most cases is just the largest normal subgroup of $P_Y$ that is contained in $Q_Y$ and does not contain the $T_Y$ root subgroup corresponding to $-\gamma$. Let $\Sigma_\gamma(Y)$ denote the set of roots in $\Sigma(Y)$ having coefficient of $\gamma$ equal to $-1$ and zero coefficient for other roots in $\Pi(Y) - \Pi(L_Y)$. Then let $K_\gamma$ be the product of those $T_Y$ root subgroups $U_\beta$ for which $\beta \in \Sigma^-(Y) - \Sigma^-(L_Y) - \Sigma_\gamma(Y)$. From the commutator relations it follows that $K_\gamma \trianglelefteq P_Y$.

<u>(2.1).</u> ((3.1) of [12])

(i) $Q_Y/K_\gamma$ is isomorphic to the direct product of those $T_Y$ root subgroups for roots $\beta \in \Sigma_\gamma(Y)$.

(ii) There is an $L_Y$-module structure on $Q_Y/K_\gamma$ such that $Z_Y$ acts by scalars and such that there is a maximal vector of weight $-\gamma$.

(iii) $Q_Y/K_\gamma$ is an irreducible $L_Y$-module, unless $\gamma$ is a long root with $(\gamma, \Sigma L_Y) \neq 0$ and $\Sigma(Y) = G_2$ or $F_4$, with p = 3 or 2, respectively.□

The above considerations apply to the parabolic subgroup $P_A$. Here we have only $\alpha \in \Pi(A) - \Pi(L_A)$ and we write $Q_A/K_\alpha = Q_A^\alpha$. If (A,p) is not special and if $\alpha$ is long when (A,p) has type $(G_2,2)$, then $Q_A{}^{\alpha} = Q_A/[Q_A,Q_A]$.

<u>(2.2).</u> ((3.2) of [12])

(i) $Q_A{}^{\alpha}$ has an $L_A$-module structure and $-\alpha$ is the high weight of a composition factor.

(ii) If (A,p) is not special, $Q_A{}^{\alpha}$ is an irreducible $L_A$ module.

(iii) $Q_A{}^{\alpha}$ has a unique maximal $L_A$-invariant subgroup, $M_\alpha$, which is a submodule having quotient module with high weight $-\alpha$.□

Notation: We will write $I_\alpha$ for the irreducible quotient, $(Q_A{}^{\alpha})/M_\alpha$, described in (iii) of the above result.

<u>(2.3).</u> ((3.3) of [12]) Assume $Z_A \le Z_Y$ and let $\gamma \in \Pi(Y) - \Pi(L_Y)$.

(i) $Q_AK_\gamma/K_\gamma$ is an $L_A$ invariant submodule of $Q_Y/K_\gamma$.

(ii) If $V_\gamma(Q_Y) \neq 0$, then $Q_A \not\leq K_\gamma$.

(iii) Suppose rank$(\Sigma(A)) > 1$ and $\langle \lambda, \gamma \rangle \neq 0$. Then $(\gamma, \Sigma(L_Y)) \neq 0$.□

<u>(2.4).</u> Assume $Z_A \le Z_Y$ and let $\gamma \in \Pi(Y) - \Pi(L_Y)$. Suppose that $\delta \in \Sigma(A)$, $\delta$ has $\alpha$-coefficient equal to $-e$ and $U_\delta \not\leq K_\gamma$. Then

(i) $\gamma|Z_A = r\alpha|Z_A$, where r is a positive integral power of p. Moreover, $r = q_i$, for some i, in case each $V_j$ is basic and $V_\gamma(Q_Y) \neq 0$.

(ii) There exist unique roots $\beta_1, \ldots, \beta_s$ in $\Sigma_\gamma(Y)$ and $c_1, \cdots, c_s$ in $k^*$ such that for $t \in k$, $x_\delta(t)K_\gamma = x_{\beta_1}(c_1t^{r/e}) \cdots x_{\beta_s}(c_st^{r/e})K_\gamma$.

(iii) $\beta_j|T_A = (r/e)\delta$, for $j = 1, \ldots, s$ as in (ii).

(iv) If $D = Q_A \cap K_\gamma$, then $U_{-\alpha} \not\leq D$, so $Q_AK_\gamma/K_\gamma$ has an $L_A$ composition factor isomorphic to $(I_\alpha)^r$.

(v) If (A,p) is not special, $Q_AK_\gamma \le \langle U_{-s} \mid U_{-s} \not\leq K_\gamma, s|T(L_A') = r\eta$, for some $\eta \in \Sigma^+(A)$ with $U_{-\eta} \not\leq Q_A'\rangle K_\gamma$. If (A,p) is special and $\eta \in \Sigma^+(A)$ with $U_{-\eta} \not\leq M_\alpha$ (see 2.2(iii)), then
$U_{-\eta}K_\gamma \le \langle U_{-s} \mid U_{-s} \not\leq K_\gamma, s|T(L_A') = r\eta\rangle K_\gamma$.

<u>Proof</u>: Statements (i) - (iii) are contained in (3.4) of [12]. As well,

if when $(A,p) = (G_2,2)$ we take $\alpha$ to be long, (iv) follows from the proof of (3.4). Suppose $(A,p) = (G_2,2)$ and $\alpha$ is short. Let $\Pi(L_A) = \{\beta\}$. Then, $Q_A' = \langle U_{-3\alpha-\beta}, U_{-3\alpha-2\beta}\rangle \leq D$. If $U_{-\alpha} \leq D$, then $(U_{-\alpha})^{s_\beta} = U_{-\alpha-\beta} \leq D$, where $s_\beta$ is the reflection corresponding to the root $\beta$. But as well, $(U_{-\alpha})^{x_{-\beta}(1)} \leq D$; and a nonidentity element from $U_{-2\alpha-\beta}$ occurs in the factorization of this last expression. Hence, if $U_{-\alpha} \leq D$, then $Q_A \leq D$. Contradiction. Thus, (iv) holds. Finally, since $Q_AK_\gamma/K_\gamma$ is an $L_A$ invariant submodule of $Q_Y/K_\gamma$, it is a sum of $T(L_A')$ weight spaces. The result of (v) then follows from (iv).□

For the following result, we will need additional notation. Recall that $V|A = V_1^{q_1}\otimes\cdots\otimes V_k^{q_k}$, where $V_i$ is a restricted irreducible kA module and $q_1,\ldots,q_k$ are distinct p-powers. Write $V_i = V_i^s \otimes V_i^\ell$, where $V_i = V_i^s$ unless (X,p) is special. If (X,p) is special, $V_i^s$ and $V_i^\ell$ are the short and long parts of $V_i$, as in (1.7). We will write $V_i^\sim$ to indicate one of $V_i$, $V_i^s$, $V_i^\ell$.

<u>(2.5).</u> ((3.5) of [12])

(i) For $i = 1,\ldots,r$, $M_i$ is restricted.

(ii) For $i = 1,\ldots,r$, $M_i$ is irreducible under the action of $L_A'$ and there is a uniquely determined subset $\{q_{i_1},\ldots,q_{i_d}\}$ of $\{q_1,\ldots,q_r\}$ such that $M_i|L_A' \cong (V_{i_1}{}^{\sim 1}(Q_A))^{q_{i_1}} \otimes\cdots\otimes (V_{i_d}{}^{\sim 1}(Q_A))^{q_{i_d}}$.□

<u>Definition</u>: Suppose one of the following holds:

(i) $L_i$ is a classical group with natural module $W_i$ and $W_i|L_A'$ is an algebraic conjugate, by a p-power q, of a nontrivial restricted module.

(ii) $\rho_i(L_A') \leq L_i$ is the natural embedding of a group of type $B_m$ in a group of type $D_{m+1}$ and taking $\Pi(L_A) = \{\gamma_1,\ldots,\gamma_m\}$, $\Pi(L_i) = \{\tau_1,\ldots,\tau_{m+1}\}$, we have $\rho_i(x_{\pm\gamma_\ell}(t)) = x_{\pm\tau_\ell}(t^q)$ for $1\leq\ell<m$ and $\rho_i(x_{\pm\gamma_m}(t)) = x_{\pm\tau_m}(t^q)x_{\pm\tau_{m+1}}(t^q)$, for all $c\in k^*$.

(iii) $L_A' \cong L_i$ and if $\Pi(L_A) = \{\gamma_1,\ldots,\gamma_m\}$ and $\Pi(L_i) = \{\tau_1,\ldots,\tau_m\}$,

$\rho_i(x_{\pm\gamma_j}(t)) = x_{\pm\tau_j}(t^q)$, for $1 \le j \le m$, for all $t \in k$.
Then we call q the *field twist on the embedding of $L_A'$ in $L_i$.*

(2.6). Suppose $L_i$ is a classical group and $M_i|L_A'$ is the $q_i$ twist of a nontrivial basic or p-basic module. Then, $q_i$ is the field twist on the embedding of $L_A'$ in $L_i$.

Proof: If $\rho_i(L_A') \le L_i$ is not the usual embedding of a group of type $B_m$ in a group of type $D_{m+1}$, this follows from (9.1) of [12]. Suppose $\rho_i(L_A') \le L_i$ is of type $B_m \le D_{m+1}$. Let $\Pi(L_A)$ and $\Pi(L_i)$ be as in (ii) above. Then if $\langle\lambda,\tau_i\rangle = c_i$ for $1 \le i < m$, then $\langle\lambda,\gamma_i\rangle = c_i q$, where q is the field twist on the embedding of $L_A'$ in $L_i$. And by (8.1) of [12], $\langle\lambda,\tau_m\rangle = 0$ or $\langle\lambda,\tau_{m+1}\rangle = 0$. So $\langle\lambda,\gamma_m\rangle = dq$, where $d = \langle\lambda,\tau_m\rangle + \langle\lambda,\tau_{m+1}\rangle$. By (2.5), $M_i|L_i$ is restricted so $c_i, d < p$ and $q_i = q$.□

Definition: For $\gamma \in \Pi(Y) - \Pi(L_Y)$ with $(\gamma,\Sigma L_Y) \ne 0$, suppose $Q_Y/K_\gamma$ is an irreducible $L_Y'$ module. Then, for $1 \le i \le r$ with $(\gamma,\Sigma L_i) \ne 0$, let $V_{L_i}(-\gamma)$ denote the irreducible $L_i$ module with high weight $-\gamma$. Suppose $V_{L_i}(-\gamma)|L_A' \cong D_1^{r_1} \otimes \cdots \otimes D_d^{r_{d_i}}$, for $D_1,\ldots,D_d$ nontrivial, restricted irreducible $L_A'$ modules and $r_1,\ldots,r_{d_i}$ distinct p-powers. Let $S_i(\gamma,L_A) = \{r_1,\ldots,r_{d_i}\}$.

(2.7). Assume $Z_A \le Z_Y$. Let $\gamma \in \Pi(Y) - \Pi(L_Y)$ such that $Q_A \not\le K_\gamma$. Suppose one of the following holds:

(i) There exists a unique pair $1 \le i,j \le r$ such that $(\Sigma L_i,\gamma) \ne 0 \ne (\Sigma L_j,\gamma)$ and $V_{L_m}(-\gamma)|L_A'$ is irreducible for $m = i,j$.

(ii) There exist distinct $1 \le i,j,\ell \le r$ such that $(\Sigma L_m,\gamma) \ne 0$ for $m = i,j,\ell$ and $V_{L_m}(-\gamma)|L_A'$ is irreducible for $m = i,j,\ell$.

If (i) holds, either $(A,p) = (G_2,2)$ with $\alpha$ long and $\mathrm{rank}(L_i) = 1 = \mathrm{rank}(L_j)$, or $S_i(\gamma,L_A) \cap S_j(\gamma,L_A) \ne \emptyset$. If (ii) holds, $S_i(\gamma,L_A) \cap (S_j(\gamma,L_A) \cup S_\ell(\gamma,L_A)) \ne \emptyset$.

Proof: Since $Q_A \not\le K_\gamma$, (2.4) implies that there exists an $L_A'$ composition factor of $Q_Y/K_\gamma$ isomorphic to a twist of $I_\alpha$. Now, as $L_A'$

modules $Q_Y/K_\gamma \cong V_{L_i}(-\gamma)|L_A' \otimes V_{L_j}(-\gamma)|L_A'$, if (i) holds, and $V_{L_i}(-\gamma)|L_A' \otimes V_{L_j}(-\gamma)|L_A' \otimes V_{L_k}(-\gamma)|L_A'$, if (ii) holds. If A has type $G_2$, p=2, and $\alpha$ is long, $I_\alpha \cong W \otimes W^2$, where W is the restricted 2-dimensional $L_A'$ irreducible. In every other case, $I_\alpha$ is a tensor indecomposable $L_A'$ module. The result then follows from the Steinberg tensor product theorem. (See [15].) □

Hypothesis. For the remainder of Chapter 2, assume $Z_A \leq Z_Y$.

(2.8). Assume (Y,p) is not special and let $\gamma,\delta \in \Pi(Y) - \Pi(L_Y)$.

(i) Suppose there exist $r,s \in \Sigma^+(Y) - \Sigma^+(L_Y)$ such that $r+s \in \Sigma^+(Y)$, $U_{-r} \nleq K_\gamma$, $U_{-s} \nleq K_\delta$ and $x_{-\alpha}(t) = x_{-r}(c_1t^q)x_{-s}(c_2t^{q_0})w$, for $c_i \in k^*$, q and $q_0$ positive integral powers of p and $w \in \langle U_{-\beta} | \beta \in \Sigma^+(Y) - \Sigma^+(L_Y) - \{r,s\}\rangle$. If $q \neq q_0$, there exists a pair of roots $\{r_0,s_0\} \subset \Sigma^+(Y) - \Sigma^+(L_Y) - \{r,s\}$ such that $r_0+s_0 = r+s$ and a nonidentity element from each of $U_{-r_0}$ and $U_{-s_0}$ occurs in the factorization of $x_{-\alpha}(t)$.

(ii) Let $1 \leq i,j \leq r$ such that $(\Sigma L_i,\gamma) \neq 0 \neq (\Sigma L_j,\delta)$ and $(\gamma,\delta) < 0$. If $Q_A \nleq K_\gamma$ and $Q_A \nleq K_\delta$, there exists a p-power q such that $\gamma|Z_A = q\alpha = \delta|Z_A$.

Proof: Let r, s, q and $q_0$ be as in (i). If there does not exist a pair $\{r_0,s_0\}$ as described, then in the expression for $1 = [x_{-\alpha}(t), x_{-\alpha}(u)]$, the contribution to the root group $U_{-r-s}$ is $c_1c_2(at^qu^{q_0} - bu^qt^{q_0})$, for some $a,b \in k^*$. (Here we have used the fact that (Y,p) is not a special pair.) Since $c_i \neq 0$, $a = -b$ and $q = q_0$. Thus (i) holds.

For (ii), let $r,s \in \Sigma^+(Y) - \Sigma^+(L_Y)$ such that $U_{-r} \nleq K_\gamma$ and $U_{-s} \nleq K_\delta$, and $x_{-\alpha}(t) = x_{-r}(c_1t^{q_1})x_{-s}(c_2t^{q_2})w$, for $c_i \in k^*$, $w \in Q_Y$, $q_i$ a positive integral power of p. Also, we may choose w so that no nonidentity element from the set $U_{-r}U_{-s}$ occurs in its factorization. Note that {r,s} is the unique pair of roots in $\Sigma^+(Y) - \Sigma^+(L_Y)$ whose sum is r+s. Thus, by part (i), $q_1 = q_2$ and (ii) then follows from (2.4).□

(2.9). Assume (Y,p) is not special, and let $\gamma,\delta \in \Pi(Y) - \Pi(L_Y)$, as in (2.8)(ii). Suppose $Q_A' = \{1\}$. Then, $Q_A \leq K_\gamma$ or $Q_A \leq K_\delta$.

Proof: Suppose false; i.e., suppose $Q_A \nleq K_\gamma$ and $Q_A \nleq K_\delta$. Then, by (2.8), $\delta|Z_A = q\alpha = \gamma|Z_A$, for some p-power q. Let $r \in \Sigma^+(Y) - \Sigma^+(L_Y)$ and $c_r \in k^*$ be such that $U_{-r} \nleq K_\gamma$ and $x_{-r}(c_r t^q)$ occurs in the factorization of $x_{-\alpha}(t)$. Let $\beta \in \Sigma^+(L_A)$ such that $U_{-\alpha-\beta} \nleq M_\alpha$. (See 2.2(iii).) Let $s \in \Sigma^+(Y) - \Sigma^+(L_Y)$ and $c_s \in k^*$ be such that $U_{-s} \nleq K_\delta$ and $x_{-s}(c_s t^q)$ occurs in the factorization of $x_{-\beta-\alpha}(t)$. (We have used (2.4) here.)

Consider the commutator $[x_{-\alpha}(t), x_{-\beta-\alpha}(t)]$. There is a nontrivial contribution to the root group $U_{-r-s}$ from $[x_{-r}(c_r t^q), x_{-s}(c_s t^q)]$. Thus, as the commutator is 1, there must be another contribution to this root group. Now, {r,s} is the unique pair of roots in $\Sigma^+(Y) - \Sigma^+(L_Y)$ whose sum is r+s. Thus, a nonidentity element from the root group $U_{-r}$ must occur in the factorization of $x_{-\beta-\alpha}(t)$, and a nonidentity element from the root group $U_{-s}$ must occur in the factorization of $x_{-\alpha}(t)$. But this contradicts (2.4). Thus, the result holds.□

For the following general lemmas we will need additional notation. Let $\gamma \in \Pi(Y) - \Pi(L_Y)$ and suppose that one of the following holds:

(a) There exists $\delta \in \Pi(Y) - \Pi(L_Y)$ with $(\delta,\gamma) < 0$.

(b) There exist $1 \leq j \leq r$ and $\delta \in \Pi(Y) - \Pi(L_Y)$ with $\gamma \neq \delta$ and $(\Sigma L_j,\gamma) \neq 0 \neq (\Sigma L_j,\delta)$.

Let $K \leq K_\delta$ be defined as follows: $K = \langle U_{-r} \mid r = \Sigma n_\beta \beta,\ \beta \in \Pi(Y),\ n_\gamma > 1$ or $n_\delta > 1$ or $n_\tau > 0$ for some $\tau \in \Pi(Y) - \Pi(L_Y)$ with $\tau \neq \gamma,\delta\rangle$. Then, $K \trianglelefteq P_Y$ and $K_\delta/K$ is an abelian group with an $L_Y$ module structure, where $L_Y$ acts by conjugation and the scalar action is defined as follows: for $c,d \in k$ and $s \in \Sigma^+(L_Y)$ such that $U_{-s} \nleq K$, $cx_{-s}(d)K = x_{-s}(cd)K$. Then $T_Y$ preserves this scalar action and the image of $U_{-s}$ in $K_\delta/K$ is a $T_Y$ weight space of weight $-s$. As $L_Y$ modules, $K_\delta/K \cong K_1/K \times K_2/K$, where $K_1/K$ is the irreducible $L_Y$ module with high weight $-\gamma$. If (a) holds, $K_2/K$ is the irreducible $L_Y$ module with high weight $-\gamma-\delta$. If (b) holds, let $r \in \Sigma^+(L_j)$ such that $\delta+r+\gamma \in \Sigma^+(Y)$ and if $s \in \Sigma^+(L_j)$ with $\delta+s+\gamma \in \Sigma^+(Y)$, then $ht(r) < ht(s)$. Then $K_2/K$ is the irreducible $L_Y$ module with high weight $-\gamma-r-\delta$. Let $Q_Y(\gamma,\delta)$ denote the $L_Y$ composition factor, $K_2/K$, of $Q_Y$.

We wish to study the action of $L_A$ on $K_\delta/K$. We note that the commutator relations imply that $K_2/K$ is an $L_A$ invariant subspace of $K_\delta/K$. However, $K_1/K$ is not necessarily $L_A$ invariant; in particular, under the conditions of the following

Definition: Let $\delta \in \Pi(Y) - \Pi(L_Y)$. We say *$-\delta$ is involved in $L_A'$* if there exists $r \in \Sigma(L_A)$ and $s \in \Sigma^+(Y) - \Sigma(L_Y)$ such that $U_{-s} \nleq K_\delta$ and a nonidentity element from the root group $U_{-s}$ occurs in the factorization of $x_r(t)$.

Consider the following example to see how we may insure, in a particular case, that $-\delta$ is not involved in $L_A'$, when $Q_A \leq K_\delta$. Suppose $(\delta,\Sigma L_j) \neq 0$ for a unique component $L_j$ and $L_j$ has type $A_k$ for some k. Suppose, in addition, that $V_{L_j}(-\delta) \cong W_j$, the natural module for $L_j$, and hence is an irreducible $\rho_j(L_A')$ module. Say, $W_j|\rho_j(L_A)$ has high weight $\nu_j$. Let $P_Y\hat{} \geq B_Y^-$ be the parabolic subgroup with Levi factor $L_Y\hat{} = \langle L_Y, U_{\pm\delta}\rangle$; so $P_A \leq P_Y\hat{}$ and $Q_A \leq Q_Y\hat{} = R_u(P_Y\hat{}) = K_\delta$. Let $\rho\hat{}: L_A \to L_Y\hat{}$ be the natural homomorphism and $\rho_j\hat{}$ be $\rho\hat{}|L_A'$ followed by the projection of $L_Y\hat{}$ onto the component $L_j\hat{} = \langle L_j, U_{\pm\delta}\rangle$. Then $\rho_j\hat{}$ is a rational morphism of $L_A'$ into a group of type $A_{k+1}$. Moreover, $W_j\hat{}$, the natural module for $L_j\hat{}$, has two $\rho_j(L_A')$ composition factors – a factor isomorphic to $W_j$ (or $W_j^*$) and a one-dimensional factor. Hence, if $\nu_j$ is not linked to the 0 weight, in the sense of (1.33), then $\rho_j(L_A')$ acts completely reducibly on $W_j\hat{}$ and we may assume, up to conjugacy by $L_j\hat{}$, that $-\delta$ is not involved in $L_A'$.

We give one additional criterion for $-\delta$ to be involved in $L_A'$.

(2.10). If $-\delta$ is involved in $L_A'$, then $\delta|Z_A = 0$. Moreover, if r and s are as in the above definition, $-s(h_r(c)) = c^k$ for some $k \in 2\mathbb{Z}^+ - \{0\}$.

Proof: Let r and s be as given. Then there exists $0 \neq f(t) \in k[t]$ such that $x_r(t) = \ell x_{-s}(f(t))u$, for some $\ell \in L_Y'$ and some $u \in Q_Y$ such that u has no nonidentity element from $U_{-s}$ in its factorization. Conjugating by $z \in Z_A$ and using uniqueness of factorization in $U_Y$, we have $-s(z) = 0$. But, if $\beta \in \Pi(L_Y)$, $\beta(z) = 0$, so the first statement holds. Also, conjugating by $h_r(c)$, we have $f(c^2t) = -s(h_r(c))f(t)$. Let $t = 1$ and the result follows.□

(2.11). Let $\gamma \in \Pi(Y) - \Pi(L_Y)$ with $V_\gamma(Q_Y) \neq 0$. Suppose one of the following holds:

(a) There exists $\delta \in \Pi(Y) - \Pi(L_Y)$ with $(\gamma,\delta) < 0$ and $Q_A \leq K_\delta$. If $p=2$, assume $(\gamma,\gamma) = (\delta,\delta)$ or $\delta$ is long.

(b) There exist $1 \leq j \leq r$ and $\delta \in \Pi(Y) - \Pi(L_Y)$ with $\gamma \neq \delta$, $(\Sigma L_j,\gamma) \neq 0 \neq (\delta,\Sigma L_j)$ and $Q_A \leq K_\delta$. If $p=2$, assume $\delta$ is long.

Then,

(i) if $-\delta$ is not involved in $L_A'$, $Q_Y(\gamma,\delta)$ has an $L_A'$ composition factor isomorphic to $(I_\alpha)^r$, for some p-power r, and $(\gamma+\delta)|Z_A = r\alpha$. Moreover, if $V|A = V_1^{q_1} \otimes \cdots V_k^{q_k}$, where each $V_j$ is basic, then $r = q_i$, for some $1 \leq i \leq k$.

(ii) Suppose in addition that if (b) holds with Y of type $E_n$ and $\delta=\beta_4$, then $\{\beta_2,\beta_3,\beta_5\} \not\subseteq \Pi(L_Y)$. Then there exists a parabolic subgroup $P_Y\hat{} \geq B_Y^-$, the opposite Borel subgroup of Y, with Levi factor $L_Y\hat{} = \langle L_Y, U_{\pm\delta}\rangle$ and such that $P_A \leq P_Y\hat{}$, $Q_A \leq R_u(P_Y\hat{}) = Q_Y\hat{}$.

(iii) Let $P_Y\hat{}$ be as in (ii). If $-\delta$ is involved in $L_A'$ or if $\gamma|Z_A = r\alpha$, where r is as in (i), then $Z_A \leq Z(L_Y\hat{})°$.

Proof: Let $K \leq K_\delta$, $K_1$ and $K_2$ be as defined prior to (2.10). Note that if $-\delta$ is not involved in $L_A'$, the commutator relations imply that $K_1/K$ is an $L_A$ invariant subgroup of $K_\delta/K$. Moreover, since $Z_A \leq Z_Y$, $L_A$ preserves the given scalar action on $K_\delta/K$. So $K_i/K$ has an $L_A$ module structure for each i. Consider the image of $Q_A$ in $K_\delta/K$. Now, $Q_AK/K \nleq K_1/K$, else $[V,K_1] = [V,Q_Y]$. But there exists $b \in \Sigma^+(L_Y) \cup \{0\}$, with $V_{T_Y}(\lambda-b-\gamma-\delta) \neq 0$.

And $V_{T_Y}(\lambda-b-\gamma-\delta) \leq [V,Q_Y] - [V,K_1]$. Thus, $Q_AK/K$ projects nontrivially into the $L_A$ submodule $K_2/K = Q_Y(\gamma,\delta)$. Moreover, $Z_A$ either acts trivially on $Q_Y(\gamma,\delta)$ or induces a full set of scalars on $Q_Y(\gamma,\delta)$; so in fact, $Q_AK/K$ is an $L_A$ submodule of $Q_Y(\gamma,\delta)$. One may now argue as in our proof of (2.4) and the proof of (3.4) of [12] that $U_{-\alpha}K/K \nleq K_1/K$. So the image of $x_{-\alpha}(1)K/K$ in $Q_Y(\gamma,\delta)$ affords the high weight space of an $L_A'$ composition factor.

Let $s \in \Sigma^+(L_Y) \cup \{0\}$ and $0 \neq f(x) \in k[x]$ such that $x_{-s-\gamma-\delta}(f(t))$ occurs in the factorization of $x_{-\alpha}(t)$ and such that s is of minimal height with this property. (Say ht(0) = 0.) Conjugating $x_{-\alpha}(t)$ by an element of $T_A$ which does not centralize $x_{-\alpha}(t)$, we obtain $f(c^kt) = c^{\ell}f(t)$, for some integers k and $\ell$, and for all $c \in k^*$, $t \in k$. Letting t=1, we have $f(x) = a_1x^{\ell/k}$, for some $a_1 \in k$. Moreover, having chosen s of minimal height, $x_{-\alpha}(t)x_{-\alpha}(u) = x_{-\alpha}(t+u)$ implies that $\ell/k$ is a positive integral power of p. Let $\ell/k = r$. Then, by (2.4), $(s+\gamma+\delta)|T_A = r\alpha$. Thus, $(\gamma+\delta)|Z_A = r\alpha$ and the $L_A'$ composition factor of $Q_Y(\gamma,\delta)$ afforded by the image of $U_{-\alpha}K/K$ is isomorphic to $(I_\alpha)^r$.

We must now show that if $V|A$ is a tensor product of basic modules, then $r = q_i$, for some i. Let $b \in \Sigma^+(L_Y) \cup \{0\}$ be as in the first paragraph. So $V_0 = V_{T_Y}(\lambda-b-\gamma-\delta) \neq 0$. Moreover, $V_0 \nleq [V,Q_A{}^2]$, since $Q_A \leq K_\delta$ implies that $[V,Q_A{}^2] \leq [V,K_\delta{}^2]$. Since $V_0 \nleq [V,Q_A{}^2]$, $(V_0 + [V,Q_A{}^2])/[V,Q_A{}^2] \leq V^2(Q_A)_{\lambda - q_i\alpha}$, a $Z_A$ weight space, for some $1 \leq i \leq k$. See (1.22) for the description of $Z_A$ weight spaces of $V^2(Q_A)$. Thus, $(\gamma+\delta)|Z_A = q_i\alpha$ and $r = q_i$ as desired. The statement of (ii) is clear and (iii) follows from (2.10).□

<u>(2.12).</u> Suppose there exist distinct $\tau_0, \ldots, \tau_m \in \Pi(Y) - \Pi(L_Y)$, with $(\tau_\ell,\tau_{\ell+1}) < 0$ for $0 \leq \ell < m$, $(\tau_\ell,\Sigma L_Y) = 0$ for $0 < \ell \leq m$, $(\tau_0,\Sigma L_Y) \neq 0$, and $V_{\tau_0}(Q_Y) \neq 0$. If m = 1 and p = 2, assume $(\tau_0,\tau_0) = (\tau_1,\tau_1)$; if m > 1 and Y has type $F_4$, assume p > 2. Suppose there exists a unique p-power q such

that $(I_\alpha)^q$ is an $L_A'$ composition factor of $Q_Y/K_{\tau_0}$.

Then $\tau_i|Z_A = 0$ for $i > 0$. Also, if $P_Y\hat{} \geq B_Y^-$ is the parabolic subgroup of $Y$ with Levi factor $L_Y\hat{} = \langle L_Y, U_{\pm\tau_1},\ldots,U_{\pm\tau_m}\rangle$, then $P_A \leq P_Y\hat{}$, $Q_A \leq R_u(P_Y\hat{}) = Q_Y\hat{}$ and $Z_A \leq Z(L_Y\hat{})°$.

If, in addition, there exists a unique $L_A'$ composition factor of $Q_Y/K_{\tau_0}$ isomorphic to $(I_\alpha)^q$, then $\tau_i|T_A = 0$ for $i > 0$.

Proof: We use induction on m. Let $m = 1$ and note that since $(\tau_1,\Sigma L_Y) = 0$, $\tau_1|T(L_A') = 0$ and (2.10) implies that $-\tau_1$ is not involved in $L_A'$. By (2.4), $\tau_0|Z_A = q\alpha$ and since $Q_A \leq K_{\tau_1}$, (2.11) implies that there is a nontrivial image of $I_\alpha$ in $Q_Y(\tau_0,\tau_1)$. But $Q_Y(\tau_0,\tau_1) \cong Q_Y/K_{\tau_0}$, as $L_Y'$ modules, so all $L_A'$ composition factors of $Q_Y(\tau_0,\tau_1)$ isomorphic to a twist of $I_\alpha$ are isomorphic to $(I_\alpha)^q$. Thus, $(\tau_0+\tau_1)|Z_A = q\alpha$. Hence, $\tau_1|Z_A = 0$. The statements about the parabolic $P_Y\hat{}$ are clear.

Now, suppose there exists a unique $L_A'$ composition factor of $Q_Y/K_{\tau_0}$ isomorphic to $(I_\alpha)^q$. Let $r_1,\ldots,r_j \in \Sigma^+(Y)$ be such that $x_{-r_1}(c_1)\cdots x_{-r_j}(c_j)K_{\tau_0}$ spans the high weight space of this composition factor. So $r_\ell|T_A = q\alpha$, for $1 \leq \ell \leq j$. Then, there exists an $L_A'$ composition series of $Q_Y(\tau_0,\tau_1)$ such that $x_{-r_1-\delta}(c_1)\cdots x_{-r_j-\delta}(c_j)$ affords the high weight space of the unique composition factor of $Q_Y(\tau_0,\tau_1)$ isomorphic to $(I_\alpha)^q$. Thus, $(r_\ell+\tau_1)|T_A = q\alpha$, for $1 \leq \ell \leq j$. Hence, $\tau_1|T_A = 0$.

Now suppose $m>1$. By induction, $\tau_1|T_A = 0 = \cdots = 0 = \tau_{m-1}|T_A$. Also, the parabolic subgroup $D_Y \geq B_Y^-$ with Levi factor $M_Y = \langle L_Y, U_{\pm\tau_1},\ldots,U_{\pm\tau_{m-1}}\rangle$ has the properties: $P_A \leq D_Y$, $Q_A \leq R_u(D_Y)$, and $Z_A \leq Z(M_Y)°$. Let $K_{\tau_m} \leq R_u(D_Y)$ be as usual. Then $Q_A \leq K_{\tau_m}$, as all $L_A'$ composition factors of $R_u(D_Y)/K_{\tau_m}$ are trivial. Also, since $(\tau_m,\Sigma L_Y) = 0$, $\tau_m|T(L_A') = 0$ and (2.10) implies that $-\tau_m$ is not involved in $L_A'$. Note that the given hypotheses on $Q_Y/K_{\tau_0}$ carry over to $R_u(D_Y)/K_{\tau_m}$; i.e., if $Q_Y/K_{\tau_0}$ has a unique $L_A'$ composition factor isomorphic to a twist of $I_\alpha$, then so does $R_u(D_Y)/K_{\tau_m}$. So we may argue as in the case $m = 1$ to obtain the result.□

In (2.11) and (2.12), the purpose of constructing the parabolic $P_Y\hat{}$ is to point out that since $Q_A \le Q_Y\hat{}$, $[V,Q_A{}^2] \le [V,(Q_Y\hat{})^2]$, and so there are many $T_Y$ weight vectors of $Q_Y$ level greater than 1, which are not contained in $[V,Q_A{}^2]$. Often, this construction will produce a $V_\gamma(Q_Y\hat{})$ which exceeds the bound in (1.25).

Hypothesis (G): (i) There exists $\gamma \in \Pi(Y) - \Pi(L_Y)$ with $Q_AK_\gamma/K_\gamma = Q_Y/K_\gamma$ and $\dim(Q_AK_\gamma/K_\gamma) = \dim(I_\alpha)$.

(ii) There exists $\delta \in \Pi(Y) - \Pi(L_Y)$ with $(\gamma,\delta) < 0$, $(\delta,\Sigma L_Y) = 0$, and $(\gamma,\gamma) = (\delta,\delta)$ when $p=2$.

(iii) $V_\gamma(Q_Y) \ne 0$.

(iv) There exists a unique $1 \le j \le r$ with $(\gamma,\Sigma L_j) \ne 0$.

(v) $L_A' \cong L_j$, and q is the field twist on the embedding of $L_A'$ in $L_j$.

(vi) $V|A = V_1{}^{q_1} \otimes \cdots \otimes V_k{}^{q_k}$, where each $V_m$ is basic.

(2.13). (a) If Hypothesis (G) (i) holds, there exists a p-power r with $\gamma|T_A = r\alpha$. If (i), (iii) and (vi) hold, $r = q_i$ for some $1 \le i \le k$. If (i), (iv), and (v) hold, $q = r$.

(b) If Hypothesis (G) (i) – (iii) hold, $\delta|T_A = 0$ and $\langle\lambda,\gamma\rangle = 0$.

Proof: Condition (i), together with (2.4), implies $Q_Y/K_\gamma \cong (I_\alpha)^r$ as $L_A'$ modules, for some p-power r. Comparing high weight vectors in the two modules, we have $\gamma|T_A = r\alpha$. If (iii) and (vi) hold as well, then (2.4) implies that $r = q_i$ for some i. If $L_A' \cong L_j$, let q be as in Hypothesis (G) (v). So if $\Pi(L_A) = \{\eta_1,\ldots\eta_m\}$ we may take $\Pi(L_j) = \{\tau_1,\ldots\tau_m\}$, such that $\rho_j(h_{\eta_\ell}(c)) = h_{\tau_\ell}(c^q)$, for $1 \le \ell \le m$ and for all $c \in k^*$. Moreover, $(\alpha,\eta_\ell) = (\gamma,\tau_\ell)$ for all $\ell$, else $Q_AK_\gamma/K_\gamma$ and $Q_Y/K_\gamma$ are non-isomorphic $\rho(L_A')$ modules. Thus, the high weight space, $U_{-\gamma}K_\gamma/K_\gamma$, of $Q_Y/K_\gamma$ affords $T(L_A')$ weight $-q\alpha$, and $Q_Y/K_\gamma \cong (I_\alpha)^q$, as $L_A'$ modules. Thus, $q = r$. Hence, (a) holds.

Let $\delta$ be as in Hypothesis (G) (ii) and r as in (a). Then (2.12) implies

$\delta|T_A = 0$. Thus $\langle\lambda,\gamma\rangle = 0$, else $f_\gamma v^+$ and $f_{\gamma+\delta}v^+$ are two linearly independent vectors in $V_{T_A}(\lambda - r\alpha)$, contradicting (1.31). Thus, (b) holds.□

<u>(2.14).</u> Assume Hypothesis (G). In the p-adic expansion of $\langle\lambda,\alpha\rangle$, $q = q_i$ has nonzero coefficient. Moreover, if $p>2$ when Y has type $F_4$, $L_Y'$ is not a simple algebraic group.

<u>Proof</u>: Note that $M_j$ is nontrivial since $V_\gamma(Q_Y) \neq 0$, $(\gamma,\Sigma L_\ell) = 0$, for all $\ell\neq j$, and $\langle\lambda,\gamma\rangle = 0$, by (2.13). Choose $\beta_0 \in \Pi L_j$ of minimal distance from $\gamma$ (on the Dynkin diagram) such that $\langle\lambda,\beta_0\rangle \neq 0$. Then there exist distinct $\beta_1,\ldots,\beta_t \in \Pi L_j$ with $(\beta_\ell,\beta_{\ell+1}) \neq 0$ for $0\leq\ell<t$ and $(\beta_t,\gamma) \neq 0$. Also, there exist distinct $\alpha_1,\ldots,\alpha_t \in \Pi(L_A)$ with $\beta_\ell|T_A = q_i\alpha_\ell$, for $0\leq\ell\leq t$, $(\alpha_\ell,\alpha_{\ell+1}) \neq 0$ for $0\leq\ell<t$, $(\alpha_t,\alpha) \neq 0$, $\langle\lambda,\alpha_0\rangle \neq 0$, and $\langle\lambda,\alpha_\ell\rangle = 0$ for $0<\ell\leq t$. Let $s = \beta_0+\cdots+\beta_t$ and $r = \alpha_0+\cdots+\alpha_t$. Then $f_{s+\gamma}v^+$ and $f_{s+\gamma+\delta}v^+$ are two linearly independent vectors in $(V_{T_A}(\lambda - q_ir - q_i\alpha) + [V,Q_A{}^2])/[V,Q_A{}^2]$. But if $q_i$ has zero coefficient in the p-adic expansion of $\langle\lambda,\alpha\rangle$, the indicated $T_A$ weight space in $V^2(Q_A)$ has dimension at most 1. (Here we use (1.22) and (1.29).) Thus, the first statement of the result holds.

Now, suppose $L_Y'$ is a simple algebraic group; then $L_Y' = L_j$. Also, assume $p > 2$ when Y has type $F_4$. If $\tau \in \Pi(Y) - \Pi(L_Y)$ with $(\tau,\Sigma L_Y) \neq 0$, then $V_\tau(Q_Y) \neq 0$ and $Q_A \nleq K_\tau$. Also, $Q_Y/K_\tau$ is an irreducible $L_Y'$ module and so an irreducible $L_A'$ module. Thus, $Q_Y/K_\tau = Q_AK_\tau/K_\tau$. Comparing high weights, we have $(\tau,\beta_t) \neq 0$ and $\tau|T_A = q_i\alpha$. Now, if $\tau \in \Pi(Y) - \Pi(L_Y)$ with $(\tau,\Sigma L_Y) = 0$, (2.12) implies $\tau|T_A = 0$. Thus, we have completely determined the action of $\Pi(Y)$ on $T_A$.

The work of the first paragraph implies $V_{T_A}(\lambda - q_i\alpha) \neq 0$. Thus, there exists $\tau_0 \in \Pi(Y) - \Pi(L_Y)$ with $(\tau_0,\Sigma L_Y) \neq 0$ and $\langle\lambda,\tau_0\rangle \neq 0$. Moreover, there exists a unique $\tau_0$ with these properties, else $\dim V_{T_A}(\lambda - q_i\alpha) > 1$, contradicting (1.31). Note that there does not exist a nonzero vector with $T_A$ weight $\lambda - q_0\alpha$ for $q_0 \neq q_i$. Thus, $V|A$ is a conjugate of a basic module, and so by (1.10), $q_i = 1$. Also, note that if $Y$ has type $F_4$, the above work implies $L_Y' = \langle U_{\pm\beta_2}\rangle$.

We now claim that $\langle\lambda,\alpha\rangle = \langle\lambda,\tau_0\rangle$. Certainly, $x = \langle\lambda,\tau_0\rangle \leq \langle\lambda,\alpha\rangle = y$, as $0 \neq ((f_{\tau_0})^x)v^+ \in V_{T_A}(\lambda - x\alpha)$. But, in fact, $x \geq y$, else there does not exist a vector in $V|Y$ with $T_A$ weight $\lambda - y\alpha$. By (1.29), $\dim V_{T_A}(\lambda-r-\alpha) \leq t+2$. By (1.34), $\dim V_{T_Y}(\lambda-s-\tau_0) \geq t+1$. So $\dim(V_{T_Y}(\lambda-s-\tau_0) \oplus V_{T_Y}(\lambda-s-\gamma) \oplus V_{T_Y}(\lambda-s-\gamma-\delta)) \geq t+3$. But each of these $T_Y$ weight spaces lies in $V_{T_A}(\lambda-r-\alpha)$, contradicting the given bound.

This completes the proof of (2.14).□

<u>(2.15).</u> Assume Hypothesis (G) and let rank $A = 2$, with $L_A' = \langle U_{\pm\beta}\rangle$. Assume also that $(A,p)$ is not special and not of type $(G_2,2)$.

(i) There does not exist $\tau \in \Pi(Y) - \Pi(L_Y)$, with $\tau \neq \gamma$, such that $(\tau,\Sigma L_j) \neq 0$, $(\tau,\Sigma L_\ell) = 0$ for $\ell \neq j$.

(ii) If $Y$ has type $F_4$, assume $p>2$. Then, there does not exist $\tau \in \Pi(Y) - \Pi(L_Y)$ such that $(\tau,\delta) < 0$, $(\tau,\Sigma L_Y') = 0$.

<u>Proof</u>: Let $L_j = \langle U_{\pm\beta_j}\rangle$, for some $\beta_j \in \Pi(L_Y)$. Then $\langle\lambda,\beta_j\rangle \neq 0$, as $V_\gamma(Q_Y) \neq 0$, $\langle\lambda,\gamma\rangle = 0$ and $(\Sigma L_\ell,\gamma) = 0$ for all $\ell \neq j$. Suppose there exists $\tau$ as in (i). Then, Hypothesis (G) implies that $V_\tau(Q_Y) \neq 0$. So by (2.13), $\tau|T_A = q_i\alpha$, for some $i$. However, $f_{\beta_j+\gamma}v^+$, $f_{\beta_j+\gamma+\delta}v^+$ and $f_{\beta_j+\tau}v^+$ are three linearly independent vectors in $(V_{T_A}(\lambda - q_i\beta - q_i\alpha) + [V,Q_A{}^2])/[V,Q_A{}^2]$, contradicting (1.37). Thus, (i) holds. If there exists $\tau$ as in (ii), (2.12) implies $\tau|T_A = 0$. But then $f_{\beta_j+\gamma}v^+$, $f_{\beta_j+\gamma+\delta}v^+$ and $f_{\beta_j+\gamma+\delta+\tau}v^+$ are again three linearly independent vectors in $(V_{T_A}(\lambda - q_i\beta - q_i\alpha) + [V,Q_A{}^2])/[V,Q_A{}^2]$. Again, this produces a contradiction. Thus, (ii) holds.□

We close this chapter with three technical results which apply only when rank A = 2. Hypothesis (G) is no longer necessary.

For (2.16) – (2.18), our only assumptions are that $\Pi(A) = \{\alpha, \beta\}$, so $L_A' = \langle U_{\pm\beta} \rangle$ and $\mu_\beta$ is the fundamental dominant weight corresponding to $\beta$.

(2.16). Suppose $\dim(Q_A/[Q_A,Q_A]) = 2$. Let $\gamma \in \Pi(Y) - \Pi(L_Y)$ such that $Q_A \nleq K_\gamma$ and $L_i$ has type $A_{k_i}$, for some $k_i \geq 1$, for all i such that $(\Sigma L_i, \gamma) \neq 0$. Let $W_i$ denote the natural module for $L_i$, and suppose $W_i|L_A'$ is tensor indecomposable, for all such i. If $\dim(Q_Y/K_\gamma) > 2$ and $-\gamma$ affords $T(L_A')$ weight $q_0\mu_\beta$, then $\gamma|Z_A \neq q_0\alpha$.

Proof: Suppose false. Let $r_i$ be the field twist on the embedding of $L_A'$ in $L_i$. Since, for each i with $(\Sigma L_i, \gamma) \neq 0$, $W_i|L_A'$ is tensor indecomposable, (1.10) implies $W_i|\rho_i(L_A')$ is restricted. Then one checks, using (1.12) for $SL_2$, that for each $s \in \Pi(L_i)$, a nonidentity element from the group $U_s$ occurs in the factorization of $\rho_i(x_\beta(t))$. So, $s|T_A = r_i\beta$, for each such s. (We use (2.5) to see that the p-power is $r_i$.) Thus $U_{-\gamma}$ affords the unique 1-space of $Q_Y/K_\gamma$ with $T(L_A')$ weight $q_0\mu_\beta$, and all other $T(L_A')$ weights in $Q_Y/K_\gamma$ are strictly less than $q_0\mu_\beta$. Also, let $a_i \in \Sigma^+(L_i)$ with $-\Sigma a_i - \gamma \in \Sigma^-(Y)$, and such that $\Sigma a_i + \gamma$ has maximum height with these properties. Set $r_0 = \Sigma a_i + \gamma$. Then, $U_{-r_0}$ affords the unique 1-space of $Q_Y/K_\gamma$ with $T(L_A')$ weight $-q_0\mu_\beta$. Thus, $x_{-\alpha}(t) = x_{-\gamma}(c_1t^{q_0})u_1$ and $x_{-\alpha-\beta}(t) = x_{-r_0}(c_2t^{q_0})u_2$, for $c_i \in k^*$ and $u_i \in K_\gamma$. Now since $Q_Y/K_\gamma$ is an irreducible $L_Y'$ module with high weight space $U_{-\gamma}K_\gamma/K_\gamma$, there exists $\delta \in \Pi(L_Y)$ with $[U_\delta, U_{-r_0}] \neq 1$. Clearly, $\delta \in \Pi(L_i)$ for some i with $(\Sigma L_i, \gamma) \neq 0$. The remarks about the factorization of $x_\beta(t)$ imply that a nonidentity element from the group $U_{-r_0+\delta}$ occurs in the expression for $[x_\beta(t), x_{-\alpha-\beta}(t)]$. Since $Q_A' \leq K_\gamma$ and $\dim(Q_A/Q_A') = 2$, a nonidentity element from the group $U_{-r_0+\delta}$ occurs in the factorization of $x_{-\alpha}(t)$. Thus, $-r_0+\delta = -\gamma$ and $\Sigma r_i = \delta$. But then, $Q_Y/K_\gamma \cong U_{-\gamma} \times U_{-\gamma-\delta}$, contradicting $\dim(Q_Y/K_\gamma) > 2$.□

<u>(2.17)</u> Assume (A,p) is not special and $(A,p) \neq (G_2,2)$. Suppose there exists $\gamma \in \Pi(Y) - \Pi(L_Y)$ such that the following hold:

(i) There exists a unique pair $1 \leq i,j \leq r$, with $(\Sigma L_i,\gamma) \neq 0 \neq (\Sigma L_j,\gamma)$.

(ii) $L_i$ is of type $A_1$, $L_j$ is of type $A_k$, for some k.

(iii) $V_{L_m}(-\gamma) \cong W_m$ or $W_m{}^*$, for m = i,j, where $W_m$ is the natural module for $L_m$.

(iv) $W_j|L_A{}'$ is tensor indecomposable.

(v) $Q_A \nleq K_\gamma$.

Then there exists q, a power of p, such that q is the field twist on the embedding of $L_A{}'$ in $L_i$ and in $L_j$ and $\gamma|Z_A = q\alpha$.

<u>Proof</u>: Note that $p > k$. By (2.7), the field twists on the embeddings of $L_A{}'$ in $L_i$ and in $L_j$ are equal. Let q be the associated power of p. Let $\mu_\beta$ be the fundamental dominant weight corresponding to the root $\beta$. Then the $L_A{}'$ composition factors of $Q_Y/K_\gamma$ have high weights $\{q(k+1)\mu_\beta, q(k-1)\mu_\beta\}$. If $k+1 < p$, there exists a unique $L_A{}'$ composition factor isomorphic to a twist of $Q_A{}^\alpha$; moreover, this twist is q and the result follows from (2.4). If $k+1 = p$, the $L_A{}'$ composition factors of $Q_Y/K_\gamma$ have high weights $\{pq\mu_\beta, q(k-1)\mu_\beta\}$. If A has type $B_2$ or $G_2$ and $\beta$ is short the prime restrictions and the above argument imply the result. If $\beta$ is long, the result holds unless $\gamma|Z_A = pq\alpha$. But since $-\gamma$ affords $T(L_A{}')$ weight $pq\mu_\beta$, (2.16) implies that this cannot occur. Again the result holds.□

<u>(2.18).</u> Assume (A,p) is not special and $(A,p) \neq (G_2,2)$. Let $\gamma,\delta \in \Pi(Y) - \Pi(L_Y)$ be such that $\delta$ is long when $p = 2$ and the following hold:

(i) $(\gamma,\delta) < 0$, $V_\gamma(Q_Y) \neq 0$ and $Q_A \leq K_\delta$.

(ii) There exists a unique i (respectively, j) such that $(\Sigma L_i,\gamma) \neq 0$ $((\Sigma L_j,\delta) \neq 0)$.

(iii) $L_i$, $L_j$ and $W_m$ are as in (2.17) (ii) and (iii).

(iv) $V_{L_i}(-\gamma) \cong W_i$ or $W_i^*$ and $V_{L_j}(-\delta) \cong W_j$ or $W_j^*$.

(v) $W_j|L_A'$ is tensor indecomposable.

Then, there exists q, a power of p, such that q is the field twist on the embedding of $L_A'$ in $L_i$ and in $L_j$ and $(\gamma+\delta)|Z_A = q\alpha$.

Proof: Note that $Q_Y/K_\gamma$ is a 2-dimensional irreducible $L_A'$ module containing a nontrivial image of $Q_A{}^\alpha$, so the prime restrictions imply that $\beta$ is long. If $-\delta$ is involved in $L_A'$, then (2.10) implies k is even and $\delta|Z_A = 0$. Let $P_Y{}^\wedge \geq B_Y{}^-$ be the parabolic of Y with Levi factor $L_Y{}^\wedge = \langle L_Y, U_{\pm\delta}\rangle$. Then $P_A \leq P_Y{}^\wedge$, $Q_A \leq R_u(P_Y{}^\wedge) = Q_Y{}^\wedge$ and $Z_A \leq Z(L_Y{}^\wedge)$. The bound on $\dim V_\gamma(Q_Y{}^\wedge)$ implies that $p > k = 2$. But since $p > 2$, $2\mu_\beta|T(L_A')$ is not linked to the zero weight in the sense of (1.33); so we may assume $-\delta$ is not involved in $L_A'$. Then (2.11) implies that there is a nontrivial image of $Q_A{}^\alpha$ in $Q_Y(\gamma,\delta)$. The field twists on the embeddings of $L_A'$ in $L_i$ and in $L_j$ are equal, else $Q_Y(\gamma,\delta)$ is a tensor decomposable irreducible $L_A'$ module of dimension greater than 2. Call this twist q. Then, $\gamma|T_A = q\alpha$. As in the previous result, we are done unless $k+1 = p$ and $(\gamma+\delta)|Z_A = pq\alpha$.

So suppose $k+1 = p$ and $(\gamma+\delta)|Z_A = pq\alpha$. Let $L_i = \langle U_{\pm\beta_i}\rangle$ and $r \in \Sigma^+(L_j)$ the root of maximal height. Examining the $T(L_A')$ weight vectors in $Q_Y(\gamma,\delta)$, we have $x_{-\alpha}(t) = x_{-\gamma}(c_1t^q)x_{-\gamma-\delta}(c_2t^{pq})w_1$ and $x_{-\alpha-\beta}(t) = x_{-\gamma-\beta_i}(c_3t^q)x_{-\beta_i-\gamma-\delta-r}(c_4t^{pq})w_2$, where $c_i \in k^*$ and $w_i \in K = \langle U_{-t} \mid t = \Sigma n_\tau\tau, \tau \in \Pi(Y), n_\gamma>1$ or $n_\delta>1$ or $n_\varepsilon>0$ for some $\varepsilon \in \Pi(Y) - \Pi(L_Y)$, $\varepsilon\neq\gamma,\delta\rangle$. We have used here the fact that there is a unique $L_A'$ composition factor of $Q_Y(\gamma,\delta)$ isomorphic to $(Q_A{}^\alpha)^{pq}$.

Now $x_{\beta_i}(t^q)$ occurs in the factorization of $x_\beta(t)$. This observation, together with the earlier assumption about the factorization of $x_\beta(t)$, implies that there is a nontrivial contribution to the root group $U_{-\gamma-\delta-r}$ in the expression for $[x_{-\alpha-\beta}(t), x_\beta(t)]$, which must occur in the factorization of $x_{-\alpha}(t)$ due to the restrictions on the characteristic. But this contradicts the given factorization of $x_{-\alpha}(t)$.□

# CHAPTER 3: $Y = F_4$ or $G_2$

In this chapter, we consider the main problem where $A \leq Y$ are simple algebraic groups, with Y simply connected, having root system of type $G_2$ or $F_4$. Let $V = V(\lambda)$ be an irreducible kY module and let $T_A$, $T_Y$, $\lambda_i$, $\mu_i$ be as in (2.0). We first note that the following results were obtained in [12]:

<u>Theorem (7.1)</u> (in [12]): If rank$A <$ rank$Y$ in Y of type $G_2$ and $V|A$ is irreducible then $A = PSL_2$, $\lambda|T_Y = \lambda_1$, $\lambda|T_A = 6\mu_1$ and $p \neq 2,3,5$.

<u>Theorem (4.1)</u> (in [12]): If rank$A =$ rank$Y$, then $V|A$ is irreducible if and only if the following conditions hold:

(i) p=2 when $Y = F_4$ or $p = 3$ when $Y = G_2$, and

(ii) $\Sigma(A)$ is a subsystem of $\Sigma(Y)$ containing all long roots (respectively, all short roots) of $\Sigma(Y)$, and $\langle\lambda,\alpha\rangle = 0$ for all $\alpha \in \Pi(Y)$ with $\alpha$ short (respectively, long).

The remaining cases are handled in the following

<u>Theorem (3.0).</u> (a) If Y has type $G_2$ and $p \neq 2,3,5$, there exists a subgroup $A \leq Y$, $A \cong PSL_2$, such that $V(\lambda_1)|A$ is a restricted 7-dimensional irreducible.

(b) If Y has type $F_4$, rank$A <$ rank$Y$, and $V|A$ is irreducible, then A has type $G_2$, p=7 and $\lambda|T_A = 2\mu_1$, $\lambda|T_Y = \lambda_4$.

(c) Let Y have type $F_4$ with $p = 7$. Then there exists a closed, connected subgroup $B < Y$, B of type $G_2$, with $V(\lambda_4)|B$ irreducible.

<u>Proof of (3.0)(a) and (c):</u> The proof of (c) is contained in [16]. For

(a), let Y be a simple algebraic group of type $G_2$ with $\Pi(Y) = \{\alpha_1,\alpha_2\}$, labelled as throughout, with irreducible module $V = V(\lambda_1)$. Assume $p \neq 2,3,5$ and consider the subgroup $A = \langle x_\alpha(t), x_{-\alpha}(t) \mid t \in k\rangle$, where $x_\alpha(t) = x_{\alpha_1}(6t)x_{\alpha_2}(10t)x_{\alpha_1+\alpha_2}(30t^2)x_{2\alpha_1+\alpha_2}(120t^3)x_{3\alpha_1+\alpha_2}(-540t^4)\cdot x_{3\alpha_1+2\alpha_2}(-2160t^5)$ and $x_{-\alpha}(t) = x_{-\alpha_1}(t)x_{-\alpha_2}(t)x_{-\alpha_1-\alpha_2}(-\frac{1}{2}t^2)\cdot x_{-2\alpha_1-\alpha_2}((1/3)t^3)x_{-3\alpha_1-\alpha_2}((1/4)t^4)x_{-3\alpha_1-2\alpha_2}((-1/10)t^5)$. Considering first the action of $L(Y)$ on V, we obtain the following description of the root groups of $G_2$ (in $SL_7$), where we use $E_{ij}$ to mean the matrix whose $k\ell$ entry is $\delta_{ik}\delta_{j\ell}$ and write I for the identity matrix:

$x_{\alpha_1}(t) = I + t(E_{12} + 2E_{34} - E_{45} + E_{67}) - t^2E_{35};$

$x_{-\alpha_1}(t) = I + t(E_{21}+ E_{43} - 2E_{54} + E_{76}) - t^2E_{53};$

$x_{\alpha_2}(t) = I - t(E_{23} + E_{56}); x_{-\alpha_2}(t) = (x_{\alpha_2}(t))^t;$

$x_{\alpha_1+\alpha_2}(t) = I + t(E_{13} - 2E_{24} - E_{46} - E_{57}) + t^2E_{26};$

$x_{-\alpha_1-\alpha_2}(t) = I + t(E_{31} - E_{42} - 2E_{64} - E_{75}) + t^2E_{62};$

$x_{2\alpha_1+\alpha_2}(t) = I + t(2E_{14} + E_{25} + E_{36} - E_{47}) - t^2E_{17};$

$x_{-2\alpha_1-\alpha_2}(t) = I + t(E_{41} + E_{52} + E_{63} - 2E_{74}) - t^2E_{71};$

$x_{3\alpha_1+\alpha_2}(t) = I + t(E_{15} - E_{37}); x_{-3\alpha_1-\alpha_2}(t) = (x_{3\alpha_1+\alpha_2}(t))^T;$

$x_{3\alpha_1+2\alpha_2}(t) = I + t(E_{16} + E_{27}); x_{-3\alpha_1-2\alpha_2}(t) = (x_{3\alpha_1+2\alpha_2}(t))^T.$

Let P be the diagonal matrix diag(360,60,−12,−3,2,−12,−1). Then one checks that $P^{-1}x_\alpha(t)P$ (respectively, $P^{-1}x_{-\alpha}(t)P$) is the $7 \times 7$ matrix obtained by considering the action of $\begin{pmatrix}1 & t\\ 0 & 1\end{pmatrix}$ (respectively, $\begin{pmatrix}1 & 0\\ t & 1\end{pmatrix}$) on the irreducible 7-dimensional module with ordered basis $\{x^6,x^5y,x^4y^2,x^3y^3, x^2y^4,xy^5,y^6\}$ and action $\begin{pmatrix}a & b\\ c & d\end{pmatrix}(x^iy^j) = (ax+cy)^i(bx+dy)^j$. Hence, $A \cong PSL_2$, $V|A$ is irreducible with the correct high weight and (3.0)(a) holds.□

For the remainder of this chapter, let $Y = F_4$. The proof of (3.0)(b) involves a straightforward reduction to the case where rank$A = 2$ and then a detailed study of the possible embeddings of a maximal parabolic subgroup of A. We adopt Notation and Hypothesis (2.0), and note that since all components of $L_Y{}'$ are necessarily of classical type, (1.5) implies

$Z_A \leq Z_Y$.

Suppose p=2. Then there is a surjection (isomorphism of abstract groups) $\varphi$: $Y \longrightarrow Y$. (See Section 10 of [14].) We may consider V as a module for $\varphi^{-1}(Y)$.

<u>(3.1).</u> Assume V is p-basic. Then $V|\varphi^{-1}(Y)$ is an algebraic conjugate of a basic module. (See (1.8) for the definition of basic and p-basic.)

<u>Proof</u>: This follows from (2.2) of [12].□

<u>Hypothesis</u>: If p=2, we will assume $\lambda|T_Y$ has short support; i.e., $V|Y$ is basic. For if $\lambda$ has both long and short support, $V|Y$ is tensor decomposable, by (1.7). If $\lambda$ has long support, the above remarks and (3.1) give rise to a configuration $(\varphi^{-1}(A),\varphi^{-1}(Y),V)$, where V is a conjugate of a basic module. But then, we reduce to $(\varphi^{-1}(A),\varphi^{-1}(Y),W)$, where W is basic.

<u>(3.2).</u> If rank A = 3 and $L_A{}'$ is of type $B_2$ ( = $C_2$), then $\dim V^1(Q_A) = 1$.

<u>Proof</u>: Since $P_Y$ is minimal, $L_Y{}' = \langle U_{\pm\beta_2}, U_{\pm\beta_3}\rangle$. Also, if p > 2, $Q_Y/K_{\beta_1}$ and $Q_Y/K_{\beta_4}$ are nonisomorphic irreducible $L_A{}'$ modules. Thus, $Q_A \leq K_{\beta_1}$ or $Q_A \leq K_{\beta_4}$ and (2.3) implies $\dim V^1(Q_Y) = \dim V^1(Q_A) = 1$. Thus, if $\dim V^1(Q_A) > 1$, p = 2 and $\lambda|T_Y = \lambda_3$ or $\lambda_3+\lambda_4$. Using induction to determine the possible labellings of $V|A$, we see that either $\lambda|T_Y = \lambda_4$, with $\lambda|T_A = q\mu_2$, $q\mu_3$ or $q_1\mu_1 + q\mu_3$, or $\lambda|T_Y = \lambda_3+\lambda_4$, with $\lambda|T_A = q\mu_1 + q\mu_2$ or $q\mu_3$. (Here q and $q_1$ are distinct p-powers.) But Table 1 of [5] implies that $\dim V|A < \dim V|Y$ in each case. Hence, $\dim V^1(Q_A) = 1$.□

<u>(3.3).</u> If rank A = 3 and $L_A{}'$ is of type $A_2$, then $\dim V^1(Q_A) = 1$.

<u>Proof</u>: Suppose false; i.e. suppose $\dim V^1(Q_A) > 1$. We first note that size restrictions imply that $L_Y{}'$ is a simple algebraic group. If p >2, (2.14) implies $L_Y{}'$ is not of type $A_2$. Thus, the Main Theorem of [12] and

the above remarks imply $p = 3$ and $L_Y'$ is of type $B_3$. However, the $L_A'$ composition factors of $Q_Y/K_{\beta_4}$ have dimensions 1 and 7, while $\dim(Q_A{}^{\alpha}) = 3$ or 6. Thus, $Q_A \leq K_{\beta_4}$. But $V_{\beta_4}(Q_Y) \neq 0$, contradicting (2.3). Now if $p = 2$, induction and the labelling of $V|Y$ imply $L_Y' = \langle U_{\pm\beta_3}, U_{\pm\beta_4}\rangle$. Using (3.2), (1.23) and Table 1 of [5], we find that $\dim V|A < \dim V|Y$ in every case. Hence, $\dim V^1(Q_A) = 1$.□

The above results imply rank $A = 2$. For the remainder of this chapter we use the following

Notation: Let $\Pi(A) = \{\alpha, \beta\}$ and $\Pi(L_A) = \{\beta\}$. Let $\mu_\alpha$ (respectively, $\mu_\beta$) denote the fundamental dominant weight corresponding to $\alpha$ (respectively, $\beta$). We will also use $\mu_\beta$ to mean $\mu_\beta|T(L_A')$.

(3.4). There are no examples $(A,Y,V)$ in the main theorem with $p = 2$, $A$ simple, $Y$ of type $F_4$ and rank $A <$ rank $Y$.

Proof: Suppose false. Then (3.2) and (3.3) imply rank $A = 2$. By [8], $\dim V|Y = 26, 246$ or $4096$ and $\dim V|A = 8^k \cdot 3^\ell, 4^m, 6^r \cdot 14^s \cdot 64^t$, where $A$ has type $A_2, B_2, G_2$ respectively and $k,\ell,m,r,s,t \in \mathbb{Z}^+$. So $\lambda|T_Y = \lambda_3 + \lambda_4$. If $A$ has type $A_2$, $\lambda|T_A = (q_1 + q_2 + q_3 + q_4)(\mu_1 + \mu_2)$ for distinct $p$-powers $q_1, q_2, q_3$ and $q_4$. Thus, for a fixed maximal parabolic $P_A$, $\dim V^1(Q_A) = 16$. But there is no parabolic $P_Y$ of $Y$ with $\dim V^1(Q_Y) = 16$, contradicting [13]. Thus, $A$ must have type $B_2$ or $G_2$.

Suppose $A$ has type $B_2$. Let $P_A$ be a maximal parabolic of $A$ with $\dim V^1(Q_A) > 1$. By the main theorem of [12], $\dim V^1(Q_A) = 2, 4$ or $8$. In fact since $\dim V|A = 4096$, we must have $\dim V^1(Q_A) = 8$, else $\dim V|A < \dim V|Y$. So $P_Y$ (as in Hypothesis and Notation (2.0)) has type $B_3$. However, $Q_Y/K_{\beta_4}$ is then an 8-dimensional irreducible $L_A'$ module and hence cannot contain a nontrivial image of $Q_A{}^{\alpha}$, contradicting (2.3).

Finally, we must consider the case where $A$ has type $G_2$. The above remarks imply $\lambda|T_A = (q_1 + q_2)(\mu_1 + \mu_2)$ for $q_1$ and $q_2$ distinct $p$-powers.

Now let $P_A$ be the maximal parabolic of $A$ with Levi factor $L_A = \langle U_{\pm\beta}\rangle T_A$ where $\beta$ is long. By induction, $P_Y$ has Levi factor of type $B_2$. However, $Q_Y/K_{\beta_4}$ is then a 4-dimensional irreducible $L_A{}'$ module which cannot contain a nontrivial image of $Q_A{}^{\alpha}$. But this contradicts (2.3).□

<u>(3.5).</u> (i) If $\langle \lambda,\beta\rangle \neq 0$ and $L_Y{}'$ is quasisimple, then $(A,p) = (G_2,3)$, $L_Y{}'$ has type $A_1$ and $\Pi(L_Y) \neq \{\beta_3\}$ or $A$ has type $B_2$, $L_Y{}' = \langle U_{\pm\beta_1}, U_{\pm\beta_2}\rangle$ and $\beta$ is short.

(ii) Assume $\beta$ is long, unless $(A,p) = (G_2,3)$, in which case $\beta$ is arbitrary. If $L_Y{}' = L_1 \times L_2$, for $L_i$ a simple algebraic group with $V^1(Q_Y) = M_1 \otimes M_2$ where $M_i$ is an irreducible $kL_i$ module, then at most one of $M_1$ and $M_2$ is nontrivial.

<u>Proof</u>: Suppose $\langle \lambda,\beta\rangle \neq 0$ and $L_Y{}'$ is a simple algebraic group. Since $p > 2$, if rank $L_Y{}' = 1$, there exists $\gamma \in \Pi(Y) - \Pi(L_Y)$ such that $V_\gamma(Q_Y) \neq 0$ and $\dim(Q_Y/K_\gamma) = 2$. So if $(A,p) \neq (G_2,3)$, $\beta$ must be long. Also, (3.4) and (2.14) imply that either $(A,p) = (G_2,3)$ or $L_Y{}' = \langle U_{\pm\beta_3}\rangle$. If $L_Y{}' = \langle U_{\pm\beta_3}\rangle$, then $V_{\beta_2}(Q_Y) \neq 0$ so $Q_A \not\leq K_{\beta_2}$. But $Q_Y/K_{\beta_2}$ is a 3-dimensional irreducible $L_A{}'$ module, while $\dim(Q_A K_{\beta_2}/K_{\beta_2}) = 2$. Thus, if $\mathrm{rank}(L_Y{}') = 1$, $(A,p) = (G_2,3)$ and $L_Y{}' \neq \langle U_{\pm\beta_3}\rangle$. If $\mathrm{rank}(L_Y{}') = 2$, we use the condition that each $Q_Y/K_\gamma$ have an $L_A{}'$ composition factor of dimension $\dim(Q_A{}^{\alpha})$, for $\gamma \in \Pi(Y) - \Pi(L_Y)$ with $V_\gamma(Q_Y) \neq 0$. This results in the second configuration of (i).

If $L_Y{}'$ has type $B_3$, the $L_A{}'$ composition factors of $Q_Y/K_{\beta_4}$ are of dimensions 1 and 7. Hence, $L_Y{}'$ is not of type $B_3$. Thus, $L_Y{}'$ has type $C_3$. Since $Q_A \not\leq K_{\beta_1}$, $Q_Y/K_{\beta_1}$ must be a reducible $L_A{}'$ module. Hence, by (7.1) of [12], $V^1(Q_Y)$ is isomorphic to the natural module for $L_Y{}'$. Moreover, $\langle \lambda,\beta_1\rangle = 0$ else the bound on $\dim V_{\beta_1}(Q_Y)$ is exceeded. Thus, $\lambda | T_Y = \lambda_4$ and $\dim V|Y \leq 26$. However, $\langle \lambda,\beta\rangle = 5\cdot q_1$ or $q_1 + 2q_2$, for $q_1$ and $q_2$ distinct $p$-powers. Applying (1.23) when $A$ has type $A_2$, we see that $\dim V|A > \dim V|Y$. Contradiction. This completes the proof of (i).

Let $L_i$ and $M_i$ be as in (ii), and suppose $M_1$ and $M_2$ are both nontrivial. Let $q_i$ be the field twist on the embedding of $L_A'$ in $L_i$ for i = 1,2. (This is well-defined as $L_i$ has type $A_1$ or $A_2$.) Then $q_1 \neq q_2$ by (2.5) and (2.6). Suppose there exists $\gamma \in \Pi(Y) - \Pi(L_Y)$ such that $(\gamma, \Sigma L_i) \neq 0$ for i = 1, 2. Examining the $L_A'$ composition factors of $Q_Y/K_\gamma$ and recalling that $p > 2$, we reduce to $\Pi(L_Y) = \{\beta_1, \beta_3, \beta_4\}$. Temporarily assume $(A,p) \neq (G_2,3)$. The $L_A'$ composition factors of $Q_Y/K_{\beta_2}$ have high weights $(q_1 + 4q_2)\mu_\beta$ and $q_1\mu_\beta$; if $p = 3$ and $q_1 = 3q_2$, the composition factors have high weights $(2q_1 + q_2)\mu_\beta$, $q_2\mu_\beta$ and $q_1\mu_\beta$. Thus, if $p \neq 3$, $\beta_2|Z_A = q_1\alpha$. If $p = 3$ and $q_1 = 3q_2$, $\beta_2|Z_A = q_1\alpha$ or $q_2\alpha$. We notice that a nonidentity element from the set $U_{-\beta_2} \cdot U_{-\beta_1-\beta_2}$ must occur in the factorization of some element in $Q_A - Q_A'$, since $\langle \lambda, \beta_1 \rangle \neq 0$. Since $-\beta_2$ (respectively, $-\beta_1-\beta_2$) affords $T(L_A')$ weight $(q_1 + 4q_2)\mu_\beta$ (respectively, $(-q_1 + 4q_2)\mu_\beta$), (2.4) implies $p = 3$, $q_1 = 3q_2$ and $\beta_2|Z_A = q_2\alpha$. We also note that a nonidentity element from the set $U_{-\beta_2} \cdot U_{-\beta_2-\beta_3} \cdot U_{-\beta_2-\beta_3-\beta_4}$ must occur in the factorization of some element in $Q_A - Q_A'$, since $M_2$ is nontrivial. However, $-\beta_2-\beta_3$ (respectively, $-\beta_2-\beta_3-\beta_4$) affords $T(L_A')$ weight $(q_1 + 2q_2)\mu_\beta = 5q_2\mu_\beta$ $(q_1\mu_\beta = 3q_2\mu_\beta)$, contradicting $\beta_2|Z_A = q_2\alpha$.

Now suppose $(A,p) = (G_2,3)$ with $\Pi(L_Y) = \{\beta_1, \beta_3, \beta_4\}$ and $q_1$, $q_2$ as above. Consider the action of $L_A'$ on the 25-dimensional restricted irreducible kY module, $V(\lambda_4)$. There is a 6-dimensional $L_A'$ composition factor with high weight $(\lambda-\beta_2-\beta_3-\beta_4)|T(L_A)$. However, there is no $L_A'$ module of dimension 25 affording such an $L_A'$ composition factor.

Now suppose $\Pi(L_Y) = \{\beta_1, \beta_4\}$ and assume $(A,p) \neq (G_2,3)$. Then $Q_A \nleq K_{\beta_k}$, for k = 1,4, and we find that $\beta_2|Z_A = q_1\alpha$ and $\beta_4|Z_A = q_2\alpha$ (or vice versa). But this contradicts (2.8). If $(A,p) = (G_2,3)$, again consider the action of $L_A'$ on $V(\lambda_4)$. There is a 4-dimensional $L_A'$ composition factor with high weight $(\lambda-\beta_2-2\beta_3-\beta_4)|T(L_A')$. Now argue as before to produce a contradiction.

(3.6). Assume $\beta$ is long unless $(A,p) = (G_2,3)$, in which case $\beta$ is arbitrary. If $\langle\lambda,\beta\rangle \neq 0$, one of the following holds:

(a) $(A,p) \neq (G_2,3)$, $\Pi(L_Y) = \{\beta_1,\beta_3\}$, $\langle\lambda,\beta_1\rangle \leq 2$, $\langle\lambda,\beta_2+\beta_3\rangle = 0$.

(b) $(A,p) \neq (G_2,3)$, $\Pi(L_Y) = \{\beta_1,\beta_3\}$, $\langle\lambda,\beta_1\rangle = 0 = \langle\lambda,\beta_2\rangle$.

(c) $\Pi(L_Y) = \{\beta_1,\beta_4\}$, $\langle\lambda,\beta_1\rangle\cdot\langle\lambda,\beta_4\rangle = 0$ and $Q_A \nleq K_{\beta_2}$ and $Q_A \nleq K_{\beta_3}$.

(d) $(A,p) = (G_2,3)$, $\Pi(L_Y) = \{\beta_1,\beta_3\}$ and $\langle\lambda,\beta_1\rangle\cdot\langle\lambda,\beta_3\rangle = 0$.

(e) $\Pi(L_Y) = \{\beta_1,\beta_3,\beta_4\}$, $\lambda|T_Y = \lambda_4$, $(A,p) \neq (G_2,3)$.

(f) $\Pi(L_Y) = \{\beta_1,\beta_2,\beta_4\}$, $\lambda|T_Y = c\lambda_4$, $c \leq 2$, $(A,p) \neq (G_2,3)$.

(g) $\Pi(L_Y) = \{\beta_1,\beta_2,\beta_4\}$, $\lambda|T_Y = \lambda_2$, $(A,p) \neq (G_2,3)$.

(h) $(A,p) = (G_2,3)$, $L_Y'$ has type $A_1$ and $\Pi(L_Y) \neq \{\beta_3\}$.

Proof: By (3.5), either (h) holds or $L_Y'$ has type $A_1 \times A_1$ or $A_1 \times A_2$ with only one component acting nontrivially on $V^1(Q_Y)$.

Case 1: Suppose $L_Y'$ is of type $A_1 \times A_1$.

Applying (1.15), we see that $\Pi(L_Y) \neq \{\beta_2,\beta_4\}$. Consider the case $\Pi(L_Y) = \{\beta_1,\beta_3\}$ and $(A,p) \neq (G_2,3)$. We claim that $\langle\lambda,\beta_2\rangle = 0$. Otherwise, a nonidentity element from the root group $U_{-\beta_2}$ occurs in the factorization of some element of $Q_A - Q_A'$. But then $\beta_2|Z_A = q\alpha$, where $-\beta_2$ affords $T(L_A')$ weight $q\mu_\beta$, for some p-power q. This contradicts (2.16). Finally, the condition $\langle\lambda,\beta_1\rangle \leq 2$ in (a) follows by considering the bound on $\dim V_{\beta_2}(Q_Y)$ given in (1.25) and the $L_Y'$ composition factor in $V_{\beta_2}(Q_Y)$ afforded by $f_{\beta_1+\beta_2}v^+$. Now, if $\Pi(L_Y) = \{\beta_1,\beta_4\}$, $p > 2$, (2.10) and (2.11) imply that $Q_A \nleq K_{\beta_k}$ for $k = 2,3$. This completes the consideration of Case 1.

Case 2: Suppose $L_Y'$ is of type $A_1 \times A_2$.

Assume for now that if $A = G_2$, then $p \neq 3$. If $\Pi(L_Y) = \{\beta_1,\beta_3,\beta_4\}$, (1.35), (1.36) and the bound on $\dim V_{\beta_2}(Q_Y)$ given in (1.25) imply that $\lambda|T_Y = \lambda_4$. If $\Pi(L_Y) = \{\beta_1,\beta_2,\beta_4\}$, (1.36) and the bound on $\dim V_{\beta_3}(Q_Y)$ imply that $\lambda|T_Y = \lambda_1$, $\lambda_2$ or $c\lambda_4$, for $c \leq 2$. Recall that $\lambda|T_Y \neq \lambda_1$.

It remains to consider the case where $A = G_2$, $p = 3$ and $L_Y'$ has type $A_1 \times A_2$. Note that when $p = 3$, $G_2$ irreducibles have dimensions $7^k\cdot 27^\ell$ for

$k, \ell \in \mathbb{Z}^+$. Suppose $\Pi(L_Y) = \{\beta_1,\beta_3,\beta_4\}$. If $\langle\lambda,\beta_1\rangle \neq 0$, (1.36) and the bound on $\dim V_{\beta_2}(Q_Y)$ of (1.25) imply $\langle\lambda,\beta_2\rangle = 0$. Since $\lambda|T_Y \neq \lambda_1$, we have $\lambda|T_Y = 2\lambda_1$. However, by [8], $\dim V|A \neq \dim V|Y$. Hence, $\langle\lambda,\beta_1\rangle = 0$. Again the bound on $\dim V_{\beta_2}(Q_Y)$ and (1.35) imply $\langle\lambda,\beta_2\rangle = 0$. So $\lambda|T_Y = \lambda_3$ or $\lambda_4$ and by [8] $\dim V|A \neq \dim V|Y$. So $\Pi(L_Y) \neq \{\beta_1,\beta_3,\beta_4\}$.

Finally, we note that if $(A,p) = (G_2,3)$, with $\Pi(L_Y) = \{\beta_1,\beta_2,\beta_4\}$, then in the action of $L_A'$ on the 25-dimensional irreducible $kY$ module, $V(\lambda_4)$, there are no 1-dimensional $L_A'$ composition factors. But every 25-dimensional $kA$ module has 1-dimensional $L_A'$ composition factors.

This completes the proof of (3.6).

<u>(3.7).</u> A is not of type $A_2$.

<u>Proof</u>: Suppose false; choose $\beta$ such that $\langle\lambda,\beta\rangle \neq 0$. Then (2.9) rules out the configuration described in (c) of (3.6). Apply (1.23) in the remaining cases to obtain a precise description of $\lambda|T_A$. Then (1.26) and the methods of (1.30) and (1.32) imply $\dim V|A < \dim V|Y$ in the configurations of (a), (f) and (g). If $\lambda|T_Y = \lambda_4$ and $\lambda|T_A = q(2\mu_\alpha + 2\mu_\beta)$ as in (e), we must use (1.33) and Table 1 of [5] to see that the $\dim V|A \neq \dim V|Y$. Thus, we have reduced to $\lambda|T_Y = c\lambda_3 + x\lambda_4$, for $0<c<p$, $0 \leq x < p$ and $\lambda|T_A = q(c\mu_\alpha + c\mu_\beta)$, the configuration implied by (b) of (3.6). However, $\dim V^2(Q_Y) > \dim V^2(Q_A)$. Contradiction.□

<u>(3.8).</u> If A is of type $B_2$ and $\beta$ is short, then $\langle\lambda,\beta\rangle = 0$.

<u>Proof</u>: Suppose false; i.e., suppose $\langle\lambda,\beta\rangle \neq 0$. Then (3.4), (3.5) and (7.1) of [12] imply that each component, $L_i$, of $L_Y'$ has type $A_{k_i}$ for some $k_i$ and if $k_i > 1$, $M_i$ is isomorphic to the natural module (or dual) for $L_i$. However, we also have $h_\beta(-1) \in Z(A) \leq Z(Y) = 1$; so in fact, $L_Y'$ is quasisimple and by (3.5), $L_Y' = \langle U_{\pm\beta_1}, U_{\pm\beta_2}\rangle$. In this case, $Q_AK_{\beta_3}/K_{\beta_3} = Q_Y/K_{\beta_3}$ and (2.3) and (2.13) imply $\langle\lambda,\beta_3+\beta_4\rangle = 0$. Thus, $\lambda|T_Y = \lambda_2$. Also, (3.6) implies $\lambda|T_A = 2q\mu_\beta + cq_0\mu_\alpha$, for some p-powers

$q$ and $q_0$ and $c = 0$ or 2. By (1.26) and [8], $\dim V|A \leq 140 < \dim V|Y$. Contradiction.□

(3.9). A is not of type $B_2$.

Proof: Suppose false. Then (3.8) implies that if $\langle\lambda,\beta\rangle \neq 0$, then $\beta$ is long. So (3.6) gives the possible Levi factors, $L_Y$, of $P_Y$. Using (1.26) to obtain an upper bound for $\dim V|A$, we see that (3.6)(b) or (c) holds. If (3.6)(b) holds, so $\lambda|T_Y = c\lambda_3 + x\lambda_4$, for $p>c>0$, $p>x\geq 0$ and $\lambda|T_A = cq\mu_\beta$ for $q$ a $p$-power, then $\dim V^2(Q_Y) > \dim V^2(Q_A)$. Thus, (3.6)(c) holds.

Suppose $\lambda|T_Y = c\lambda_1 + x\lambda_2 + y\lambda_3$, for $c > 0$, $x,y \geq 0$ and $\lambda|T_A = cq\mu_\beta$, for $q$ a $p$-power. We first claim that $x = 0 = y$. For otherwise, applying (2.3) and (2.13), we find that $\beta_2|T_A = q_0\alpha$ or $\beta_3|T_A = q_0\alpha$, for $q_0$ some $p$-power. Thus, $f_{\beta_2}v^+$ or $f_{\beta_3}v^+$ is a nonzero vector in $V_{T_A}(\lambda - q_0\alpha)$. But $\langle\lambda,\alpha\rangle = 0$. So we now have $\lambda|T_Y = c\lambda_1$. It requires an easy check to see that $\dim V^2(Q_Y) = \dim V^2(Q_A)$ in this case. Thus $[V,Q_A{}^2] = [V,Q_Y{}^2]$. Also, $\dim V^3(Q_A) \leq 2c$. But if $c > 1$, $f_{2\beta_1+2\beta_2}v^+$ and $f_{\beta_1+\beta_2+\beta_3}v^+$ afford $L_Y'$ composition factors in $V^3(Q_Y)$ of dimensions $c-1$ and $2c$, respectively. Thus, $c = 1$. But then $\dim V|A < \dim V|Y$.

Thus, it remains to consider $L_Y' = \langle U_{\pm\beta_1}\rangle \times \langle U_{\pm\beta_4}\rangle$, $\lambda|T_Y = x\lambda_2 + y\lambda_3 + c\lambda_4$, $\lambda|T_A$ as above. But the same argument as above implies $x = 0 = y$ and $c = 1$. So $\dim V|A < \dim V|Y$. Contradiction.

(3.10). Let A be of type $G_2$. Then $p \neq 3$.

Proof: Suppose false; choose $\beta \in \Pi(A)$ such that $\langle\lambda,\beta\rangle \neq 0$. Then (3.6) implies $L_Y' = \langle U_{\pm\beta_k}\rangle$ for $k = 1, 2$ or 4 or $L_Y' = \langle U_{\pm\beta_i}\rangle \times \langle U_{\pm\beta_j}\rangle$ for $\{i,j\} = \{1,3\}$ or $\{1,4\}$. In the latter case, $\langle\lambda,\beta_i\rangle = 0$ or $\langle\lambda,\beta_j\rangle = 0$. In particular, if $\langle\lambda,\beta_\ell\rangle \leq 1$ for all $\ell$, then $\dim V|A = 7$ or 49 and $\dim V|A \leq 27^2$ in any case. Now [8] implies $\lambda|T_Y \neq \lambda_k$ for any $k$. Also, $\lambda|T_Y \neq 2\lambda_k$ for any $k$. For otherwise, [8] and (3.5) imply $\lambda|T_Y = 2\lambda_2$. But then (1.38) implies $\dim V|A < \dim V|Y$. Using (1.32) we see that $\dim V|Y > 49$. So there exists $1 \le j \le 4$ with $\langle\lambda,\beta_j\rangle = 2$. Let $W_0$ be the stabilizer in $W$ of $\lambda$. If rank $W_0 = 1$, counting only the conjugates of $V_{T_Y}(\lambda)$ and $V_{T_Y}(\lambda-\beta_j)$ (where $\langle\lambda,\beta_j\rangle = 2$) we see that $\dim V|Y > 27^2$. Thus, $\mathrm{rank}(W_0) = 2$. It is now a check to see that $\dim V|Y > 27^2$ in every case, completing the proof of (3.10).□

<u>(3.11).</u> Let $A$ be of type $G_2$ with short root $\beta$. If $\langle\lambda,\beta\rangle \neq 0$, then $L_Y' = \langle U_{\pm\beta_1}\rangle \times \langle U_{\pm\beta_3}, U_{\pm\beta_4}\rangle$, $p = 7$ and $\lambda|T_Y = \lambda_4$ or $\lambda|T_Y = 2\lambda_1$.

<u>Proof</u>: By (3.5) and (3.10), $L_Y'$ is not a simple algebraic group. Consider first the case where $L_Y'$ has type $A_1 \times A_2$. If $\gamma \in \Pi(Y) - \Pi(L_Y)$, $Q_Y/K_\gamma$ contains an $L_A'$ composition factor isomorphic to a twist of $Q_A{}^\alpha$ only if the field twists on the embeddings of $L_A'$ in the two components are equal. Also $p > 2$ implies $V_\gamma(Q_Y) \neq 0$ for $\gamma \in \Pi(Y) - \Pi(L_Y)$. Thus, (2.3), (2.5) and (2.6) imply that only one component acts nontrivially on $V^1(Q_Y)$.

Suppose $\Pi(L_Y) = \{\beta_1,\beta_3,\beta_4\}$. In the action of $L_A'$ on the 26-dimensional $kY$-module with high weight $\lambda_4$, there will be a 5-dimensional composition factor. The only 26-dimensional $kA$ module affording this is a conjugate of the irreducible $kA$ module with high weight $2\mu_\beta$ when $p = 7$; hence, the prime restriction of the result. Continuing with $L_Y'$ as above, using (1.36) and the bound on $\dim V_{\beta_2}(Q_Y)$ given in (1.25), we reduce to the configurations of the result.

Consider now the case where $\Pi(L_Y) = \{\beta_1,\beta_2,\beta_4\}$. Since $Q_A \nleq K_{\beta_3}$, the field twists on the embeddings of $L_A'$ in $\langle U_{\pm\beta_4}\rangle$ and in $\langle U_{\pm\beta_1}, U_{\pm\beta_2}\rangle$ are equal. Consider the action of $L_A'$ on the 26- dimensional irreducible $kY$ module, $V(\lambda_4)$. One checks that there are two 4-dimensional, three 3-dimensional, four 2-dimensional and one 1- dimensional $L_A'$ composition factors. But there is no 26-dimensional $kA$ module affording such an $L_A'$ composition series. Hence $\Pi(L_Y) \neq \{\beta_1,\beta_2,\beta_4\}$.

It remains to show that $L_Y'$ does not have type $A_1 \times A_1$. Since

$\dim(Q_A{}^{\alpha}) = 4$, $\Pi(L_Y) \neq \{\beta_1,\beta_4\}$. If $\Pi(L_Y) = \{\beta_2,\beta_4\}$ (respectively, $\{\beta_1,\beta_3\}$), $Q_A \leq K_{\beta_1}$ (respectively, $K_{\beta_4}$). But then $p > 2$, (2.10) and (2.11) produce a contradiction. This completes the proof of (3.11).

(3.12). Let A be of type $G_2$ with $\beta$ long and $\alpha$ short. Then $\langle\lambda,\alpha\rangle \neq 0$.

Proof: Suppose $\langle\lambda,\alpha\rangle = 0$. Then $\langle\lambda,\beta\rangle \neq 0$ for $\beta$ the long root. Recall, by (3.4) and (3.10), $p > 3$. Thus, (3.6) and (1.10) imply that $V|A$ is basic. Consider the configuration of (3.6)(c); the above remarks imply that $\beta_2|T_A = \alpha = \beta_3|T_A$, while $\beta_1|T_A = \beta = \beta_4|T_A$. So $\langle\lambda,\beta_2+\beta_3\rangle = 0$ since $\langle\lambda,\alpha\rangle = 0$. Now, $V_{T_A}(\lambda-2\beta_1-2\beta_2)\oplus V_{T_A}(\lambda-\beta_1-\beta_2-\beta_3-\beta_4) \oplus V_{T_A}(\lambda-\beta_2-\beta_3-2\beta_4)\oplus V_{T_A}(\lambda-2\beta_3-2\beta_4)\oplus V_{T_A}(\lambda-2\beta_1-\beta_2-\beta_3) \leq V_{T_A}(\lambda-2\alpha-2\beta)$. If $\langle\lambda,\beta\rangle > 1$ (so $\langle\lambda,\beta_1\rangle > 1$ or $\langle\lambda,\beta_4\rangle > 1$), the last weight space has dimension 2 while the sum of weight spaces in $V|Y$ has dimension 3. So $\langle\lambda,\beta\rangle = 1$ and $\dim V|A < \dim V|Y$. Thus, (3.6)(c) does not hold.

Using (1.30), (1.26) and [8], we may argue that $\dim V|A \neq \dim V|Y$ in ther configurations of (3.6) (a), (e), (f) and (g). Thus, the only possible configuration is as described in (3.6) (b); $L_Y{}' = \langle U_{\pm\beta_1}\rangle \times \langle U_{\pm\beta_3}\rangle$, $\lambda|T_Y = c\lambda_3 + x\lambda_4$, $c > 0$, $x \geq 0$ and $\lambda|T_A = c\mu_\beta$. However, $\dim V_{\beta_2}(Q_Y) + \dim V_{\beta_4}(Q_Y)$ exceeds the bound on $\dim V^2(Q_A)$. Contradiction.

Proof of (3.0)(b): Under the hypotheses of (3.0)(b), results (3.2) - (3.12) imply that $A = G_2$ and $p > 3$. Let $\beta \in \Pi(A)$ be short, so by (3.12), $\langle\lambda,\beta\rangle \neq 0$. As well, by (3.11), $\Pi(L_Y) = \{\beta_1,\beta_3,\beta_4\}$, $p = 7$ and $\lambda|T_Y = \lambda_4$ or $\lambda|T_Y = 2\lambda_1$. In each case, $\langle\lambda,\beta\rangle = 2q$, for q a p-power. If $\lambda|T_Y = \lambda_4$, $\dim V|Y = 26$, so $\langle\lambda,\alpha\rangle = 0$, else the methods of (1.30), (1.32) and (1.35) imply $\dim V|A > \dim V|Y$. Then (1.10) implies the result of (3.0)(b). If $\lambda|T_Y = 2\lambda_1$, then $\langle\lambda,\alpha\rangle \neq 0$, else by (1.26) and [8] $\dim V|A < \dim V|Y$. Thus (3.6) implies that $\langle\lambda,\alpha\rangle = 2q_0$, for a p-power $q_0$. But then $\dim V|A \neq \dim V|Y$, by [8]. This completes the proof of (3.0)(b).□

# CHAPTER 4: THE ONE COMPONENT THEOREM

Let $A < Y$ be simple algebraic groups, with $Y$ simply connected, having root system of type $E_n$ and with rank $Y >$ rank $A > 2$. In this chapter, we prove a result which is a useful tool in finding the configurations of the Main Theorem when rank $A > 2$. Throughout this chapter, we adopt Hypothesis and Notation (2.0), with the following additional restrictions: If $A$ has type $B_m$, $C_m$ or $F_4$, then $p>2$. The result we prove is the following:

Theorem 4.0. If, in addition to the above hypotheses, $L_Y'$ is a simple algebraic group of type $A_k$ or $D_k$, for some $k$, and if $\dim V^1(Q_Y) > 1$, then the pair $(A,L_A')$, $(Y,L_Y')$ is one of the following:

(i) $(C_4,C_3)$, $(E_n,A_5)$;

(ii) $(C_4,A_3)$, $(E_6,A_5)$;

(iii) $(F_4,C_3)$, $(E_6,A_5)$;

(iv) $(F_4,B_3)$, $(E_6,D_4)$;

(v) $(C_5,C_4)$, $(E_8,A_7)$; or

(vi) $(C_3,C_2)$, $(E_8,D_5)$, $p=5$, $L_Y' = \langle U_{\pm\beta_i} | 1 \le i \le 5\rangle$, and $\lambda | T_Y = \lambda_1$.

Moreover, in each case, if $\rho: L_A \to L_Y$ is the natural homomorphism, the pairs $\rho(L_A') \le L_Y'$ occur as natural embeddings of classical groups.

Remarks. 1. Under the hypotheses of Theorem (4.0), we see that $V_\gamma(Q_Y) \ne 0$ for all $\gamma \in \Pi(Y) - \Pi(L_Y)$ such that $(\gamma, \Sigma L_Y) \ne 0$. (See (1.24).)

2. The hypotheses of Theorem (4.0) and (1.5) imply that $Z_A \le Z_Y$.

(4.1). Under the hypotheses of Theorem (4.0), $L_A' \not\cong L_Y'$.

Proof: Suppose false. Let q be the field twist on the embedding of $L_A'$ in $L_Y'$. Then, since distinct $L_Y'$-irreducibles restrict to distinct $L_A'$ irreducibles, $Q_Y/K_\gamma \cong (Q_A{}^\alpha)^q$ as $L_A'$-modules, for all $\gamma \in \Pi(Y)-\Pi(L_Y)$ such that $(\gamma,\Sigma L_Y) \neq 0$. In particular, if $\gamma_1$, $\gamma_2$ are two such roots, $-\gamma_1|(T_Y \cap L_Y') = -\gamma_2|(T_Y \cap L_Y')$. Also, (2.14) implies that there does not exist $\delta \in \Pi(Y)-\Pi(L_Y)$ with $(\delta,\Sigma L_Y) \neq 0$. Thus, $L_Y'$ must be a maximal parabolic of Y. So rank A = rank Y. Contradiction.□

(4.2). Assume $L_Y'$ has type $A_k$ for some k, W is the natural module for $L_Y'$, and $W|L_A' \cong (Q_A{}^\alpha)^q$ or $W^*|L_A' \cong (Q_A{}^\alpha)^q$ for some p-power q. If $\dim V^1(Q_Y) > 1$, then (4.0) (i) or (v) holds.

Proof: By (4.1), A does not have type $A_m$, for any m and if A is of type $B_m$, $L_A'$ does not have type $A_{m-1}$.

(4.2.1) $(A,L_A')$ is not of type $(B_3,B_2)$.

Proof: Suppose false. By assumption, p>2 and $L_Y' = A_4$. Theorem (8.1) of [12] implies that there does not exist $\gamma \in \Pi(Y)-\Pi(L_Y)$ with $(\gamma,\Sigma L_Y) \neq 0$ such that $Q_Y/K_\gamma \cong W\wedge W$ or $W^*\wedge W^*$. Hence $Y = E_7$ or $E_8$ and $Q_Y/K_\gamma \cong W$ or $W^*$ for all $\gamma \in \Pi(Y)-\Pi(L_Y)$ with $(\gamma,\Sigma L_Y) \neq 0$. So in particular, $Q_A K_\gamma/K_\gamma = Q_Y/K_\gamma$ for such $\gamma$. Thus, if q is the field twist on the embedding of $L_A'$ in $L_Y'$, (2.13) implies $\gamma|T_A = q\alpha$ for all $\gamma \in \Pi(Y)-\Pi(L_Y)$ such that $(\gamma,\Sigma L_Y) \neq 0$. In addition, (2.12) implies $\tau|T_A = 0$ for all $\tau \in \Pi(Y)-\Pi(L_Y)$ such that $(\tau,\Sigma L_Y) = 0$. Finally, note that for all $\beta \in \Pi(L_Y)$, $\beta|T_A = q\eta$ for some $\eta \in \Pi(L_A)$.

By Theorem (8.1) of [12], $V^1(Q_Y) \cong W$ (or $W^*$) or $V^1(Q_Y) \cong W\wedge W$ (or $W^*\wedge W^*$). Thus, $\langle\lambda,\alpha\rangle \neq 0$, else by (1.26) and (1.32), $\dim V|A < \dim V|Y$. So there exists $\gamma \in \Pi(Y)-\Pi(L_Y)$ such that $(\gamma,\Sigma L_Y) \neq 0$ and $\langle\lambda,\gamma\rangle \neq 0$, else there is no vector in $V|Y$ with $T_A$ weight $\lambda - q_0\alpha$. (See the preceeding work describing the restriction of $\beta_i$ to $T_A$ for $1 \le i \le n$.) These remarks, together with (2.13), imply that $\Pi(L_Y) = \{\beta_4,\beta_5,\beta_6,\beta_7\}$ with $\langle\lambda,\beta_2\rangle \neq 0$ if

$Y = E_7$ or $\langle\lambda,\beta_2+\beta_8\rangle \neq 0$ if $Y = E_8$. We now argue that $V^1(Q_Y) \cong W$ or $W^*$. For if $V^1(Q_Y) \cong W \wedge W$ or $W^* \wedge W^*$, we first pass to the parabolic $P_Y\hat{}$ of (2.12), where $P_Y\hat{}$ has Levi factor $L_Y\hat{} = \langle L_Y, U_{\pm\beta_1}\rangle$. If $\langle\lambda,\beta_2\rangle \neq 0$, then $\langle\lambda,\beta_5\rangle = 1$, else $f_{\beta_2}v^+$ and $f_{3456}v^+$ afford $(L_Y\hat{})'$ composition factors of $V^2(Q_A)_{\lambda-q\alpha}$, exceeding the bound of (1.22). With $\langle\lambda,\beta_5\rangle = 1$, $f_{\beta_2}v^+$ affords an $(L_Y\hat{})'$ composition factor of $V^2(Q_A)_{\lambda-q\alpha}$ of dimension 40 unless p=3, in which case the composition factor has dimension 30. (See (1.34).) As well, $f_{345}v^+$ affords an $(L_Y')\hat{}$ composition factor of dimension 20. Hence, p=3. But then (1.34) implies that $\dim V_{T_Y}(\lambda-\beta_2-\beta_4-\beta_5) = 3$, while the multiplicity of this weight in the first composition factor mentioned is only 1. So the bound on $\dim V^2(Q_A)_{\lambda-q\alpha}$ is exceeded. We may argue similarly if $Y = E_8$ with $\langle\lambda,\beta_8\rangle \neq 0$. Hence, $V^1(Q_Y) \cong W$ or $W^*$.

Using (1.34) carefully, as above, we find that the bound on $\dim V^2(Q_A)_{\lambda-q\alpha}$ is again exceeded unless $Y = E_8$ with $\lambda|T_Y = \lambda_7+y\lambda_8$, for some $y>0$. Now there does not exist a vector in $V|Y$ with weight $\lambda-q_0\alpha$, for $q_0\neq q$, so $V|A$ is a conjugate of a restricted module and hence by (1.10), $\lambda|T_A = a\mu_1+\mu_2$ for some $0<a<p$. Since $((f_{\beta_8})^y)v^+ \in V_{T_A}(\lambda-y\alpha_1)$, $y \leq a$. But if $y<a$, there is no vector in $V|Y$ with $T_A$ weight $\lambda-a\alpha_1$. So $y = a$. Now, let $L_0 = \langle U_{\pm\beta_k} \mid 3\leq k\leq 8\rangle$, a group of type $A_6$ with natural subgroup, B, of type $B_3$. We note that $v^+$ affords an $L_0$ composition factor of V which restricts to B to produce a composition factor with the same high weight as $B_3$ module as $V|A$. But $L_0$ lies in a proper parabolic of Y and hence acts reducibly on V. Thus, $\dim V|A < \dim V|Y$. Contradiction.

<u>(4.2.2).</u> $(A,L_A')$ is not of type $(D_4,A_3)$.

<u>Proof</u>: Suppose false; then $L_Y' = A_5$. Let $\Pi(L_A) = \{\alpha_1,\alpha_2,\alpha_3\}$, so $\Pi(A) - \Pi(L_A) = \{\alpha_4\}$. Then there does not exist $\gamma \in \Pi(Y)-\Pi(L_Y)$ with $(\gamma,\Sigma L_Y) \neq 0$, $Q_Y/K_\gamma \cong W\wedge W$ (or $W^*\wedge W^*$). This follows from Theorem (8.1) of [12] if p > 2. If p=2, though $(W\wedge W)|L_A'$ is reducible, one checks that there is no $L_A'$ composition factor of $W\wedge W$ isomorphic to a twist of $Q_A{}^\alpha$. Let q be the field twist on the embedding of $L_A'$ in $L_Y'$. Consider

first the case where $\Pi(L_Y) = \{\beta_1, \beta_3, \beta_4, \beta_5, \beta_6\}$. Then, $Q_Y/K_{\beta_2} \cong W\wedge W\wedge W$ as $L_Y'$-modules. One checks that the $L_A'$ composition factors of $W\wedge W\wedge W$ have high weights $2q\mu_1$ and $2q\mu_3$. Thus, p=2, as $Q_A \nleq K_{\beta_2}$. So by the Main Theorem of [12], $V^1(Q_Y) \cong W$ or $W^*$. By (1.25), dim $V_{\beta_2}(Q_Y) \le 36$, so $\langle\lambda,\beta_2\rangle = 0$. Moreover, (2.13) implies that $\langle\lambda,\beta_7\rangle = 0 = \langle\lambda,\beta_8\rangle$ if $Y = E_8$; and recall that $\lambda|T_Y \neq \lambda_1$ if $Y = E_7$. Thus, one of the following holds:

(a) $Y = E_6$, $\lambda|T_Y = \lambda_1$ (or $\lambda_6$).

(b) $Y = E_7$, $\lambda|T_Y = \lambda_1 + \lambda_7$ or $\lambda|T_Y = \lambda_6 + x\lambda_7$. (c) $Y = E_8$ and $\lambda|T_Y = \lambda_1$ or $\lambda_6$.

Referring to Table 1 of [5], we see that $\dim V|A = 8^k \cdot 26^\ell \cdot 160^m$, for $k,\ell,m \in \mathbb{Z}^+$. So [8] implies that (a) does not hold, if (b) holds $\lambda|T_Y = \lambda_6+\lambda_7$, and if (c) holds $\lambda|T_Y = \lambda_6$. By induction and Theorem (7.1) of [12], $\langle\lambda,\alpha_4\rangle = 0$, $q_1$, or $q_1+q_2$, for $q_1$ and $q_2$ distinct p-powers. But then $\dim V|A < \dim V|Y$, by (1.32) and (1.38). Thus, $L_Y' \neq \langle U_{\pm\beta_i} \mid i = 1,3,4,5,6\rangle$.

It remains to consider the case where $L_Y' = \langle U_{\pm\beta_i} \mid 4 \le i \le 8\rangle$. Let q be as above. Then (2.12) and (2.4) imply $\beta_1|T_A = 0$ and $\beta_3|T_A = q\alpha_4 = \beta_2|T_A$. One checks that $\beta_4|T_A = q\alpha_2 = \beta_8|T_A$, $\beta_5|T_A = q\alpha_i = \beta_7|T_A$, for i = 1 or 3, respectively, and $\beta_6|T_A = q(\alpha_3 - \alpha_1)$ or $q(\alpha_1 - \alpha_3)$, respectively. Also, by (2.13), $\langle\lambda,\beta_3\rangle = 0$.

Now, Theorem (8.1) of [12] implies $V^1(Q_Y) \cong W$, $W^*$, $W\wedge W$, or $W^*\wedge W^*$ (the latter two only if $p\neq 2$). Thus, $\langle\lambda,\alpha_4\rangle \neq 0$, else (1.26) and (1.32) imply that dim $V|A <$ dim $V|Y$. So, in particular, $\lambda - q_0\alpha_4$ is a $T_A$ weight in $V|A$, for some p-power $q_0$. The above remarks imply that $\langle\lambda,\beta_2\rangle \neq 0$, else there is no vector in $V|Y$ with $T_A$ weight $\lambda - q_0\alpha_4$.

We now argue carefully, using (1.34) and the parabolic $P_Y\hat{}$ of (2.12) (as in the proof of (4.3)), to see that the bound on $\dim V^2(Q_A)_{\lambda - q\alpha_4}$ is exceeded in every configuration. This completes the proof of (4.2.2).

<u>(4.2.3).</u> $(A, L_A')$ is not of type $(C_3,C_2)$.

<u>Proof</u>: Suppose false; then $L_Y' = A_3$. Examining $L_Y'$ composition

factors of $W\wedge W$, we see that there does not exist $\gamma \in \Pi(Y)-\Pi(L_Y)$ such that $(\gamma,\Sigma L_Y) \neq 0$ and $Q_Y/K_\gamma \cong W\wedge W$. Thus, $Q_Y/K_\gamma \cong W$ or $W^*$ for all $\gamma \in \Pi(Y)-\Pi(L_Y)$ with $(\gamma,\Sigma L_Y) \neq 0$. So $Q_Y/K_\gamma = Q_AK_\gamma/K_\gamma$ for all such $\gamma$. Let $\Pi(A) = \{\alpha_1,\alpha_2,\alpha_3\}$. Thus, if q is the field twist on the embedding of $L_A'$ in $L_Y'$, (2.13) implies $\gamma|T_A = q\alpha_1$ for all $\gamma \in \Pi(Y)-\Pi(L_Y)$ such that $(\gamma,\Sigma L_Y) \neq 0$. In addition, (2.12) implies $\tau|T_A = 0$ for all $\tau \in \Pi(Y)-\Pi(L_Y)$ such that $(\tau,\Sigma L_Y) = 0$. If $\Pi(L_Y) = \{r_1,r_2,r_3\}$ with $(r_i,r_{i+1}) < 0$, for $i = 1,2$, then $r_1|T_A = q\alpha_2 = r_3|T_A$ and $r_2|T_A = q\alpha_3$. Now, there exists at most one $\gamma \in \Pi(Y)-\Pi(L_Y)$ with $(\gamma,\Sigma L_Y) \neq 0$ and $\langle\lambda,\gamma\rangle \neq 0$, else $\dim V_{T_A}(\lambda-q\alpha_1) > 1$, contradicting (1.31). So, in fact, $\langle\lambda,\alpha_1\rangle = cq$ for some $0 \le c < p$, as there does not exist a $T_Y$ weight restricting to $\lambda - q_0\alpha_1$ for $q_0 \neq q$. So by (1.10), $q=1$.

The restriction $Q_Y/K_\gamma \not\cong W\wedge W$ implies $\Pi(L_Y)$ is (a) $\{\beta_1,\beta_3,\beta_4\}$, (b) $\{\beta_4,\beta_5,\beta_6\}$, (c) $\{\beta_5,\beta_6,\beta_7\}$, or (d) $\{\beta_6,\beta_7,\beta_8\}$. In (a), (b), (c), (d), respectively, $\langle\lambda,\beta_4\rangle = 0$, $\langle\lambda,\beta_4\rangle = 0$, $\langle\lambda,\beta_5\rangle = 0$, $\langle\lambda,\beta_6\rangle = 0$, respectively. Otherwise, $\dim V_{T_A}(\lambda-\alpha_1-\alpha_2) > 2$, contradicting (1.29).

If $\langle\lambda,\alpha_1\rangle = 0$, then Theorem (8.1) of [12] implies $\lambda|T_A = k\mu_2$, for $0<k<p$ or $\lambda|T_A = a\mu_2 + b\mu_3$, where $a\neq 0 \neq b$ and $a+b = p-1$. In the first case, $\dim V_{T_A}(\lambda-\alpha_1-2\alpha_2-\alpha_3) \le 3$, and in the second case, $\dim V_{T_A}(\lambda-\alpha_1-\alpha_2-\alpha_3) \le 2$. (See (1.29).) Thus, the information in the above 2 paragraphs implies that $Y = E_6$, $L_Y'$ is as in (a) or (b) and $\lambda|T_Y = k\lambda_1$ or $k\lambda_6$, respectively. However, this configuration is ruled out by a direct application of (1.23). Thus, $\langle\lambda,\alpha_1\rangle \neq 0$.

Now, $V_{T_A}(\lambda-\alpha_1) \neq 0$ implies there exists $\gamma \in \Pi(Y)-\Pi(L_Y)$, with $(\gamma,\Sigma L_Y) \neq 0$, and $\langle\lambda,\gamma\rangle \neq 0$. Recall that there exists only one such $\gamma$. Applying these remarks, (2.13) and symmetry, we restrict still further to:

(A) $\Pi(L_Y) = \{\beta_1,\beta_3,\beta_4\}$, $\langle\lambda,\beta_2\rangle \neq 0$, $\langle\lambda,\beta_i\rangle = 0$, $i \ge 4$.

(B) $\Pi(L_Y) = \{\beta_4,\beta_5,\beta_6\}$, $Y = E_7$, $\langle\lambda,\beta_2+\beta_7\rangle \neq 0$, $\langle\lambda,\beta_1+\beta_3+\beta_4\rangle = 0$.

(C) $\Pi(L_Y) = \{\beta_5,\beta_6,\beta_7\}$, $Y = E_8$, $\langle\lambda,\beta_8\rangle \neq 0$, $\langle\lambda,\beta_i\rangle = 0$, $1 \le i \le 4$.

(D) $\Pi(L_Y) = \{\beta_4,\beta_5,\beta_6\}$, $Y = E_8$, $\langle\lambda,\beta_2\rangle \neq 0$, $\langle\lambda,\beta_1+\beta_3+\beta_4+\beta_7+\beta_8\rangle = 0$.

Note, (1.23) rules out Y of type $E_6$, entirely.

By Theorem (8.1) of [12], and the above work, $\lambda|T_A = y\mu_1 + k\mu_2$, for $0<y,k<p$, or $\lambda|T_A = y\mu_1 + a\mu_2 + b\mu_3$, for $0<y,a,b<p$, $a+b = p-1$. We use (1.29) to check that $\dim V_{T_A}(\lambda-\alpha_1-\alpha_2) \leq 2$, $\dim V_{T_A}(\lambda-2\alpha_1-\alpha_2) \leq 2$, and $\dim V_{T_A}(\lambda-\alpha_1-\alpha_2-\alpha_3) \leq 4$. Also, $\dim V_{T_A}(\lambda-\alpha_1-2\alpha_2-\alpha_3) \leq 5$, if $\lambda|T_A = y\mu_1 + k\mu_2$.

Consider the configuration of (B) when $\lambda|T_Y = b\lambda_5 + a\lambda_6 + x\lambda_7$, $0<a,b,x<p$, $a+b = p-1$. Then $f_{245}v^+$, $f_{345}v^+$, $f_{1345}v^+$, $f_{\beta_7}f_{45}v^+$ and $f_{567}v^+$ are five linearly independent vectors in $V_{T_A}(\lambda-\alpha_1-\alpha_2-\alpha_3)$, contradicting the given bound. Using (1.34) to argue in this manner for each case, we reduce to Y of type $E_7$, $\lambda|T_Y = k\lambda_6 + x\lambda_7$, $\lambda|T_A = y\mu_1 + k\mu_2$, for $0<k,y,x<p$.

We now argue that $x=y$. Since $0 \neq (f_{\beta_7})^x v^+ \in V_{T_A}(\lambda- x\alpha_1)$, $x\leq y$. Moreover, there does not exist a vector in $V|Y$ with $T_A$ weight $\lambda - z\alpha_1$ where $z>x$. For if $0\neq w$ is a $T_Y$ weight vector with weight $\lambda - \Sigma c_i\beta_i$ and $c_7 > x$, then $c_6 > 0$. So $y\leq x$.

Let $X = \langle U_{\pm\beta_3},U_{\pm\beta_4},U_{\pm\beta_5},U_{\pm\beta_6},U_{\pm\beta_7}\rangle$. Then X contains a natural subgroup, $C \leq X$, of type $C_3$. The X composition factor of V afforded by $v^+$ has dimension strictly less than $\dim V|Y$, since X is contained in the Levi factor of a proper parabolic of Y. But the C composition factor of V afforded by $v^+$ has the same high weight as $V|A$, as $C_3$ module. Thus, $\dim V|A < \dim V|Y$. Contradiction. This completes the proof of (4.2.3).

<u>(4.2.4).</u> $(A, L_A')$ is not of type $(C_3,A_2)$.

<u>Proof</u>: Suppose false. By assumption, $p>2$ and $L_Y' = A_5$. By the Main Theorem of [12], $L_A'$ acts irreducibly on $W\wedge W$ and on $W^*\wedge W^*$. One checks that $W\wedge W\wedge W$ has no 6-dimensional $L_A'$ composition factor. Thus, $Y = E_8$ and $L_Y' = \langle U_{\pm\beta_i} \mid 4\leq i\leq 8\rangle$. Let q be the field twist on the embedding of $L_A'$ in $L_Y'$.

Now, one checks that $Q_AK_{\beta_2}/K_{\beta_2} = Q_Y/K_{\beta_2}$ forces an embedding of $L_A'$ in $L_Y'$ which gives the following: $\beta_4|T_A = q\alpha_2 = \beta_5|T_A = \beta_7|T_A$, $\beta_6|T_A = q(\alpha_1 - \alpha_2)$ and $\beta_8|T_A = q\alpha_1$. Also, by (2.13), $\beta_2|T_A = q\alpha_3 = \beta_3|T_A$ and $\beta_1|T_A = 0$. Moreover, (2.13) also implies that $\langle\lambda,\beta_1+\beta_3\rangle = 0$. Thus, either $\langle\lambda,\alpha_3\rangle = 0$ or $\langle\lambda,\beta_2\rangle \neq 0$. If $\langle\lambda,\beta_2\rangle \neq 0$, we argue carefully using (1.34) (as in the proof of (4.2.1)) that the bound on $\dim V^2(Q_A)_{\lambda-q\alpha_3}$ is exceeded in every configuration. Thus, $\langle\lambda,\alpha_3\rangle = 0$. But it is a straightforward check, using induction, (1.26) and (1.32), to see that $\langle\lambda,\alpha_3\rangle = 0$ implies $\dim V|A < \dim V|Y$. This completes the proof of (4.2.4).

It remains to consider the case where $L_Y' = A_7$ and $(A,L_A')$ has type $(D_5,D_4)$ or $(F_4,B_3)$. In the first case, consideration of the quotient $Q_Y/K_{\beta_2}$, in view of Theorem (8.1) of [12], implies that $p = 2$ and $V^1(Q_Y) \cong W$ or $W^*$. But now in each case, the bound on $\dim V_{\beta_2}(Q_Y)$ implies $\lambda|T_Y = \lambda_8$, a contradiction.□

<u>(4.7).</u> Suppose $L_Y'$ is of type $A_k$, with natural module W, and $(Q_A{}^{\alpha})^q \not\cong W$ or $W^*$, as $L_A'$ modules, for any p-power q. If $\dim V^1(Q_Y) > 1$, then Theorem (4.0) (ii) or (iii) holds.

<u>Proof</u>: Since $L_A'$ acts irreducibly on $W \not\cong (Q_A{}^{\alpha})^q$, (2.3) implies that there does not exist $\gamma \in \Pi(Y) - \Pi(L_Y)$, with $(\gamma,\Sigma L_Y) \neq 0$, such that $Q_Y/K_\gamma \cong W$ or $W^*$. In particular, we have $k \geq 4$. In fact, $k>4$. For otherwise, since rank$A > 2$, $L_A'$ must be of type $B_2$ ($= C_2$) in order to have a 5-dimensional irreducible representation. Moreover, since $W \not\cong (Q_A{}^{\alpha})^q$, $A = C_3$. However, there exists $\gamma \in \Pi(Y)-\Pi(L_Y)$, with $(\gamma,\Sigma L_Y) \neq 0$, such that $Q_Y/K_\gamma \cong W\wedge W$ or $W^*\wedge W^*$, a 10-dimensional irreducible $L_A'$ module. (Recall $p \neq 2$ when $A = C_3$.) Thus $Q_A \leq K_\gamma$, contradicting (2.3).

Consider the case where $L_Y' = A_5$. The existence of a 6-dimensional irreducible $L_A'$-module not isomorphic to a twist of $Q_A{}^{\alpha}$ implies $(A,L_A')$ is of type $(A_3,A_2)$, $(B_3,A_2)$, $(A_4,A_3)$, $(B_4,A_3)$, $(C_4,A_3)$, or $(F_4,C_3)$. If $L_A' = A_2$, then $p>2$. For the pairs $(A_3,A_2)$, $(A_4,A_3)$, and $(B_3,A_2)$,

$\dim Q_A{}^{\alpha} < \dim W$. But the Main Theorem of [12] implies that $L_A{}'$ acts irreducibly on $W \wedge W$ and on $W^* \wedge W^*$, unless p=2 and $L_A{}' = A_3$. Even in the latter case, the $L_A{}'$ composition factors of $W \wedge W$ have dimensions 14 and 1. Hence, there does not exist $\gamma \in \Pi(Y) - \Pi(L_Y)$ with $(\gamma, \Sigma L_Y) \neq 0$ and $Q_Y/K_\gamma \cong W \wedge W$ or $W^* \wedge W^*$. Also, by the Main Theorem of [12], $V^1(Q_Y) \cong W$, $W^*$, $W \wedge W$ or $W^* \wedge W^*$. But now a direct application of (1.23) rules out all possible configurations.

For $(A, L_A{}')$ of type $(B_4, A_3)$, the preceeding remarks imply $L_Y{}' = \langle U_{\pm\beta_i} \mid i = 1,3,4,5,6\rangle$. Then, $Q_Y/K_{\beta_2} \cong W \wedge W \wedge W$, as $L_Y{}'$-modules. The $L_A{}'$ composition factors of $W \wedge W \wedge W$ have high weights $2\mu_1$ and $2\mu_3$. But since p>2, this implies $Q_A \leq K_{\beta_2}$, contradicting (2.3). In the $(F_4, C_3)$ and $(C_4, A_3)$ cases, there does not exist an $L_A{}'$ composition factor of $W \wedge W$ isomorphic to a twist of $Q_A{}^{\alpha}$. Thus, there does not exist $\gamma \in \Pi(Y) - \Pi(L_Y)$ with $(\gamma, \Sigma L_Y) \neq 0$ and $Q_Y/K_\gamma \cong W \wedge W$ or $W^* \wedge W^*$. So $Y = E_6$ and we have the result.

Consider now the case where $L_Y{}' = A_6$. Since W is a 7-dimensional irreducible $L_A{}'$-module ( $\not\cong (Q_A{}^{\alpha})^q$ ), either $L_A{}' = A_2$ and p=3, or $(A, L_A{}')$ is of type $(F_4, B_3)$. In the latter case, Theorem (8.1) of [12] produces a contradiction to (2.3). So we consider the case where $L_A{}' = A_2$ and p=3. Let $\Pi(L_A) = \{\alpha_1, \alpha_2\}$. By (1.25), for $\gamma \in \Pi(Y) - \Pi(L_Y)$, $\dim V_\gamma(Q_Y) \leq 21$, if $A = A_3$ or $B_3$, and $\dim V_\gamma(Q_Y) \leq 42$, if $A = C_3$. This restriction implies that either (a) $Y = E_7$ with $\lambda|T_Y = \lambda_7$ or (b) $Y = E_8$ with $\lambda|T_Y = \lambda_2$ and $A = C_3$. (Note that we used the methods of (1.30), (1.32), and (1.35) to find a lower bound for $\dim V_\gamma(Q_Y)$.) In each case, $\lambda|T_A = q(\mu_1 + \mu_2) + x\mu_3$, for some $x \geq 0$ and some p-power q. If $A = A_3$, (1.23) implies that $x = q \cdot 1$. In the configuration of (a), $\dim V|Y = 56$. However, referring to Table 1 of [5], and using the methods of (1.32) and (1.30), we see that $\dim V|A \neq 56$. In the configuration of (b), $Q_Y/K_{\beta_3} \cong W \wedge W$. But $W \wedge W$ has $L_A{}'$ composition factors with high weights $3q\mu_1$, $3q\mu_2$, $q(\mu_1+\mu_2)$ and 0. (We have used (2.6) to identify q with the field twist on the embedding of $L_A{}'$ in $L_Y{}'$.)

Thus, there is no $L_A'$ composition factor of $Q_Y/K_{\beta_3}$ isomorphic to a twist of $Q_A^{\alpha_3}$. But this contradicts (2.3) and (2.4). Thus, $L_Y'$ does not have type $A_6$.

It remains to consider the case where $L_Y' = A_7$. Since W is an 8-dimensional irreducible $L_A'$-module ( $\not\cong (Q_A^{\alpha})^q$ ), $L_A' = A_2$ or $D_4$, or $(A,L_A)$ has type $(B_4,B_3)$. Note that $V^1(Q_Y) \not\cong W$ or $W^*$. For otherwise the bound on $\dim V_{\beta_2}(Q_Y)$ of (1.25) implies that $\lambda|T_Y = \lambda_8$. Thus, the Main Theorem of [12] implies that $A = D_5$, $L_A' = D_4$, and $p>2$. However, then $L_A'$ acts irreducibly on the 56-dimensional $L_Y'$ module $Q_Y/K_{\beta_2}$. But this implies $Q_A \le K_{\beta_2}$, contradicting (2.3). Thus, $L_Y'$ does not have type $A_7$.

This completes the proof of (4.7).□

<u>(4.8).</u> If $L_Y'$ is of type $D_k$, for some $k\ge 4$ and $\dim V^1(Q_Y) > 1$, then Theorem (4.0) (iv) or (vi) holds.

<u>Proof</u>: Let $\mu_1, \ldots, \mu_k$ be the fundamental dominant weights for $D_k$. Then we will call $W = V(\mu_1)$ the natural module for $L_Y'$.

Suppose $L_A'$ acts irreducibly on some $L_Y'$ module other than W. Then the Main Theorem of [12] implies that $L_A'$ acts irreducibly on the $L_Y'$-modules with high weights $\mu_{k-1}$ and $\mu_k$. In every case, there exists $\gamma \in \Pi(Y)-\Pi(L_Y)$ with $(\gamma,\Sigma L_Y) \ne 0$ such that the irreducible $L_Y'$-module $Q_Y/K_\gamma$ has high weight $\mu_{k-1}$ or $\mu_k$. Thus, $Q_A^{\alpha}$ is isomorphic to the $L_A'$ irreducible afforded by $\mu_{k-1}$ or $\mu_k$. The Main Theorem of [12] then implies that $A = F_4$, $L_A' = B_3$, and $L_Y' = D_4$.

With $(A,L_A',L_Y')$ of type $(F_4,B_3,D_4)$, suppose $Y = E_7$ or $E_8$. Let q be the field twist on the embedding of $L_A'$ in $L_Y'$. Then, by (2.13) and (2.6), $\beta_1|T_A = q\alpha_4 = \beta_6|T_A$, $\beta_k|T_A = 0$, and $\langle\lambda,\beta_k\rangle = 0$ for $k>6$. Also, since the embedding of $L_A'$ in $L_Y'$ is the natural embedding of classical groups, $\beta_2|T_A = q\alpha_1$, $\beta_4|T_A = q\alpha_2$ and $\beta_3|T_A = q\alpha_3 = \beta_5|T_A$. We claim that $\langle\lambda,\alpha_4\rangle = \langle\lambda,\beta_1\rangle$. Certainly, $x = \langle\lambda,\beta_1\rangle \le \langle\lambda,\alpha_4\rangle$, as $0 \ne (f_{\beta_1})^x v^+ \in V_{T_Y}(\lambda - xq\alpha_4)$. Also, $\langle\lambda,\beta_1\rangle \ge \langle\lambda,\alpha_4\rangle = y$. For otherwise, there is no vector in $V|Y$ with $T_A$ weight $\lambda - yq\alpha_4$. Now, let $X = \langle U_{\pm\beta_i} | 1 \le i \le 6\rangle$. Then, X contains a natural subgroup of type $F_4$, say $A_0 \le X$. Moreover, the X composition factor of V afforded by $v^+$ is not all of V. But the $A_0$ composition factor of V afforded by $v^+$ has the same dimension as $V|A$. Thus, $\dim V|A < \dim V|Y$. Hence, if $(A, L_A', L_Y')$ has type $(F_4, B_3, D_4)$, $Y = E_6$ and the result holds.

Now consider the case where $V^1(Q_Y) \cong W$ and $L_A'$ acts reducibly on every nontrivial, restricted $L_Y'$ module other than W. The Main Theorem of [12] then implies that the triple $(L_A', L_Y', p)$ is one of $(A_2, D_4, p)$, $(B_2, D_5, 5)$, $(B_2, D_7, 3)$, $(C_3, D_7, 3)$ or $(C_3, D_7, 7)$. We first note that $L_Y'$ does not have type $D_7$. For otherwise, the bound on $\dim V_{\beta_1}(Q_Y)$ implies $\lambda|T_Y = \lambda_8$. As well, $(L_A', L_Y')$ does not have type $(A_2, D_4)$; for one checks that $L_A'$ acts irreducibly on each of the 8-dimensional irreducible $L_Y'$-modules. So $Q_A \le K_{\beta_1}$, contradicting (2.3). Thus, $L_A' = B_2$ and $L_Y' = D_5$, with p=5. For $\gamma \in \Pi(Y) - \Pi(L_Y)$ such that $Q_Y/K_\gamma$ is one of the restricted irreducible spin modules for $L_Y'$, $Q_Y/K_\gamma$ has $L_A'$ composition factors of dimensions 12 and 4. Hence, (2.3) implies that $A = C_3$. Also, by (1.25), $\dim V_\gamma(Q_Y) \le 40$, so $\langle \lambda, \gamma \rangle = 0$. An application of (1.23) then rules out Y of type $E_6$. Moreover, $L_Y' = \langle U_{\pm\beta_i} | 1 \le i \le 5\rangle$, else $Q_Y/K_{\beta_7}$ is a 10-dimensional irreducible $L_A'$ module containing a nontrivial image of $Q_A{}^\alpha$. The above remarks and (2.3) imply $\lambda|T_Y = \lambda_1$. Thus, $Y = E_8$ and the result holds.

This completes the proof of (4.8).□

The proof of Theorem (4.0) is now complete.

## CHAPTER 5: RANK(A) $\geq 3$

In this chapter we establish the Main Theorem under the following conditions: Y has type $E_n$ and rank $A > 2$. We adopt Notation and Hypothesis (2.0) throughout the chapter. Note that Theorem (4.1) of [12] implies rank A < rank Y. Our result is the following.

<u>Theorem (5.0).</u> If $V|A$ is irreducible, then $Y = E_6$ and one of the following holds:

(i) A is the fixed point subgroup of the graph automorphism of Y, so A has type $F_4$, and $\lambda|T_Y = (p-3)\lambda_1$ or $(p-3)\lambda_6$, for $p>3$, or $\lambda|T_Y = \lambda_1 + (p-2)\lambda_3$ or $\lambda_6 + (p-2)\lambda_5$, for $p>2$. Moreover, with A and V as described, $V|A$ is irreducible.

(ii) $p>2$, A has type $C_4$, $\lambda|T_A = \mu_2$ and $\lambda|T_Y = \lambda_1$ or $\lambda_6$.

Moreover, if $p>2$, $Y = E_6$ and B is the fixed point subgroup of the automorphism $\tau i_x$, where $i_x$ is the inner automorphism associated with $x = h_{\beta_1+2\beta_2+2\beta_3+3\beta_4+2\beta_5+\beta_6}(-1)$, then B has type $C_4$, and $V(\lambda_1)|B$ and $V(\lambda_6)|B$ are irreducible with high weight $\mu_2$.

<u>Proof of existence statement in (ii):</u> (due to G. Seitz) It is a check to see that the fixed point subgroup of the automorphism $\tau i_x$ is $B = \langle x_{\pm\beta_1}(t)x_{\pm\beta_6}(t), x_{\pm\beta_3}(t)x_{\pm\beta_5}(t), x_{\pm\beta_4}(t),$ $x_{\pm(\beta_2+\beta_3+\beta_4)}(t)x_{\pm(\beta_2+\beta_4+\beta_5)}(-t) \mid t \in k\rangle$, that this group has type $C_4$ and that $V(\lambda_1)$ and $V(\lambda_6)$ have a B composition factor with high weight $\mu_2$. But then, $p>2$ and Table 1 of [5] imply that this composition factor has dimension 27 and so $V(\lambda_1)|B$ and $V(\lambda_6)|B$ are irreducible.□

(5.1) Assume there are no examples (B,Y,W) satisfying the hypotheses of the Main Theorem with Rank(B) = 3. Also, assume (A,p) is not special. If $L_A'$ is of type $A_3$ and $\dim V^1(Q_A) > 1$, then $L_Y'$ is a simple algebraic group.

Proof. Suppose false; i.e., suppose $L_Y'$ has more than one component. Let $\Pi(L_A) = \{\alpha_1,\alpha_2,\alpha_3\}$, so $\alpha = \alpha_4$. Then by rank restrictions $L_Y'$ has 2 components. In fact $L_Y'$ has type $A_3 \times A_3$, since $P_Y$ is minimal and $A_3$ has no 5-dimensional irreducible representation. Thus, (1.5) implies $Z_A \le Z_Y$. By (2.7), we may assume the field twist on the embedding of $L_A'$ in each component of $L_Y'$ is q, for some p-power q.

Consider the case where $\Pi(L_Y) = \{\beta_1,\beta_3,\beta_4,\beta_6,\beta_7,\beta_8\}$. The $L_A'$ composition factors of $Q_Y/K_{\beta_5}$ have high weights among $\{q(\eta_1+\eta_3),0,2q\eta_1, 2q\eta_3, q\eta_2\}$, where $\eta_1$, $\eta_2$, $\eta_3$ represent the fundamental dominant weights of $A_3$, labelled as throughout. Since there is a nontrivial image of $Q_A{}^\alpha$ in $Q_Y/K_{\beta_5}$, $A = D_4$ or $C_4$, or p=2 and $A = A_4$. Then (1.23) forces $\dim V|A < \dim V|Y$ if p=2 and A has type $A_4$. Thus, $A = D_4$ or $C_4$. Now, $\dim(Q_Y/K_{\beta_2}) < \dim Q_A{}^\alpha$, so $Q_A \le K_{\beta_2}$. But (1.33) implies that $q\eta_i$ and 0 cannot be high weights of an indecomposable $L_A'$ module, for i = 1 or 3. Hence, we may assume that $-\beta_2$ is not involved in $L_A'$. So (2.11) implies that there is a nontrivial image of $Q_A{}^\alpha$ in $Q_Y(\beta_5,\beta_2)$. However, the $L_A'$ composition factors of this $L_Y'$ irreducible have high weights among $\{q(\eta_1+\eta_2), q(\eta_2+\eta_3), q\eta_1, q\eta_3\}$. Thus, $\Pi(L_Y) \ne \{\beta_1,\beta_3,\beta_4,\beta_6,\beta_7,\beta_8\}$.

We now have $\Pi(L_Y) = \{\beta_2,\beta_3,\beta_4,\beta_6,\beta_7,\beta_8\}$. The $L_A'$ composition factors of $Q_Y/K_{\beta_5}$ have high weights among $\{q(\eta_1+\eta_2), \eta_1, q(\eta_2+\eta_3), q\eta_3\}$. Since (2.3) implies that there is a nontrivial image of $Q_A$ in $Q_Y/K_{\beta_5}$, $A = A_4$ or $B_4$. As above, if $Q_A \le K_{\beta_1}$ then $-\beta_1$ is not involved in $L_A'$. But there is no 4-dimensional $L_A'$ composition factor of $Q_Y(\beta_5,\beta_1)$. Hence (2.11) implies $Q_A \not\le K_{\beta_1}$. Since each of the irreducible $L_Y'$ modules $Q_Y/K_{\beta_5}$ and $Q_Y/K_{\beta_1}$ must have an $L_A'$ composition factor isomorphic to a twist of $Q_A{}^{\alpha_4}$, the embedding of $T(L_A')$ in $T(L_Y')$ is as follows: $h_{\alpha_1}(c) = h_{\beta_2}(c^q)\cdot h_{\beta_6}(c^q)$, $h_{\alpha_2}(c) = h_{\beta_4}(c^q)\cdot h_{\beta_7}(c^q)$ and $h_{\alpha_3}(c) = h_{\beta_3}(c^q)\cdot h_{\beta_8}(c^q)$. Considering the $T(L_A')$ weight vectors in $Q_Y/K_{\beta_5}$, we see that $x_{-\alpha_4}(t) = x_{-245}(c_1t^q)x_{-567}(c_2t^q)x_{-456}(c_3t^q)w$, where $c_i \in k$, some $c_i$ nonzero, and $w \in K_{\beta_5}$. So $\beta_5|T_A = q(\alpha_4-\alpha_1-\alpha_2)$. This implies $\langle\lambda,\beta_k\rangle = 0$ for $3\le k\le 6$, else $(V_{T_Y}(\lambda-q\alpha_4+q\alpha_1+q\alpha_2) \oplus V_{T_Y}(\lambda-q\alpha_4+q\alpha_1) \oplus V_{T_A}(\lambda-q\alpha_4+q\alpha_1) \oplus V_{T_A}(\lambda-q\alpha_4-q\alpha_3+q\alpha_1)) \ne 0$. Also, (2.13) implies $\beta_1|T_A = q\alpha_4$.

Suppose A is of type $B_4$, so $p>2$. Then $\langle\lambda,\alpha_i\rangle = 0$ for $i=2,3,4$, else rank restrictions imply there is a parabolic subgroup of Y, $P_0$, containing the $B_3$ parabolic of A with containment of unipotent radicals and such that the Levi factor of $P_0$ is a simple algebraic group. However, Theorem (4.0) and the induction hypothesis imply that no such configuration occurs. Thus $\lambda|T_A = q(a\mu_1)$ for some $0<a<p$. Comparing this with the information in the preceeding paragraph, we have $\lambda|T_Y = x\lambda_1+a\lambda_2$, for some $x\ge 0$. However, $0 \ne f_{\beta_2+\beta_4+\beta_5}v^+ \in V_{T_A}(\lambda-q\alpha_4)$. Contradiction.

So we have reduced to the case where $A = A_4$. Applying (1.23) and (1.10) and the above remarks we find that either $\lambda|T_A = a\mu_1+a\mu_4$ and $\lambda|T_Y = x\lambda_1+a\lambda_2$, for some $a>0$, $x\ge 0$ or $\lambda|T_A = b\mu_2+b\mu_3$ and $\lambda|T_Y = y\lambda_1+b\lambda_7+b\lambda_8$, for some $b>0$, $y\ge 0$. In the first case, $a > 1$, else $\dim V|A < \dim V|Y$. But then $f_{245}f_2v^+$ and $f_{2456}v^+$ are linearly independent vectors in $V_{T_A}(\lambda-\alpha_1-\alpha_4)$, which is a 1-dimensional weight space. In the second case, $0 \ne f_{567}v^+ \in V_{T_A}(\lambda-\alpha_4)$, contradicting $\langle\lambda,\alpha_4\rangle = 0$.

This completes the proof of (5.1).□

<u>(5.2).</u> Let (A,Y,V) be as in the Main Theorem with Y of type $E_n$ and rank$A > 3$. Assume there are no examples (B,Y,W) satisfying the hypotheses of the Main Theorem with rank(B) = 3. Also assume (A,p) is not special. Then Y has type $E_6$ and one of the following holds:

(i) A is of type $C_4$, so $p>2$, $\lambda|T_A = \mu_2$, and $\lambda|T_Y = \lambda_1$ (or $\lambda_6$).

(ii) A is of type $F_4$.

Proof: Choose $P_A$ such that $\dim V^1(Q_A) > 1$. Consider first the case where rankA = 4. By (5.1) and size restrictions, $L_Y'$ has one component. Then, the induction hypothesis implies that Theorem (4.0) applies. In particular, (1.5) implies $Z_A \leq Z_Y$.

Consider the case where $A = C_4$ and $L_A' = A_3$, as in (4.0) (ii). Then $Y = E_6$ and $L_Y' = A_5$. The embedding of $L_A'$ in $L_Y'$ is the natural embedding of $A_3$ in $A_5$. The Main Theorem of [12] implies that one of the following holds:

(a) $\lambda|T_Y = \lambda_1+x\lambda_2$ or $\lambda_6+x\lambda_2$, for $0 \leq x<p$ and $\lambda|T_A = q\mu_2+y\mu_4$, where q is some p-power and $y \geq 0$.

(b) $\lambda|T_Y = x\lambda_2+\lambda_3$ or $x\lambda_2+\lambda_5$ and $\lambda|T_A = q\mu_1+q\mu_3+y\mu_4$, where x, q and y are as in (a).

In either case, the $C_3$ parabolic of A acts nontrivially on $\langle v^+ \rangle$. Thus, Theorem (4.0) and the induction hypothesis imply that the $C_3$ parabolic of A is contained in a conjugate of $P_Y$, with containment of unipotent radicals. Theorem (8.1) in [12] rules out the configuration of (b) and forces $y = 0$ in the configuration of (a). Thus $V|A$ is a conjugate of a basic module (recall that $p>2$ when $A = C_4$), and so by (1.10), $V|A$ has high weight $\mu_2$. Moreover $x=0$, else (1.26) and (1.32) imply that $\dim V|A < \dim V|Y$. Thus $\lambda|T_Y = \lambda_1$ and we have the configuration of (i) in the statement of the result.

Consider now the case where $A = C_4$, $L_A' = C_3$, $Y = E_n$ and $L_Y' = A_5$, as in (4.0) (i). The embedding of $L_A'$ in $L_Y'$ is the natural embedding of $C_3$ in $A_5$. Theorem (8.1) of [12] implies that the $A_3$ parabolic of A acts nontrivially on $\langle v^+ \rangle$, so we may reduce to the first case. Thus, if rankA=4, (i) or (ii) holds.

Suppose rankA = 5 and consider the case where $L_Y'$ is a simple algebraic group of classical type; so by (4.0), $A = C_5$, $L_A' = C_4$, and

$L_Y' = A_7$, as in (4.0) (v). The embedding of $L_A'$ in $L_Y'$ is the natural embedding of classical groups. By (8.1) of [12], the $A_4$ parabolic of A acts nontrivially on $\langle v^+ \rangle$. Thus, size restrictions and the preceeding two paragraphs imply that Theorem (4.0) applies. But there are no examples $(C_4,A_4)$, $(Y,L_Y')$ in Theorem (4.0). Thus, if rankA = 5, $L_Y' = E_6$ or $E_7$. The only possible configuration, given inductively by the preceeding two paragraphs would be: $A = C_5$, $L_A' = C_4$, and $L_Y' = E_6$. However, $Q_Y/K_{\beta_7}$ is a 27-dimensional irreducible $L_A'$ module on which $Z_A$ induces scalars. But then, $Q_AK_{\beta_7}/K_{\beta_7}$ is an 8-dimensional $L_A'$ submodule of $Q_Y/K_{\beta_7}$, by (2.4). Thus rankA $\neq$ 5. Also, by induction and Theorem (4.0), rank A $\neq$ 6,7.

This completes the proof of (5.2).□

<u>(5.3).</u> Suppose (A,Y,V) are as in the Main Theorem and the pair (A,Y) has type $(F_4,E_6)$, with $p>2$. Then V is a basic module for A. Moreover, A is the fixed point subgroup under the graph automorphism of Y.

<u>Proof</u>: Let $P_1$ (respectively, $P_2$) be the maximal parabolic of A, containing $B_A^-$, corresponding to the simple root $\alpha_4$ (respectively, $\alpha_1$). If $P_i = L_iQ_i$ is the Levi decomposition for $P_i$, then $L_1' = B_3$ and $L_2' = C_3$. Rank restrictions imply that Theorem (4.0) applies whenever $\dim V^1(Q_i) > 1$. In particular, Theorem (8.1) in [12] implies that $V^1(Q_i)$ is a tensor indecomposable $L_i'$ module. Thus, if V|A is tensor decomposable, the above remarks and (1.7) imply that $\lambda|T_A = q_1a\mu_1+q_2b\mu_4$, where $p>a,b>0$ and $q_1$ and $q_2$ are distinct p-powers. However, Theorem (8.1) of [12] implies that $V^1(Q_1)$ cannot have high weight $(q_1a\mu_1)|T(L_1')$. Thus, V|A is a tensor indecomposable module. Then $p>2$ and (1.10) imply that V|A is basic. Hence Proposition (2.8) of [12] holds.

Let $P_Y^1$ and $P_Y^2$ be as in Proposition (2.8) of [12]. That is, if $P_Y^i = L_Y^iQ_Y^i$ is the Levi decomposition of $P_Y^i$, then $P_i \leq P_Y^i$, $Q_i \leq Q_Y^i$, $L_i \leq L_Y^i$ and $Z_i = Z(L_i)^\circ \leq Z_Y^i = Z(L_Y^i)^\circ$ for i=1,2. Moreover, the fixed maximal torus $T_Y$ is contained in $L_Y^i$. Now V|Y nontrivial implies that

$\dim V^1(Q_i) > 1$ for $i = 1$ or $2$. Induction and the Main Theorem of [12] imply $\dim V^1(Q_2) > 1$ in every possible configuration, and that $(L_Y{}^2)'$ has type $A_5$.

Choose a base $\Pi(Y) = \{\beta_1, \beta_2, \beta_3, \beta_4, \beta_5, \beta_6\}$ of $\Sigma(Y)$, labelled as throughout, such that $\Pi((L_Y{}^2)) = \{\beta_1, \beta_3, \beta_4, \beta_5, \beta_6\}$ and $Q_Y{}^2 = \langle U_r \mid r \in \Sigma^-(Y) - \Sigma((L_Y{}^2)')\rangle$. Then, $x_{\alpha_2}(t) = x_{\beta_4}(t)$, $x_{\alpha_3}(t) = x_{\beta_3}(t)x_{\beta_5}(t)$ and $x_{\alpha_4}(t) = x_{\beta_1}(t)x_{\beta_6}(t)$. Also, examining the $T(L_2')$ weights in $Q_Y{}^2/K_{\beta_2}$, we see that $x_{-\alpha_1}(t) = x_{-\beta_2}(at)u$, where $a \in k^*$, $u \in K_{\beta_2}$.

We claim that $(L_Y{}^1)'$ has type $D_4$. This follows from the Main Theorem and induction, if $\dim V^1(Q_1) > 1$. So suppose $\dim V^1(Q_1) = 1$. Then, $\lambda|T_Y = a\lambda_1 + x\lambda_2$ and $\lambda|T_A = a\mu_4$. In fact, $x=0$ as there is no nontrivial embedding of $B_3$ in $A_4$. If $(L_Y{}^1)'$ is not of type $D_4$, $(L_Y{}^1)'$ is a conjugate of $\langle U_{\pm\beta_2}, U_{\pm\beta_3}, U_{\pm\beta_4}, U_{\pm\beta_5}, U_{\pm\beta_6}\rangle$, and $\dim V^2(Q_Y{}^1) \geq 16$. However, (1.25) implies $\dim V^2(Q_1) \leq 8$. Thus, $(L_Y{}^1)'$ has type $D_4$, as claimed. Also, $L_1' \leq (L_Y{}^1)'$ must be the natural embedding of classical groups.

So there exists $\{\gamma_1, \gamma_2, \gamma_3, \gamma_4\} \subseteq \Sigma(Y)$ such that $(L_Y{}^1)' = \langle U_{\pm\gamma_1}, U_{\pm\gamma_2}, U_{\pm\gamma_3}, U_{\pm\gamma_4}\rangle$ is of type $D_4$ and such that $x_{-\alpha_1}(t) = x_{-\gamma_1}(b_1 t)$, $x_{\alpha_2}(b_2 t) = x_{\gamma_2}(t)$ and $x_{\alpha_3}(t) = x_{\gamma_3}(b_4 t)x_{\gamma_4}(b_5 t)$, for some $b_i \in k^*$. Comparing this with the known information about the factorization of these elements, we see that $\gamma_1 = \beta_2$, $\gamma_2 = \beta_4$, $\{\gamma_3, \gamma_4\} = \{\beta_3, \beta_5\}$ and $b_4 = b_5$. Then, $A$ is in fact the fixed point subgroup under the graph automorphism of $Y$.□

In the following few results, we will determine on which modules $V$, the fixed point subgroup of the graph automorphism of $Y = E_6$ acts irreducibly. We are forced to do tedious calculations within the universal enveloping algebra of $L(Y)$. Thus, it is necessary to know the structure constants of $L(Y)$; i.e., we need a set of consistent signs for the commutator relations among elements of a Chevalley basis of $L(Y)$. (Since all root strings in $\Sigma(Y)$ have length 1, the constants involved are either 1 or −1.) In Section 4.2 of [6], a method for constructing a set of structure constants is described. It involves choosing a set of "extraspecial" pairs (r,s) of roots. The structure constants, $N_{r,s}$, where $[e_r,e_s] = N_{r,s}e_{r+s}$, may be chosen arbitrarly for these pairs. Then, using Theorem 4.1.2 of [6], one generates the remaining structure constants.

The set of extraspecial pairs is chosen by first fixing a total ordering on the space $\mathcal{V}$ containing the roots. We do this as follows. Let $v_1 = \beta_1$, $v_2 = \beta_3$, $v_3 = \beta_4$, $v_4 = \beta_2$, $v_5 = \beta_5$, $v_6 = \beta_6$. Then, we say $0 \prec \Sigma c_i v_i$ if and only if the first nonzero coefficient $c_i$ is positive. An ordered pair (r,s) is said to be special if $r+s \in \Sigma(Y)$ and $0 \prec r \prec s$. An ordered pair (r,s) is said to be extraspecial if (r,s) is special and if for all special pairs $(r_1,s_1)$ with $r+s = r_1+s_1$, we have $r \preceq r_1$. Then, every root in $\Sigma^+(Y)$ which is the sum of two roots in $\Sigma^+(Y)$ can be uniquely expressed as a sum of an extraspecial pair. We choose $N_{r,s} = 1$ for all extraspecial pairs.

(5.4). Let (A,Y,V) be as in (5.3), so p>2. Then one of the following holds: (i) $\lambda|T_Y = (p-3)\lambda_1$ or $(p-3)\lambda_6$, for p>3.

(ii) $\lambda|T_Y = \lambda_1 + (p-2)\lambda_3$ or $\lambda_6 + (p-2)\lambda_5$, for p>2.

Proof: In view of (5.3) and (1.1), it will suffice to work with the Lie algebras $L(A) \leq L(Y)$. Actually, we do all computations inside of the universal enveloping algebra of L(Y), where we view L(Y) as a subalgebra of its universal enveloping algebra. (See Section 17 of [9].) Let $\Pi(A) = \{\alpha_1,\alpha_2,\alpha_3,\alpha_4\}$ and $\Pi(Y) = \{\beta_1,\beta_2,\beta_3,\beta_4,\beta_5,\beta_6\}$ be labelled as throughout. Let $\{e_{\beta_i},f_{\beta_i}|1\leq i\leq 6\}$ be the corresponding elements of L(Y). Let $\langle v^+\rangle$ be the unique 1-space of V such that $e_{\beta_i}v^+ = 0$ for $1\leq i\leq 6$. Then we have the elements $e_{\alpha_i}$ of L(A) as follows:

(5.4.1). $e_{\alpha_1} = e_{\beta_2}$, $\quad e_{\alpha_2} = e_{\beta_4}$,

$e_{\alpha_3} = e_{\beta_3} + e_{\beta_5}$, $\quad e_{\alpha_4} = e_{\beta_1} + e_{\beta_6}$

In the following, we will consider the possible modules V|Y given inductively by repeated applications of Theorem (8.1) of [12]. In all cases except the two in the statement of the result, we will produce a vector $w \in V - \langle v^+ \rangle$, which is a maximal vector for $L^+(A)$, the Lie subalgebra of $L(A)$ generated by $e_{\alpha_i}$, for $1 \le i \le 4$. Thus, $L(A)$ acts reducibly on $V|Y$ except in these two configurations. (Note that (1.34) is frequently used in the check that w is a maximal vector.)

By Theorem (8.1) of [12], the $C_3$ parabolic of A will act nontrivially on $\langle v^+ \rangle$ in every possible configuration. (For convenience, we will refer to (8.1) of [12] simply as (8.1) for the next few results.) Suppose $\langle \lambda, \beta_k \rangle = 0$ for k=3,4,5,6, as in the first configuration of (8.1) (c). Then, (8.1)(d) implies that $\lambda|T_Y = a\lambda_1$, for some a>0. Consider the vector $w = (-f_{(1,1,2,2,1,0)}f_1 - f_{12345}f_{134} + f_{(1,1,1,2,1,0)}f_{13} + f_{1234}f_{1345} -(a+3)f_{(1,1,2,2,1,1)} + (a+3)f_{(1,1,1,2,2,1)})v^+$. Then w is a maximal vector for $L^+(A)$. Moreover, if $a \ne p-3$, then $w \ne 0$, as $e_{(1,1,1,2,2,1)}w = (a+3)h_{(1,1,1,2,2,1)}v^+ = a(a+3)v^+$. Thus, if $\lambda|T_Y = a\lambda_1$, (i) holds.

In the configuration where $\lambda|T_Y = b\lambda_3 + x\lambda_2$, for $1 \ne b = p-1$ and $p > x \ge 0$, (8.1)(d) implies that x=0 or $x \ne 0$ and $x + b + 2 \equiv 0 \pmod p$. Suppose x = 0. Consider the vector $w = (f_{(1,1,2,2,1,0)} + f_{(1,1,1,2,1,0)}f_3 + f_{12345}f_{34} + f_{1345}f_{234} - 2f_{2345}f_{134} + f_{(0,1,1,2,1,1)}f_3 - f_{23456}f_{34} + f_{3456}f_{234} + f_{(0,1,1,2,2,1)})v^+$. (Recall, p>2.) Using the fact that b=p−1, we can show that $f_{13}f_{34}v^+ = f_{134}f_3v^+$. Then, applying other elements of L(Y) to this equation and using commutator relations, we obtain other such dependence relations. These are necessary to show that, in fact, $L^+(A)w = 0$. Moreover, $w \ne 0$ since $e_{(0,1,1,2,2,1)}w = h_{(0,1,1,2,2,1)}v^+ = bv^+$. Thus, L(A) does not act irreducibly on V. Suppose now that $\lambda|T_Y = b\lambda_3 + x\lambda_2$ and $x \ne 0$. Then, since $x + b + 2 \equiv 0 \pmod p$, and b = p−1, we have x = b = p−1. Let $w = (f_{1234}f_3 - f_{234}f_{13} - f_{12345} + f_{1345}f_2 - f_{23456} + f_{3456}f_2)v^+$. Then $L^+(A)w = 0$ and $w \ne 0$ as $e_{12345}w = -bv^+$. Thus, L(A) does not act irreducibly on V.

Consider next the configuration where $\lambda|T_Y = c\lambda_1 + b\lambda_3 + x\lambda_2$, for $c\neq 0\neq b$, $c+b = p-1$ and $p>x\geq 0$. If $x=0$, let $w = ((b+2)f_{(0,1,1,2,2,1)} + f_{(0,1,1,2,1,1)}f_3 + f_{3456}f_{234} - f_{23456}f_{34} + f_{(0,1,1,2,1,0)}f_{13} + (b+2)f_{(1,1,2,2,1,0)} + f_{1234}f_{345} - f_{2345}f_{134})v^+$. Then $L^+(A)w = 0$ and $e_{(0,1,1,2,2,1)}w = (b+2)h_{(0,1,1,2,2,1)}v^+ = (b+2)bv^+$; so if $b \neq p-2$, then $w\neq 0$ and (ii) holds. Suppose $\lambda$ is as above with $x>0$. Then (8.1)(d) implies that $b+x+2 \equiv 0 \pmod p$. Let $w = (f_{234}f_1 + (b+1)f_{1234} + f_{245}f_1 + cf_{2456})v^+$. Then $L^+(A)w = 0$ and $w \neq 0$, as $e_{34}w = f_2f_1v^+ \neq 0$.

Finally, we must consider the case where $\lambda|T_Y = b\lambda_3 + a\lambda_4 + x\lambda_2$, for $a\neq 0\neq b$, $a+b = p-1$ and $p>x\geq 0$. Theorem (8.1) implies that $x = 0$. Let $w = (f_{2456}f_{45} - f_{456}f_{245} + af_{(1,1,1,2,1,0)} - (a+1)f_{1234}f_{45} + af_{1345}f_{24} + f_{134}f_{245} + (a+1)f_{23456}f_4 - f_{3456}f_{24} - af_{2456}f_{34} + f_{1234}f_{34} - f_{134}f_{234})v^+$. Then $L^+(A)w = 0$ and $w \neq 0$, as $e_{(0,1,1,2,1,1)}w = a(a+1)v^+$.

This completes the proof of (5.4).□

Definition: If $\mu = \lambda - \Sigma c_i\beta_i$ is a $T_Y$ weight in V, then the <u>level</u> of $\mu$ is $\Sigma c_i$. We define the level of a $T_A$ weight $\nu = \lambda - \Sigma d_i\alpha_i$ similarly. For each $1\leq i\leq 6$, $\beta_i|T_A = \alpha_j$ for some $1\leq j\leq 4$, so level is preserved under restriction.

<u>(5.5).</u> Let Y have type $E_6$ and let $A<Y$ be the fixed point subgroup of the graph automorphism of Y and assume $p>2$. Suppose V is a restricted irreducible, rational kY-module with high weight $\lambda$. Let $\langle v^+\rangle$ be the unique 1-space of $V|Y$ invariant under $B_Y$. If $V|A$ is reducible, there exists a maximal vector for $B_A$, $w\in V - \langle v^+\rangle$, such that one of the following holds:

(i) $\lambda|T_Y = (p-3)\lambda_1$, for $p>3$, and $e_{(1,1,2,2,1,1)}w\in\langle v^+\rangle$.

(ii) $\lambda|T_Y = \lambda_1 + (p-2)\lambda_3$, for $p>2$, and $e_{(1,1,2,2,1,0)}w\in\langle v^+\rangle$ or $e_{(1,1,2,2,1,1)}w\in\langle v^+\rangle$.

<u>Proof</u>: We first prove the following

Claim: There exists a maximal vector for $B_A$, $w\in V - \langle v^+\rangle$.

Note that the long word for the Weyl group of $E_6$, $w_0$, lies in A. That is, there exists a coset representative, $n_0$, such that $w_0 = n_0 N_Y(T_Y)$ and $n_0 \in A$. Indeed, one checks that $w_0 = (s_{\beta_1+2\beta_2+2\beta_3+3\beta_4+2\beta_5+\beta_6} \cdot s_{\beta_4} \cdot s_{\beta_3+\beta_4+\beta_5} \cdot s_{\beta_1+\beta_3+\beta_4+\beta_5+\beta_6}) N_Y(T_Y)$, where $s_r$ is the reflection corresponding to the root r. Now, since $V|A$ is reducible, there exists $0<W<V$, an irreducible A-submodule of V. If $v^+ \notin W$, the result follows. So suppose $v^+ \in W$. Since $n_0 \in A$, $n_0 v^+ \in W$. Also, $n_0 v^+$ is a maximal vector for $B_Y^- = \langle U_{-r} | r \in \Sigma^-(Y)\rangle T_Y$.

Consider now the kY-module $V^*$, with high weight $-w_0\lambda$, by (1.11). Write $V = \langle n_0 v^+\rangle \oplus V_0$ and define a vector $f^+ \in V^*$ as follows: $f^+(n_0 v^+) = 1$ and $f^+(v_0) = 0$ for $v_0 \in V_0$. One checks that $f^+$ is a maximal vector for $B_Y^+$ and has weight $-w_0\lambda$. Thus, $\langle f^+\rangle$ is the unique 1-space of $V^*$ with these properties. Moreover, $f^+ \notin \mathrm{Ann}(W)$ as $n_0 v^+ \in W$. Hence, in $V^*|A$, Ann(W) is an invariant submodule, not containing the maximal vector $f^+$. So there exists a vector $g^+ \in V^* - \langle f^+\rangle$, such that $g^+$ is a maximal vector for $B_A$. So the claim holds for $V^*$. But $V^*|A \cong V|A$, and so the claim holds in general.

We now pass to the level of Lie algebras. Thus, there exists a vector $w \in V - \langle v^+\rangle$ such that $L^+(A)w = 0$, where $L^+(A)$ is the Lie algebra span of the elements $e_{\alpha_i}$, for $1 \le i \le 4$. And we may assume, $w \in V_{T_A}(\nu)$, for some dominant weight $\nu$. However, $w \notin V_{T_Y}(\mu)$, for any weight $\mu = \lambda - \Sigma d_i \beta_i$. For otherwise, (5.4.1) and the linear independence of weight vectors with distinct weights would imply $w \in \langle v^+\rangle$. Choose w to have minimal level. Since $w \notin \langle v^+\rangle$, w is not a maximal vector for $L^+(Y)$, the Lie algebra span of the elements $e_{\beta_i}$, for $1 \le i \le 6$. Thus, (5.4.1) implies that $e_{\beta_1} w \neq 0$ or $e_{\beta_3} w \neq 0$.

Case I: $e_{\beta_1} w \neq 0$.

Note that $e_{\beta_1} w \notin \langle v^+\rangle$, else $w \in V_{T_A}(\lambda-\alpha_4) = V_{T_Y}(\lambda-\beta_1)$, contradicting the opening remarks of the proof. Thus, by minimality, $e_{\beta_1} w$ is not a maximal vector for $L^+(A)$. Since $e_{\alpha_i} e_1 w = 0$ for i=1, 2 and 4,

$e_{\alpha_3}e_1w = (e_3+e_5)e_1w = e_{13}w \neq 0$. Here also $e_{13}w \notin \langle v^+\rangle$, else $w \in V_{T_Y}(\lambda-\beta_1-\beta_3)$. So by minimality, $e_{13}w$ is not a maximal vector for $L^+(A)$. Now, $e_{\alpha_i}e_{13}w = 0$ for i=1, 3 and 4, so $e_{\alpha_2}e_{13}w = e_{134}w \neq 0$. As above, $e_{134}w \notin \langle v^+\rangle$, and so is not a maximal vector for $L^+(A)$. But, $e_{\alpha_i}e_{134}w = 0$ for i = 2 and 4. Thus, $e_{\alpha_1}e_{134}w \neq 0$ or $e_{\alpha_3}e_{134}w \neq 0$.

Suppose $e_{\alpha_1}e_{134}w = e_{1234}w \neq 0$. Since the only $T_Y$ weight restricting to $\lambda-\alpha_1-\alpha_2-\alpha_3-\alpha_4$ is $\lambda-\beta_1-\beta_2-\beta_3-\beta_4$, $e_{1234}w \notin \langle v^+\rangle$ and so is not a maximal vector for $L^+(A)$. Since $e_{\alpha_i}e_{1234}w = 0$ for i=1, 2 and 4, $e_{\alpha_3}e_{1234}w = e_{12345}w \neq 0$. Suppose now that $e_{\alpha_3}e_{134}w = e_{1345}w \neq 0$. Note that $e_{1345}w \notin \langle v^+\rangle$, else the opening remarks of the proof imply that $\lambda|T_Y = \lambda_1+(p-2)\lambda_3$ and $w \in V_{T_Y}(\lambda-\beta_1-\beta_3-\beta_4-\beta_5) \oplus V_{T_Y}(\lambda-\beta_3-\beta_4-\beta_5-\beta_6) \oplus V_{T_Y}(\lambda-\beta_1-2\beta_3-\beta_4)$. So $w = af_{1345}v^+ + bf_{3456}v^+ + cf_{134}f_3v^+$, for some $a,b,c \in k$. But $e_{\alpha_4}w = 0$ implies that b = 0, $e_{\alpha_2}w = 0$ implies that c = 0 and $e_{\alpha_3}w = 0$ implies that a = 0, contradicting the original choice of w. So by minimality $e_{1345}w$ is not a maximal vector for $L^+(A)$. Now $e_{\alpha_i}e_{1345}w = 0$ for i=2, 3. In fact $e_{\alpha_4}e_{1345}w = e_{13456}w = 0$ , as $e_{13456} \in L^+(A)$. (We use here the fact that p>2.) Thus, $e_{\alpha_1}e_{1345}w = e_{12345}w \neq 0$, in this case also.

Now, if $e_{12345}w \in \langle v^+\rangle$, $w \in V_{T_A}(\lambda-\alpha_1-\alpha_2-2\alpha_3-\alpha_4)$. But, $\lambda-\alpha_1-\alpha_2-2\alpha_3-\alpha_4$ is not a dominant weight, with $\lambda$ as given. Thus, $e_{12345}w \notin \langle v^+\rangle$, and so is not a maximal vector for $L^+(A)$. Now, $e_{\alpha_i}e_{12345}w = 0$ for i = 1 and 3. In fact, $e_{\alpha_4}e_{12345}w = e_{123456}w = 0$, as $e_{123456} \in L^+(A)$. Hence $e_{\alpha_2}e_{12345}w = e_{(1,1,1,2,1,0)}w \neq 0$. Note that $e_{(1,1,1,2,1,0)}w \notin \langle v^+\rangle$, else $w \in V_{T_A}(\lambda-\alpha_1-2\alpha_2-2\alpha_3-\alpha_4)$. But $\lambda-\alpha_1-2\alpha_2-2\alpha_3-\alpha_4$ is not a dominant weight. Thus, by minimality, $e_{(1,1,1,2,1,0)}w$ is not a maximal vector for $L^+(A)$. Now, $e_{\alpha_i}e_{(1,1,1,2,1,0)}w = 0$ for i=1 and 2. In fact $e_{\alpha_4}e_{(1,1,1,2,1,0)}w = 0$, as $e_{(1,1,1,2,1,1)} \in L^+(A)$. Hence $e_{\alpha_3}e_{(1,1,1,2,1,0)}w = e_{(1,1,2,2,1,0)}w \neq 0$. Suppose $e_{(1,1,2,2,1,0)}w \in \langle v^+\rangle$. Then $w \in V_{T_A}(\lambda-\alpha_1-2\alpha_2-3\alpha_3-\alpha_4)$. Since $\lambda-\alpha_1-2\alpha_2-3\alpha_3-\alpha_4$ is dominant only if $\lambda|T_Y = \lambda_1 + (p-2)\lambda_3$, we have one of the configurations of the result.

Suppose now that $e_{(1,1,2,2,1,0)}w \notin \langle v^+\rangle$ and so is not a maximal vector for $L^+(A)$. Since $e_{\alpha_i}e_{(1,1,2,2,1,0)}w = 0$ for i=1, 2 and 3, we have $e_{\alpha_4}e_{(1,1,2,2,1,0)}w = e_{(1,1,2,2,1,1)}w \neq 0$. Note that $e_{\alpha_i}e_{(1,1,2,2,1,1)}w = 0$ for $1\le i\le 4$, as $e_{(1,1,2,2,2,1)} \in L^+(A)$. So by minimality, $e_{(1,1,2,2,1,1)}w \in \langle v^+\rangle$.

This completes the consideration of Case I.

Case II: $e_{\beta_3}w \neq 0$.

Note that $e_{\beta_3}w \notin \langle v^+\rangle$, else $w \in V_{T_A}(\lambda-\alpha_3) = V_{T_Y}(\lambda-\beta_3)$, contradicting the opening remarks of the proof. Thus, by minimality, $e_{\beta_3}w$ is not a maximal vector for $L^+(A)$. Since, $e_{\alpha_i}e_{\beta_3}w = 0$ for i = 1,3, we have $e_{\alpha_2}e_{\beta_3}w = e_{34}w \neq 0$ or $e_{\alpha_4}e_{\beta_3}w = -e_{13}w \neq 0$. If $e_{13}w \neq 0$, we may refer to Case I. So suppose $e_{34}w \neq 0$. Note that $e_{34}w \notin \langle v^+\rangle$, else $w \in V_{T_Y}(\lambda-\beta_3-\beta_4)$, contradicting the opening remarks of the proof. Hence, by minimality, $e_{34}w$ is not a maximal vector for $L^+(A)$.

Now $e_{\alpha_2}e_{34}w = 0$, and in fact, $e_{\alpha_3}e_{34}w = 0$, as $e_{345} \in L^+(A)$. Thus, $e_{\alpha_1}e_{34}w = e_{234}w \neq 0$ or $e_{\alpha_4}e_{34}w = -e_{134}w \neq 0$. If $e_{134}w \neq 0$ we may refer to Case I. So suppose $e_{234}w \neq 0$. Then $e_{234}w \notin \langle v^+\rangle$, as above, and so is not a maximal vector for $L^+(A)$. But $e_{\alpha_i}e_{234}w = 0$ for i = 1,2. In fact, $e_{\alpha_3}e_{234}w = 0$, as $e_{2345} \in L^+(A)$. Hence, $e_{\alpha_4}e_{234}w = -e_{1234}w \neq 0$. But now we may refer to Case I.

This completes the proof of (5.5).□

<u>(5.6).</u> Let (A,Y) be as in (5.5), so p>2. If $\lambda|T_Y = (p-3)\lambda_1$, for p>3, V|A is irreducible.

<u>Proof</u>: Suppose false; i.e., suppose V|A is reducible. Then (5.5) implies that there exists $w \in V - \langle v^+\rangle$, a maximal vector for $B_A$, such that $e_{(1,1,2,2,1,1)}w \in \langle v^+\rangle$. Hence, $e_{\alpha_i}w = 0$ for $1\le i\le 4$, where we now view V as a module for L(A). Now, $w \in V_{T_A}(\lambda-\alpha_1-2\alpha_2-3\alpha_3-2\alpha_4)$. The nontrivial $T_Y$ weights restricting to $\lambda-\alpha_1-2\alpha_2-3\alpha_3-2\alpha_4$ are

$\lambda-2\beta_1-\beta_2-2\beta_3-2\beta_4-\beta_5$, $\lambda-\beta_1-\beta_2-2\beta_3-2\beta_4-\beta_5-\beta_6$, and $\lambda-\beta_1-\beta_2-\beta_3-2\beta_4-2\beta_5-\beta_6$. The last two weights are conjugate to $\lambda-\beta_1$ and so have multiplicity 1. A spanning set for the weight space $V_{T_Y}(\lambda-2\beta_1-\beta_2-2\beta_3-2\beta_4-\beta_5)$ is $w_1 = f_{(1,1,2,2,1,0)}f_{\beta_1}v^+$, $w_2 = f_{12345}f_{134}v^+$, $w_3 = f_{(1,1,1,2,1,0)}f_{13}v^+$ and $w_4 = f_{1234}f_{1345}v^+$. Hence $w = \Sigma c_i w_i$, $1\le i\le 6$, for some $c_i \in k$ and $w_5 = f_{(1,1,2,2,1,1)}v^+$ and $w_6 = f_{(1,1,1,2,2,1)}v^+$.

Applying $e_{\alpha_i}$, $1\le i\le 4$, and $e_{2345}$ (an element of $L^+(A)$) to $w$, we find that $L^+(A)w = 0$ only if $c_5 = 0 = c_6$. So $w \in V_{T_Y}(\lambda-2\beta_1-\beta_2-2\beta_3-2\beta_4-\beta_5)$. But, the linear independence of weight vectors with distinct weights and (5.4.1) imply that $e_{\beta_i}w = 0$, for $1\le i\le 6$. Since $w \notin \langle v^+\rangle$, this is a contradiction.

This completes the proof of (5.6).□

<u>(5.7).</u> Let $(A,Y)$ be as in (5.5), so $p>2$. If $\lambda|T_Y = \lambda_1 + (p-2)\lambda_3$, for $p>2$, then $V|A$ is irreducible.

<u>Proof</u>: Suppose false. Then (5.5) implies that there exists $w \in V - \langle v^+\rangle$ a maximal vector for $B_A$ such that $e_{(1,1,2,2,1,0)}w \in \langle v^+\rangle$ or $e_{(1,1,2,2,1,1)}w \in \langle v^+\rangle$. In particular, $L^+(A)w = 0$.

Case I: $e_{(1,1,2,2,1,0)}w \in \langle v^+\rangle$.

Then $w \in V_{T_A}(\lambda-\alpha_1-2\alpha_2-3\alpha_3-\alpha_4)$. The nontrivial $T_Y$ weights restricting to $\lambda-\alpha_1-2\alpha_2-3\alpha_3-\alpha_4$ are $\mu_1 = \lambda-\beta_2-\beta_3-2\beta_4-2\beta_5-\beta_6$, $\mu_2 = \lambda-\beta_2-2\beta_3-2\beta_4-\beta_5-\beta_6$, $\mu_3 = \lambda-\beta_1-\beta_2-3\beta_3-2\beta_4$, and $\mu_4 = \lambda-\beta_1-\beta_2-2\beta_3-2\beta_4-\beta_5$. Let $w_1 = f_{(0,1,1,2,2,1)}v^+$, $w_2 = f_{(0,1,1,2,1,1)}f_3v^+$, $w_3 = f_{3456}f_{234}v^+$, $w_4 = f_{23456}f_{34}v^+$, $w_5 = f_{1234}f_{34}f_3v^+$, $w_6 = f_{(1,1,1,2,1,0)}f_3v^+$, $w_7 = f_{(1,1,2,2,1,0)}v^+$, $w_8 = f_{1234}f_{345}v^+$, and $w_9 = f_{2345}f_{134}v^+$. Then $V_{T_Y}(\mu_1) = \langle w_1\rangle$, $V_{T_Y}(\mu_2) = \langle w_2,w_3,w_4\rangle$, $V_{T_Y}(\mu_3) = \langle w_5\rangle$ and $V_{T_Y}(\mu_4) = \langle w_i | 6 \le i \le 9\rangle$. Thus, $w = \Sigma c_i w_i$, for some $c_i \in k$, $1\le i\le 9$.

Applying $e_{2345}$, $e_{345}$, $e_{\beta_2}$, $e_{34}- e_{45}$ and $e_{134}+ e_{456}$ (all elements of

$L^+(A)$) to w, we find that $L^+(A)w = 0$ only if $c_6 = -c_7 = c_8 = -c_9$, $c_1 = 0 = c_5$ and $c_2 = c_3 = -c_4$.

We now claim that $w \in \langle v^+ \rangle$, which will contradict the choice of w. It suffices to show that $e_{\beta_i}w = 0$ for $1 \le i \le 6$. By hypothesis, $e_{\beta_2}w = 0 = e_{\beta_4}w$. Consider now $e_{\beta_1}w = c_6(f_{(0,1,1,2,1,0)}f_{\beta_3} + f_{234}f_{345} - f_{2345}f_{34})v^+$. It is a straightforward check that $e_{\beta_i}(e_{\beta_1}w) = 0$ for $1 \le i \le 6$. Thus $e_{\beta_1}w = 0$; so $e_{\beta_6}w = 0$.

Now $e_{\alpha_3}w = e_{\beta_3}w + e_{\beta_5}w = 0$ and since $e_{\beta_3}w \in V_{T_Y}(\lambda-\beta_2-\beta_3-2\beta_4-\beta_5-\beta_6) \oplus V_{T_Y}(\lambda-\beta_1-\beta_2-\beta_3-2\beta_4-\beta_5)$ and $e_{\beta_5}w \in V_{T_Y}(\lambda-\beta_1-\beta_2-2\beta_3-2\beta_4)$, we must have $e_{\beta_3}w = 0 = e_{\beta_5}w$. Hence, $e_{\beta_i}w = 0$ for $1 \le i \le 6$, and $w \in \langle v^+ \rangle$ as claimed. This completes the consideration of Case I.

Case II: $e_{(1,1,2,2,1,1)}w \in \langle v^+ \rangle$.

Then, $w \in V_{T_A}(\lambda-\alpha_1-2\alpha_2-3\alpha_3-2\alpha_4)$. The nontrivial $T_Y$ weights restricting to $\lambda-\alpha_1-2\alpha_2-3\alpha_3-2\alpha_4$ are $\mu_1 = \lambda-2\beta_1-\beta_2-3\beta_3-2\beta_4$, $\mu_2 = \lambda-\beta_1-\beta_2-\beta_3-2\beta_4-2\beta_5-\beta_6$, $\mu_3 = \lambda-\beta_1-\beta_2-2\beta_3-2\beta_4-\beta_5-\beta_6$ and $\mu_4 = \lambda-2\beta_1-\beta_2-2\beta_3-2\beta_4-\beta_5$. Also, $V_{T_Y}(\mu_1) = \langle w_1 \rangle$, $V_{T_Y}(\mu_2) = \langle w_2 \rangle$, $V_{T_Y}(\mu_3) = \langle w_i | 3 \le i \le 7 \rangle$ and $V_{T_Y}(\mu_4) = \langle w_k | 8 \le k \le 10 \rangle$, where
$w_1 = f_{1234}f_{13}f_{34}v^+$, $w_2 = f_{(1,1,1,2,2,1)}v^+$, $w_3 = f_{(1,1,2,2,1,1)}v^+$,
$w_4 = f_{13456}f_{234}v^+$, $w_5 = f_{23456}f_{134}v^+$, $w_6 = f_{123456}f_{34}v^+$,
$w_7 = f_{(0,1,1,2,1,1)}f_{13}v^+$, $w_8 = f_{(1,1,2,2,1,0)}f_1v^+$, $w_9 = f_{12345}f_{134}v^+$,
$w_{10} = f_{(1,1,1,2,1,0)}f_{13}v^+$.

Applying $e_{\alpha_4}$, $e_{24}$, $e_{2345}$, $e_{1234}+e_{2456}$ and $e_{134}+e_{456}$ (all elements of $L^+(A)$), we see that $L^+(A)w = 0$ only if $c_i = 0$ for $i = 1,2,8,9,10$. So $w \in V_{T_Y}(\mu_3)$. But (5.4.1) and the linear independence of weight vectors with distinct weights implies $e_{\beta_i}w = 0$ for $1 \le i \le 6$. But this implies $w \in \langle v^+ \rangle$. Contradiction.

This completes the proof of (5.7).□

The remainder of this chapter is devoted to proving that there are no examples (A,Y,V) in the main theorem with A simple, of rank 3 and Y of type $E_n$, nor with (A,p) special and rankA > 2. This will complete the proof of Theorem (5.0).

<u>(5.8).</u> Suppose rank A = 3 with (A,p) not special, $L_A'$ is of type $A_2$ with $\dim V^1(Q_A) > 1$. Then $L_Y' = L_1 \times L_2$, with $L_i$ of type $A_2$ or $D_4$. Moreover, if $L_i$ has type $D_4$, then $\langle \lambda, \beta_j \rangle = 0$ for $1 \le j \le 5$.

<u>Proof</u>: If $L_Y'$ is quasisimple, Theorem (4.0) implies that $L_Y'$ is of exceptional type. Then (6.0) implies that $L_Y' = E_6$ and $Q_Y/K_{\beta_7}$ is a 27-dimensional irreducible $L_A'$ module. Thus, $Z_A$ induces scalars on $Q_Y/K_{\beta_7}$, forcing $Q_A K_{\beta_7}/K_{\beta_7}$ to be an $L_A'$ submodule of $Q_Y/K_{\beta_7}$. Since $\dim Q_A < 27$, $Q_A \le K_{\beta_7}$, contradicting (2.3). Thus, $L_Y'$ is not a quasisimple.

By size retrictions, $L_Y' = L_1 \times L_2$, where $L_i$ is a simple algebraic group of classical type. Hence, (1.5) implies $Z_A \le Z_Y$. Notice, if $A_2$ is embedded in $D_5$, acting reducibly on the natural module for $D_5$, $A_2$ lies in a proper parabolic of $D_5$. Thus, the minimality of $P_Y$ and the Main Theorem of [12] imply that $L_i$ has type $A_2$ or $D_4$. Suppose $L_i = D_4$. Since $P_Y$ is minimal, $\rho_i(L_A')$ is either irreducible on $Q_Y/K_{\beta_1}$, or p=3 and $L_A'$ acts on $Q_Y/K_{\beta_1}$ with composition factors of dimensions 1 and 7. Thus, $Q_A \le K_{\beta_1}$ and by (2.3), $\langle \lambda, \beta_j \rangle = 0$ for $1 \le j \le 5$.□

<u>(5.9).</u> If $(A, L_A')$ is of type $(B_3, B_2)$ and p>2, then $\dim V^1(Q_A) = 1$.

<u>Proof</u>: Suppose $\dim V^1(Q_A) > 1$. Then $L_Y'$ is not a simple algebraic group. For otherwise, Theorem (4.0) implies $L_Y'$ is of exceptional type. But the result of Chapter 7 indicates that there is no such embedding. Thus $Y = E_8$ and $L_Y'$ is of type $A_3 \times A_3$ or $A_3 \times A_4$. Also, the projection of $L_A'$ into each component of $L_Y'$ must act irreducibly on the natural module for that component, as $P_Y$ is minimal. Note that $h_{\alpha_3}(-1) \in Z(A) \le Z(Y) = \{1\}$. Since Y is simply connected, $h_{\alpha_3}(-1)$ must be in the kernel of the 4-dimensional representation of $L_A'$. But it is easy to check that this is not the case. Contradiction.□

(5.10). If A is of type $A_3$ or $B_3$, with (A,p) not special, and $L_A'$ is of type $A_2$, then $\dim V^1(Q_A) = 1$.

Proof: Suppose false. Let $\Pi(L_A) = \{\alpha_1,\alpha_2\}$, so $\Pi(A) - \Pi(L_A) = \{\alpha_3\}$. By (5.8), $L_Y' = L_1 \times L_2$, with $L_i$ of type $A_2$ or $D_4$. Let $q_i$ be the field twist on the embedding of $L_A'$ in $L_i$ for i = 1,2. Thus, (1.5) implies $Z_A \le Z_Y$. Suppose $L_1 = D_4$. Then $Y = E_8$, $\Pi(L_1) = \{\beta_2,\beta_3,\beta_4,\beta_5\}$, $\Pi(L_2) = \{\beta_7,\beta_8\}$, and $\langle\lambda,\beta_j\rangle = 0$ for $1 \le j \le 5$. By (1.23) and (5.9), $\lambda|T_A = q(c\mu_1+d\mu_2+c\mu_3)$ if $A = A_3$ or $\lambda|T_A = qc\mu_1$ if $A = B_3$, with $\lambda|T_Y = x\lambda_6+c\lambda_j+d\lambda_k$ where $\{j,k\} = 7,8$ and q is some p-power. If p=3, $0 \le c,d \le 2$ and (1.26) and (1.32) imply $\dim V|A < \dim V|Y$. Hence, $p \ne 3$.

Since $p \ne 3$, $\rho_1(L_A')$ acts irreducibly on $V_{L_1}(-\beta_6)$ and (2.7) implies that $q_1=q_2$. One checks that if $\rho_2(h_{\alpha_1}(c)) = h_{\beta_8}(c^{q_2})$ and $\rho_2(h_{\alpha_2}(c)) = h_{\beta_7}(c^{q_2})$, the $L_A'$ composition factors of $Q_Y/K_{\beta_6}$ have high weights $q_2(\mu_1+2\mu_2)$, $q_2(2\mu_1)$ and $q_2\mu_2$. By symmetry, if $\rho_2(h_{\alpha_1}(c)) = h_{\beta_7}(c^{q_2})$ and $\rho_2(h_{\alpha_2}(c)) = h_{\beta_8}(c^{q_2})$, there is no $L_A'$ composition factor of $Q_Y/K_{\beta_6}$ isomorphic to a twist of $Q_A{}^{\alpha_3}$. Thus, $L_A'$ must project into $L_2$ in the first way described.

We claim $\langle\lambda,\beta_6+\beta_7\rangle = 0$. For otherwise, since $V_{T_Y}(\lambda-\beta_6) \oplus V_{T_Y}(\lambda-\beta_6-\beta_7) \ne 0$, some nonidentity element from the set $U_{-\beta_6} \cdot U_{-\beta_6-\beta_7}$ must occur in the factorization of an element in $Q_A - Q_A'$. However, $-\beta_6$ (respectively, $-\beta_6-\beta_7$) affords $T(L_A')$ weight $q_2(\mu_1+2\mu_2)$ (respectively, $2q_2\mu_1$). But no such weight vectors occur in $Q_AK_{\beta_6}/K_{\beta_6}$. Thus, if $L_Y'$ has type $D_4 \times A_2$, $\lambda|T_Y = c\lambda_8$, for some $1<c<p$. Then (1.23) implies $\lambda|T_A = q_2(c\mu_1+c\mu_3)$ if $A = A_3$, and (5.9) implies $\lambda|T_A = q_2c\mu_1$, if $A = B_3$. In any case, $\dim V_{\beta_6}(Q_Y) \le 3/2(c+1)(c+2)$, by (1.25) and (1.12). However, $f_{\beta_6+\beta_7+\beta_8}v^+$ affords an $L_Y'$ composition factor in $V_{\beta_6}(Q_Y)$ of dimension $4c(c+1)$. Contradiction.

Now consider the case where $L_i$ has type $A_2$ for i = 1,2, and $L_1$ and $L_2$ are separated by exactly two nodes of the Dynkin diagram. For convenience, temporarily label as follows:

$L_Y' = \langle U_{\pm\gamma_1}, U_{\pm\gamma_2}\rangle \times \langle U_{\pm\gamma_5}, U_{\pm\gamma_6}\rangle = L_1 \times L_2$, $\gamma_3,\gamma_4 \in \Pi(Y) - \Pi(L_Y)$, with $(\gamma_i,\gamma_{i+1}) < 0$ for $1 \le i \le 5$. Then (2.8), (2.5) and (2.6) imply that only one of $L_1$ and $L_2$ act nontrivially on $V^1(Q_Y)$. Say $L_1v^+ \neq v^+$, $L_2v^+ = v^+$. Since $Q_A \nleq K_{\gamma_3}$, we compare high weights of the $L_A'$ modules $Q_AK_{\gamma_3}/K_{\gamma_3}$ and $Q_Y/K_{\gamma_3}$ to see that $\rho_1(h_{\alpha_1}(c)) = h_{\gamma_1}(c^{q_1})$ and $\rho_1(h_{\alpha_2}(c)) = h_{\gamma_2}(c^{q_1})$. It is a check to see that the $L_Y'$ composition factor of $Q_Y$ afforded by $U_{-\gamma_3-\gamma_4}$ has no 3-dimensional $L_A'$ submodule isomorphic to a twist of $Q_A{}^{\alpha_3}$. But by (1.33), we may assume that if $Q_A \le K_{\gamma_4}$, $-\gamma_4$ is not involved in $L_A'$. Hence, (2.11) implies that $Q_A \nleq K_{\gamma_4}$.

Now, $Q_A \nleq K_{\gamma_3}$ and $Q_A \nleq K_{\gamma_4}$ and (2.9) imply that $A = B_3$. Moreover, (2.13) and (2.8) imply $\gamma_3|T_A = q_1\alpha_3 = \gamma_4|T_A$. By (5.9), $\langle\lambda,\gamma_i\rangle = 0$ for $2 \le i \le 6$. Also, by (1.10), $q_1 = 1$ and so, if $\langle\lambda,\gamma_1\rangle = c$, for $0<c<p$, then $\langle\lambda,\alpha_1\rangle = c$. Now, the subgroup $X = \langle U_{\pm\gamma_1}, U_{\pm\gamma_2}, U_{\pm\gamma_3}, U_{\pm\gamma_4}, U_{\pm\gamma_5}, U_{\pm\gamma_6}\rangle$ of type $A_6$ has a natural subgroup, B, of type $B_3$. Moreover, the X composition factor of $V|Y$ afforded by $\langle v^+\rangle$ is not all of $V|Y$ as X is contained in the Levi factor of a proper parabolic of Y. But the B composition factor of $V|Y$ afforded by $\langle v^+\rangle$ has the same high weight, as $B_3$ module, as does $V|A$. Thus, $\dim V|A < \dim V|Y$. Hence, this configuration does not occur.

Consider now the case where $L_1$ and $L_2$ are separated by more than two nodes of the Dynkin diagram. Thus $Y = E_8$, $\Pi(L_1) = \{\beta_1,\beta_3\}$ and $\Pi(L_2) = \{\beta_7,\beta_8\}$. Then (2.13) and (2.3) imply that $\langle\lambda,\beta_k\rangle = 0$ for $k = 2,4,5,6$. We first note that only one of $L_1$ and $L_2$ acts nontrivially on $V^1(Q_Y)$. For otherwise (2.13) implies $\beta_4|T_A = q_1\alpha_3$, $\beta_6|T_A = q_2\alpha_3$, and $\beta_2|T_A = 0 = \beta_5|T_A$. As well, we have $\beta_1|T_A = q_1\alpha_1$, $\beta_3|T_A = q_1\alpha_2$, $\beta_8|T_A = q_2\alpha_1$, $\beta_7|T_A = q_2\alpha_2$. Thus, there is no vector in $V|Y$ with $T_A$ weight $\lambda - q\alpha_3$, for any p-power q, contradicting (2.14).

Note that A does not have type $B_3$. For otherwise, $\lambda|T_A = cq_i\mu_1$, for $i = 1$ or 2, and $\lambda|T_Y = c\lambda_1$ or $c\lambda_8$. And as above, considering the usual embedding of $B_3$ in $A_6$ we have $\dim V|A < \dim V|Y$. Also, (1.23) and (1.10) imply $V|A$ is restricted; say $\lambda|T_A = c\mu_1+d\mu_2+c\mu_3$ Let $P_Y\hat{} > B_Y^-$ be the parabolic subgroup of $Y$ with Levi factor $L_Y\hat{} = \langle L_Y, U_{\pm\beta_5}\rangle$. Then $Q_A \leq Q_Y\hat{} = R_u(P_Y\hat{})$ and $Z_A \leq Z(L_Y\hat{})^\circ$ by (2.12). For the argument which follows, we may assume $\langle\lambda,\beta_1+\beta_3\rangle = 0$ and $\langle\lambda,\beta_7+\beta_8\rangle \neq 0$. So $\lambda|T_Y = d\lambda_7+c\lambda_8$, $\beta_5|T_A = 0$, $\beta_6|T_A = \alpha_3$, $\beta_7|T_A = \alpha_2$ and $\beta_8|T_A = \alpha_1$. If $Q_A \leq K_{\beta_4}$, by (1.33) we may assume that $-\beta_4$ is not involved in $L_A'$. Hence, by (2.11), there is a nontrivial image of $Q_A^{\alpha_3}$ in $Q_Y\hat{}(\beta_6,\beta_4)$. But $Q_Y\hat{}(\beta_6,\beta_4)$ has no $L_A'$ composition factor isomorphic to a twist of $Q_A^{\alpha_3}$. Thus, $Q_A \nleq K_{\beta_4}$ and $\beta_4|T_A = q_1\alpha_3$. Now, if $V_{T_A}(\lambda-\alpha_2-x\alpha_3) \neq 0$, then conjugating by $s_{\alpha_3}$, we see that $V_{T_A}(\lambda-\alpha_2-(c+1-x)\alpha_3) \neq 0$. So $x \leq c+1$. In particular, $q_1 = 1$, else $0 \neq f_{4567}v^+ \in V_{T_A}(\lambda-\alpha_2-(q+1)\alpha_3)$. So $\beta_1|T_A = \alpha_1$, $\beta_3|T_A = \alpha_2$, and $\beta_4|T_A = \alpha_3$. Now, if $\langle\lambda,\beta_7\rangle \geq 2$, $(f_{67})^2v^+$, $(f_{567})^2v^+$, $f_{4567}f_7v^+$ and $f_{34567}v^+$ are 4 linearly independent vectors in $V_{T_A}(\lambda-2\alpha_2-2\alpha_3)$. But (1.28) implies $\dim V_{T_A}(\lambda-2\alpha_2-2\alpha_3) \leq 3$. Thus, $\langle\lambda,\beta_7\rangle = 1$. Also, $\langle\lambda,\beta_8\rangle > 1$, else (1.23) implies $\dim V|A < \dim V|Y$. If $\langle\lambda,\beta_8\rangle = c = \langle\lambda,\alpha_1\rangle = \langle\lambda,\alpha_3\rangle$, we may assume $c \neq p-2$, else by (1.35), $\dim V_{T_A}(\lambda-\alpha_2-\alpha_3) = 1$, but $f_{67}v^+$ and $f_{567}v^+$ are 2 linearly independent vectors in this weight space. Now by Theorem 3 of [3] and (1.28), if $c+1 < p-1$, $\dim V_{T_Y}(\lambda-\beta_7-\beta_8) = \dim V_{T_Y}(\lambda-2\beta_7-2\beta_8) = \dim V_{T_Y}(\lambda-\beta_7-2\beta_8) = 2$. Thus, $\dim[V_{T_Y}(\lambda-2\beta_6-2\beta_7-2\beta_8) + V_{T_Y}(\lambda-2\beta_5-2\beta_6-2\beta_7-2\beta_8) + V_{T_Y}(\lambda-\beta_5-2\beta_6-2\beta_7-2\beta_8) + V_{T_Y}(\lambda-\beta_4-\beta_5-\beta_6-2\beta_7-2\beta_8) + V_{T_Y}(\lambda-\beta_3-\beta_4-\beta_5-\beta_6-\beta_7-2\beta_8) + V_{T_Y}(\lambda-\beta_1-\beta_3-\beta_4-\beta_5-\beta_6-\beta_7-\beta_8)] > 9$. But each of these weight spaces lies in $V_{T_A}(\lambda-2\alpha_1-2\alpha_2-2\alpha_3)$, which has dimension at most 9, by (1.28). Hence, we may assume $c+1 > p-1$, so $c = p-1$. Now, (1.33) implies $\dim V_{T_Y}(\lambda-\beta_7-2\beta_8) = 2$ and since $\lambda-2\beta_7-2\beta_8$ is conjugate to $\lambda-\beta_7-2\beta_8$, $\dim V_{T_Y}(\lambda-2\beta_7-2\beta_8) = 2$. And again the dimension of the given sum of $T_Y$ weight spaces exceeds 9. Thus, $\Pi(L_Y) \neq \{\beta_1,\beta_3,\beta_7,\beta_8\}$.

It remains to consider the case where $L_i$ is of type $A_2$ for $i=1,2$ and $L_1$ and $L_2$ are separated by exactly one node of the Dynkin diagram. For convenience, temporarily label as follows: $\Pi(L_1) = \{\gamma_1,\gamma_2\}$, $\Pi(L_2) = \{\gamma_4,\gamma_5\}$, $\gamma_3 \in \Pi(Y) - \Pi(L_Y)$ and $(\gamma_i,\gamma_{i+1}) < 0$ for $1\le i\le 4$. Then (2.7) implies that the field twists on the embeddings of $L_A'$ in $L_1$ and in $L_2$ are equal. Call this twist q. Thus, by (2.5) and (2.6), only one of $L_1$ and $L_2$ acts nontrivially on $V^1(Q_Y)$. Say, $L_1v^+ \neq v^+$ and $L_2v^+ = v^+$.

Now we may assume $p>2$. For otherwise, $A = A_3$ and (1.23), (1.26) and (1.32) imply $\dim V|A < \dim V|Y$. Then, we find that $Q_Y/K_{\gamma_3}$ has an $L_A'$ composition isomorphic to a twist of $Q_A{}^{\alpha_3}$ if and only if $h_{\alpha_1}(c) = h_{\gamma_2}(c^q)\cdot h_{\gamma_4}(c^q)$ and $h_{\alpha_2}(c) = h_{\gamma_1}(c^q)h_{\gamma_5}(c^q)$. Then, considering the $T(L_A')$ weight vectors in the module $Q_Y/K_{\gamma_3}$, we see that $x_{-\alpha_3}(t) = x_{-\gamma_2-\gamma_3}(c_1t^q)\cdot x_{-\gamma_3-\gamma_4}(c_2t^q)w$, where $c_i \in k$, with $c_1$ and $c_2$ not both zero, and $w \in K_{\gamma_3}$. Thus, $\gamma_3|T_A = q(\alpha_3 - \alpha_1)$. In particular, $\langle\lambda,\gamma_3\rangle = 0$. So if $\langle\lambda,\gamma_1\rangle = d$, $\langle\lambda,\gamma_2\rangle = c$, for some $0\le c,d<p$, then $\langle\lambda,\alpha_1\rangle = cq$ and $\langle\lambda,\alpha_2\rangle = dq$. If $A = B_3$, $d=0$ by (5.9). In fact, $c=0$ also, else $f_{\gamma_2+\gamma_3}v^+$ is a nonzero vector in $V_{T_A}(\lambda - q\alpha_3)$. Thus, $A = A_3$ and (1.23) implies that $\langle\lambda,\alpha_3\rangle = cq$. Let $P \ge B_Y^-$ be the proper parabolic subgroup of Y with Levi factor $\langle U_{\pm\gamma_i} | 1\le i\le 5\rangle T_Y$. Then, considering the usual embedding of $A_3$ in $A_5$, we see that $\dim V|A \le \dim V^1(R_u(P)) < \dim V|Y$. This completes the proof of (5.10).□

<u>(5.11).</u> If A has type $C_3$, with $p>2$, and $L_A'$ has type $A_2$ with $\dim V^1(Q_A) > 1$, then Y has type $E_8$, $L_Y'$ has type $A_2 \times D_4$ and $\lambda|T_Y = c\lambda_7 + d\lambda_8$, $d \neq 0$, $\langle\lambda,\alpha_1\rangle = cq$, $\langle\lambda,\alpha_2\rangle = dq$, for some p-power q.

<u>Proof</u>: By (5.8), $L_Y' = L_1 \times L_2$, where $L_i$ has type $A_2$ or $D_4$. Thus, (1.5) implies $Z_A \le Z_Y$. Consider first the case where $L_1 = D_4$. Then $Y = E_8$, $L_Y'$ has type $D_4 \times A_2$ and $\langle\lambda,\beta_i\rangle = 0$ for $1 \le i \le 5$.

Let $q_i$ be the field twist on the embedding of $L_A'$ in $L_i$, for i=1,2. If $q_1\neq q_2$, $L_A'$ acts irreducibly on $Q_Y/K_{\beta_6}$ or p=3 and $Q_Y/K_{\beta_6}$ has $L_A'$ composition factors of dimensions 21 and 3. Thus, since $Q_A \not\le K_{\beta_6}$, $q_1=q_2$. If $\rho_2(h_{\alpha_1}(c)) = h_{\beta_7}(c^{q_1})$ and $\rho_2(h_{\alpha_2}(c)) = h_{\beta_8}(c^{q_1})$, the $L_A'$ composition factors of $Q_Y/K_{\beta_6}$ have high weights $\{q_1(2\mu_1+\mu_2), q_1\mu_1, q_12\mu_2\}$. By symmetry, if $\rho_2(h_{\alpha_1}(c)) = h_{\beta_8}(c^{q_1})$ and $\rho_2(h_{\alpha_2}(c)) = h_{\beta_7}(c^{q_1})$, there is no $L_A'$ composition factor of $Q_Y/K_{\beta_6}$ isomorphic to a twist of $Q_A^{\alpha_3}$. Thus, we see that $L_A'$ must project onto $L_2$ in the first way described and $\beta_6|Z_A = q_1\alpha_3$. Note that $\langle\lambda,\beta_6\rangle = 0$ else a nonidentity element from $U_{-\beta_6}$ must occur in the factorization of some element of $Q_A^{\alpha_3}$; but $-\beta_6$ does not afford a $T(L_A')$ weight in $(Q_A^{\alpha_3})^{q_1}$. (See (2.4).) Also, if $\lambda|T_Y = c\lambda_7+d\lambda_8$, then $\langle\lambda,\alpha_1\rangle = cq_1$ and $\langle\lambda,\alpha_2\rangle = dq_1$. Suppose $d = 0 \neq c$. Then (1.26) and (1.25) imply $\dim V_{\beta_6}(Q_Y) \leq 3(c+1)(c+2)$. Now, $f_{\beta_6+\beta_7}v^+$ affords an $L_Y'$ composition factor of $V_{\beta_6}(Q_Y)$ of dimension $8\cdot\dim V_0$, where $V_0$ is the $A_2$ module with high weight $(c-1)\eta_1+\eta_2$ and $\eta_1$, $\eta_2$ are the fundamental dominant weights for $A_2$. The corollary to Theorems 3 and 4 in [3] implies $\dim V_0 = c(c+2)$ if $c < p-1$ or $\frac{1}{2}(c+1)(c+2) + (c-1)$ if $c= p-1$. But $8\cdot\dim V_0 > 3(c+1)(c+2)$ in each case, contradicting the bound on $\dim V_{\beta_6}(Q_Y)$. Thus $d \neq 0$.

Now consider the case where $L_i$ is of type $A_2$ for $i=1,2$. Say $L_1v^+ \neq v^+$. Since $p>2$, $Q_A^{\alpha_3}$ is a 6-dimensional irreducible $L_A'$ module. So there does not exist $\gamma \in \Pi(Y) - \Pi(L_Y)$ with $(\gamma,\Sigma L_1) \neq 0$ and such that $Q_Y/K_\gamma$ is isomorphic to a 3-dimensional irreducible $L_1$ module. Thus, there exists $\gamma_0 \in \Pi(Y) - \Pi(L_Y)$ with $(\gamma_0,\Sigma L_i) \neq 0$ for $i=1,2$. For convenience, temporarily label as follows: $\Pi(L_1) = \{\gamma_1,\gamma_2\}$, $(\gamma_2,\gamma_0) \neq 0$, $\Pi(L_2) = \{\gamma_3,\gamma_4\}$, $(\gamma_3,\gamma_0) \neq 0$. Then (2.7) implies that the field twists on the embeddings of $L_A'$ in $L_1$ and in $L_2$ are equal. Call this twist q. Considering the $L_A'$ composition factors of $Q_Y/K_{\gamma_0}$, we have $h_{\alpha_1}(c) = h_{\gamma_1}(c^q)h_{\gamma_4}(c^q)$ and $h_{\alpha_2}(c)= h_{\gamma_2}(c^q)\, h_{\gamma_3}(c^q)$.

We first claim that there does not exist $\tau \in \Pi(Y) - \Pi(L_Y)$ with $(\tau,\gamma_4) \neq 0$. For if there exists such a $\tau$, $Q_Y/K_\tau$ is a 3-dimensional irreducible $L_A'$ module and so $Q_A \leq K_\tau$. By (2.10), $-\tau$ is not involved in $L_A'$, so (2.11) implies that there is a nontrivial image of $Q_A^{\alpha_3}$ in $Q_Y(\gamma_0,\tau)$, which is also a 3-dimensional irreducible $L_A'$ module. Contradiction. As well, there does not exist $\tau \in \Pi(Y) - \Pi(L_Y)$ with $(\tau,\gamma_3) \neq 0$. For if there exists such a $\tau$, as above, $Q_A \leq K_\tau$, $-\tau$ is not involved in $L_A'$ and (2.11) applies. But $Q_Y(\gamma_0,\tau)$ has $L_A'$ composition factors of dimensions 8 and 1 (7 and 1, if p=3).

The above work implies $Y = E_6$ and $\Pi(L_Y) = \{\beta_1,\beta_3,\beta_5,\beta_6\}$. We may assume, by symmetry, that $\langle\lambda,\beta_5+\beta_6\rangle = 0$, and by (2.3), $\langle\lambda,\beta_2\rangle = 0$. Also, one checks that $x_{-\alpha_3}(t) = x_{-\beta_4}(c_1t^q)w$, where $c_1 \in k^*$ and $w \in K_{\beta_4}$. Thus, by (2.4), $\beta_4|T_A = q\alpha_3$; and (2.12) implies $\beta_2|T_A = 0$. Thus, $\langle\lambda,\beta_4\rangle = 0$, else $f_4v^+$ and $f_{24}v^+$ are two linearly independent vectors in $V_{T_A}(\lambda-q\alpha_3)$.

We claim that $\langle\lambda,\alpha_3\rangle = 0$. For since $\beta_1|T_A = q\alpha_1 = \beta_6|T_A$, $\beta_3|T_A = q\alpha_2 = \beta_5|T_A$, $\beta_4|T_A = q\alpha_3$, and $\langle\lambda,\beta_4\rangle = 0$, there does not exist a vector in $V|Y$ with $T_A$ weight $\lambda - q_0\alpha_3$ for any p-power $q_0$. So we have $\lambda|T_A = q(c\mu_1 + d\mu_2)$ and $\lambda|T_Y = c\lambda_1 + d\lambda_3$ for some $0 \leq c,d < p$.

Now let $X = \langle U_{\pm\beta_1},U_{\pm\beta_3},U_{\pm\beta_4},U_{\pm\beta_5},U_{\pm\beta_6}\rangle$. Then X has a natural subgroup, C, of type $C_3$. Moreover, $v^+$ affords an X composition factor of $V|Y$ with dimension strictly less than $\dim V|Y$, as X is contained in the Levi factor of a proper parabolic of Y. But $v^+$ affords a C composition factor of $V|Y$ with the same high weight as $V|A$, as $C_3$ module. Thus, $\dim V|A < \dim V|Y$. Contradiction.□

<u>(5.12).</u> If $(A,L_A')$ is of type $(C_3,C_2)$ with $p>2$, then $\dim V^1(Q_A) = 1$.

<u>Proof</u>: Suppose false; i.e., suppose $\dim V^1(Q_A) > 1$. We first claim that $L_Y'$ is not simple. The work of Chapter 7 indicates that $L_Y'$ is not a simple algebraic group of exceptional type. Hence, if $L_Y'$ is simple, Theorem (4.0) implies $Y = E_8$, $\lambda|T_Y = \lambda_1$ and $\langle\lambda,\alpha_2\rangle = 2q$, for some p-power q. However, this contradicts (5.11). Thus, $Y = E_8$ and $L_Y'$ is of type $A_3 \times A_3$ or $A_3 \times A_4$.

We first claim that $\langle\lambda,\alpha_2\rangle \neq 0$. For otherwise, (5.11) and the Main Theorem of [12] imply $\lambda|T_A = q\mu_3$ and $\dim V|A < \dim V|Y$. Hence, (5.11) implies $\lambda|T_Y = c\lambda_7+d\lambda_8$ with $d \neq 0$. Now $\langle U_{\pm\beta_j} \mid j = 6, 7, 8\rangle$ is a component of $L_Y'$ with the embedding of $L_A'$ in $L_Y'$ the natural embedding of classical groups, up to some twist. Thus, $\langle\lambda,\alpha_2\rangle = dq_0$ and

$\langle\lambda,\alpha_3\rangle = cq_0$ for some p-power $q_0$. In fact, since $d \neq 0$, (5.11) implies $\langle\lambda,\alpha_1\rangle = cq_0$, so by (1.10), $q_0 = 1$. Now, let $X = \langle U_{\pm\beta_j} \mid 2 \leq j \leq 8\rangle$. Then X has a natural subgroup of type $C_3$, call it C. (See $IV_8$ on Table 1 of [12].) Moreover, $v^+$ affords an X composition factor of V|Y of dimension strictly less than dimV|Y. But $v^+$ affords a C composition factor of V|Y with the same high weight as V|A, as $C_3$ modules. Thus, dimV|A < dimV|Y. Contradiction.□

<u>(5.13).</u> If (A,Y,V) is an example in the main theorem, with Y of type $E_n$ and rankA > 2, then (A,p) is not special.

<u>Proof:</u> Suppose false

<u>Case I:</u> Suppose p = 2 and A has type $B_k$ or $C_k$.

We first claim that $\langle\lambda,\alpha_j\rangle > 0$ for some j > 1. For otherwise, applying induction to a maximal parabolic of A corresponding to $\alpha_k$, rank restrictions, (5.1) – (5.11), (9.4) and the Main Theorem of [12] imply that $\lambda|T_A = q\mu_1$ or $(q+q_0)\mu_1$ and $\dim V|A = 2k$ or $(2k)^2$, respectively. Also, k < rankY, by Theorem (4.1) of [12]. But then (1.32) and [8] imply dimV|Y ≠ dimV|A. So $\langle\lambda,\alpha_j\rangle > 0$ for some j > 1, as claimed.

For the remainder of Case I considerations, let $\alpha = \alpha_1$; so $L_A'$ has type $B_{k-1}$ or $C_{k-1}$, and $\dim V^1(Q_A) > 1$. Suppose $L_Y'$ has type $D_\ell$, for some $\ell \geq 4$. Then by (1.5), $Z_A \leq Z_Y$ and by (4.1) of [12], rank $L_A' = k-1 <$ rank$L_Y'$. If $L_A'$ acts irreducibly on some module other than W, the natural module for $L_Y'$, the Main Theorem of [12] implies that either $L_A'$ acts irreducibly on the 2 fundamental spin modules for $L_Y'$, or $L_A' = B_3$, $L_Y' = D_4$ and $L_A'$ acts irreducibly on 2 of the 3 restricted 8-dimensional irreducible $L_Y'$ modules. But there exists $\gamma \in \Pi(Y) - \Pi(L_Y)$ with $V_\gamma(Q_Y) \neq 0$ and $Q_Y/K_\gamma$ isomorphic to one of these two spin modules; while $I_\alpha$ is never isomorphic to one of these modules. Thus, W is the only $L_Y'$ module on which $L_A'$ acts irreducibly. So $V^1(Q_Y) \cong W$. If $L_Y' = D_7$, then $\dim V^1(Q_Y) = 14$, so k = 4. However, the bound on $\dim V_{\beta_1}(Q_Y)$ implies $\lambda|T_Y = \lambda_8$. By (4.1) of [12], the only remaining possibility is that

$L_Y' = D_4$, $k = 4$ and $\lambda|T_A = x\mu_1+q\mu_4$, for some p-power q. Induction, applied to the maximal parabolic of A corresponding to $\alpha_k$, implies that x = 0, $q_1$ or $q_1+q_2$, for $q_1$ and $q_2$ distinct p-powers. So dimV|A = 16, 128, or 1024. As well, $\langle\lambda,\beta_2+\beta_3+\beta_5\rangle = 1$ and $\langle\lambda,\beta_4\rangle = 0$. Finally, using (2.3) to obtain more information about $\lambda|T_Y$, and (1.32) and [8], we see that dimV|A $\neq$ dimV|Y. Thus, no component of $L_Y'$ has type $D_\ell$.

Suppose $A = B_3$ or $C_3$. The above work and (9.4) imply that $L_i$ has type $A_3$ for all i. Note that $\lambda|T_Y \neq \lambda_\ell$ for any $\ell$. For otherwise, $\lambda|T_A = q_2x\mu_1+q_2\mu_2$ or $q_1x\mu_1+q_2\mu_3$ for x = 0 or 1 and for some p-powers $q_1$ and $q_2$. Then Table 1 of [5] implies dimV|A = 8, 14, 48 or 64; so dimV|A $\neq$ dimV|Y by [8]. Now, in general, induction and rank restrictions imply dimV|A $\leq 6^2{\cdot}8$ if $Y = E_6$ or $E_7$ and dimV|A $\leq 6{\cdot}8^3$ if $Y = E_8$. But then the above remarks, (1.32) and [8] imply dimV|A < dimV|Y. Hence, $A \neq B_3$ or $C_3$.

Now, suppose $A = C_4$ or $B_4$. A straightforward argument shows that $L_Y' = A_5$ and $\lambda|T_A = x\mu_1+q\mu_2$ for some p-power q. If $Y = E_6$ or $E_7$, induction (applied to the $A_3$ maximal parabolic of A) implies that x = 0 or q. So dimV|A = 26 or 112. But then (1.32) and [8] imply dimV|A $\neq$ dimV|Y. So $Y = E_8$. By induction, x = 0, $q_1$, q, or $q_1+q$, for some p-power $q_1 \neq q$. Then, dimV|A = 26, 112, 208, or 896. But [8] and (1.32) imply dimV|A < dimV|Y. So $Y \neq B_4$ or $C_4$.

Suppose $A = B_5$ or $C_5$. The previous work of this result and the Main Theorem of [12] imply that $L_Y' = A_7$. The bound on $\dim V_{\beta_2}(Q_Y)$ implies that $\lambda|T_Y = \lambda_1$. Also, $\lambda|T_A = x\mu_1+q\mu_2$. However, induction (applied to the $A_4$ maximal parabolic of A) provides a contradiction. Thus, A does not have type $B_5$ or $C_5$. But this fact, together with rank restrictions implies that A does not have type $B_6$ or $C_6$, and consequently neither can A have type $B_7$ or $C_7$.

Case II: A has type $F_4$ and p = 2.

First note that $\langle\lambda,\alpha_j\rangle > 0$ for some $1 \leq j \leq 3$, else by induction, the

Main Theorem of [12] and the previous work of this result, $\lambda|T_A = q\mu_4$, for some p-power q, and $\dim V|A = 26 < \dim V|Y$. Let $\alpha = \alpha_4$, so $L_A' = B_3$ and $\dim V^1(Q_A) > 1$. If $L_Y' = D_7$, $V^1(Q_Y) \cong W$, the natural module for $L_Y'$. For otherwise, by the Main Theorem of [12], $Q_Y/K_{\beta_1}$ is a tensor decomposable, irreducible $L_A'$ module containing a nontrivial image of $Q_A{}^{\alpha}$. But now, the bound on $\dim V_{\beta_1}(Q_Y)$ implies $\lambda|T_Y = \lambda_8$. If $L_Y' = A_7$, the bound on $\dim V_{\beta_2}(Q_Y)$ implies $\lambda|T_Y = \lambda_1$, while $\lambda|T_A = q\mu_3 + x\mu_4$. Hence, $\dim V|A = 246 \cdot 26^k$ or $4096 \cdot 26^k$ for some $k \geq 0$. (See [8].) But then $\dim V|A \neq \dim V|Y$ by [8].

Suppose $L_Y' = A_5$. Since $L_A'$ acts irreducibly on W, the natural module for $L_Y'$, (2.3) implies that there does not exist $\gamma \in \Pi(Y) - \Pi(L_Y)$ with $Q_Y/K_{\gamma} \cong W$ or $W^*$. Also, by induction $V^1(Q_Y) \cong W$ or $W^*$. Hence, if $Y = E_6$, the bound on $\dim V_{\beta_2}(Q_Y)$ implies $\lambda|T_Y = \lambda_1$ or $\lambda_6$. But then [8] implies $\dim V|A \neq \dim V|Y$. So $Y = E_7$, $\Pi(L_Y) = \{\beta_j \mid j = 2,4,5,6,7\}$, and $\lambda|T_Y = x\lambda_3 + \lambda_\ell$ where $\ell = 2$ or $7$. Also, $\lambda|T_A = q\mu_1 + y\mu_4$ and by [8], $\dim V|A = 26^k$; so $x \neq 0$. By induction, $\langle \lambda, \alpha_4 \rangle = 0$ or $q_1$; so $\dim V|A \leq 26^2 < \dim V|Y$, by (1.32). So $L_Y' \neq A_5$.

It remains to consider the case where $L_Y'$ has type $D_4$ and by induction $\lambda|T_A = xq\mu_1 + q\mu_3 + y\mu_4$ for some p-power q and $x = 0$ or $1$. Also, since $Q_Y/K_{\beta_i}$ must contain a nontrivial image of $Q_A{}^{\alpha}$, for $i = 1$ and $5$, $\langle \lambda, \beta_3 + \beta_5 \rangle = 1$, $\langle \lambda, \beta_4 \rangle = 0$ and $\langle \lambda, \beta_2 \rangle = x$. As well, (2.3) implies $\langle \lambda, \beta_\ell \rangle = 0$ for $\ell \geq 7$. Applying induction to the $C_3$ maximal parabolic of A, we find that $Y = E_8$, $\lambda|T_Y = a\lambda_1 + \lambda_3$, and $\lambda|T_A = q\mu_3 + q\mu_4$. However, [8] and (1.32) imply $\dim V|A < \dim V|Y$. This completes the proof of (5.13).□

# CHAPTER 6: INITIAL RANK TWO RESULTS

In this chapter, we will prove the Main Theorem in case $A = A_2$, $p>2$ and $Y = E_n$. The method of proof depends almost entirely on restricting our attention to the embedding of one of the two maximal parabolics of A. In fact, we will actually be studying the embedding of the maximal parabolic of any rank 2 group whose Levi factor is $\langle U_{\pm\beta} \rangle$, for $\beta$ long. (Though we must assume $p > 3$, if $A = G_2$.) We establish a reasonably short list of possible such embeddings. (See (6.9).) For $A = A_2$, repeated applications of (1.23) usually enable us to determine the structure of $V|A$, by knowing this one embedding. The $A_2$ result is the following.

<u>Theorem 6.0.</u> (a) Let A be a simple algebraic group of type $A_2$, Y a simply connected, simple algebraic group of type $E_n$. Suppose $p>2$, $A < Y$ and $V|A$ is irreducible, for $V = V(\lambda)$ a nontrivial, restricted irreducible kY-module. Then, $p = 5$, $Y = E_6$ and $\lambda|T_Y = \lambda_1$ (or $\lambda_6$). Moreover, $\lambda|T_A = 2\mu_1 + 2\mu_2$, where $\mu_1$ and $\mu_2$ are the fundamental dominant weights corresponding to a fixed set of simple roots for $\Sigma(A)$.

(b) If $p \neq 2,5$ and $Y = E_6$, there exists a closed subgroup $B < Y$, $B \cong PSL_3(k)$ such that $V(\lambda_1)|B$ is irreducible.

<u>Remark:</u> The proof of (6.0)(b) is given in [16].

<u>Hypothesis:</u> For the remainder of this chapter we adopt Notation and Hypothesis (2.0) with rank $A = 2$, $\Sigma(Y)$ of type $E_n$, $p > 2$ and $p \neq 3$ when $A = G_2$. So (A,p) is not special. However, we will write $\Pi(A) = \{\alpha,\beta\}$, so $L_A = \langle U_{\pm\beta} \rangle T_A$ and write $\mu_\alpha$ and $\mu_\beta$ for the corresponding fundamental

dominant weights. We take $\beta$ to be the long root. Finally, assume $V|A$ is irreducible and $\langle\lambda,\beta\rangle \neq 0$.

(6.1). $L_Y'$ is not a simple algebraic group.

Proof: Suppose false. First note that Theorem (7.1) of [12] implies $L_Y'$ is not of type $E_k$. Thus, by (1.5), $Z_A \leq Z_Y$. Consider first the case $L_Y' = A_k$ for some k. By (2.14), k>1. Let W be the natural module for $L_Y'$. Then by (7.1) of [12], $V^1(Q_Y) \cong W$ or $W^*$. Thus, there does not exist $\gamma \in \Pi(Y)-\Pi(L_Y)$ such that $(\gamma,\Sigma L_Y) \neq 0$ and $Q_Y/K_\gamma \cong W$ or $W^*$. Moreover, since $W\wedge W$ (or $W^*\wedge W^*$) has all even weights as an $L_A'$ module and $p > 2$, there does not exist $\gamma \in \Pi(Y)-\Pi(L_Y)$ such that $(\gamma,\Sigma L_Y) \neq 0$ and $Q_Y/K_\gamma \cong W\wedge W$ or $W^*\wedge W^*$. These remarks imply that $L_Y' = A_{n-1}$, where $Y = E_n$, $n = 6, 7, 8$. Moreover, by (1.25), $\dim V_\gamma(Q_Y) \leq 2n$ in each case. However, it is an easy check to see that $\dim V_{\beta_2}(Q_Y) \geq \frac{1}{2}n(n-1)$ in each case. Contradiction.

Suppose $L_Y' = D_5$ and $V^1(Q_A)$ has high weight $(3q_1 + 3q_2)\mu_\beta$, as described in Theorem (7.1)(c) of [12]. Then there exists $\gamma \in \Pi(Y)-\Pi(L_Y)$ such that $(\gamma,\Sigma L_Y) \neq 0$ and $Q_Y/K_\gamma$ is a 16-dimensional irreducible $L_A'$-module. Thus $Q_A \leq K_\gamma$, contradicting $\langle\lambda,\beta\rangle \neq 0$.

Since p>2, it remains to consider the case where $L_Y' = D_k$ for some k and $V^1(Q_Y) \cong W$, the natural module for $L_Y'$. Note that by (1.14), W is a tensor decomposable $L_A'$ module and $L_Y' = D_4$ or $D_6$. If $L_Y' = D_6$, $f_{134567}v^+$ affords an $L_Y'$ composition factor of $V_{\beta_1}(Q_Y)$ of dimension 32. However this exceeds the bound of (1.25). Hence, $L_Y' = D_4$ and $V^1(Q_A)$ has high weight $(q_1+3q_2)\mu_\beta$, for $q_1$ and $q_2$ distinct p-powers. One checks that $L_A'$ acts irreducibly on two of the three fundamental 8-dimensional irreducible $L_Y'$-modules. Hence, $L_A'$ acts irreducibly on $Q_Y/K_{\beta_1}$ or on $Q_Y/K_{\beta_6}$, forcing $Q_A \leq K_{\beta_1}$ or $K_{\beta_6}$. But this contradicts (2.3).

This completes the proof of (6.1).□

Remark (6.2). (1) If there exists $1 \le i \le r$ with $L_i$ of exceptional type, then $\dim V^1(Q_Y) > 1$, induction and Theorem (7.1) of [12] imply that $Y = E_8$, $L_Y' = L_1 \times L_2$, where $L_1 = E_6$ and $L_2 = \langle U_{\pm\beta_8}\rangle$ and $\lambda|T_Y = x\lambda_7 + c\lambda_8$, for some $p>x\ge 0$ and $p>c>0$. Now, $\dim V^2(Q_Y) \ge 27c$. Thus, the bound on $\dim V^2(Q_A)_{\lambda - q\alpha}$ and the description of $V^2(Q_A)$ in (1.22) imply that $k\ge 7$, where $V|A = V_1{}^{q_1} \otimes \cdots \otimes V_k{}^{q_k}$, as in (2.0). In particular, if $V|A$ is tensor indecomposable and $V^1(Q_A)$ is nontrivial, all components of $L_Y'$ have classical type.

(2) We will often use without reference the fact that $U_{-2\alpha-\beta}\cdot U_{-3\alpha-\beta}\cdot U_{-3\alpha-2\beta} \le Q_A'$. This follows from the stated prime restrictions.

(6.3). If $L_i$ has type $D_{k_i}$ for some $i$ and $k_i$, then $Y$ has type $E_8$, $\lambda|T_Y = x\lambda_7 + c\lambda_8$, for $c>0$ and $\Pi(L_Y) = \{\beta_2,\beta_3,\beta_4,\beta_5,\beta_8\}$ or $\{\beta_1,\beta_2,\beta_3,\beta_4,\beta_5,\beta_8\}$.

Proof: Since all components of $L_Y'$ are necessarily of classical type, (1.5) implies $Z_A \le Z_Y$. Moreover, by (6.1), $L_Y'$ is not simple, so size restrictions imply $L_i = D_4$ or $D_5$ and $Y$ has type $E_7$ or $E_8$.

Case I: Suppose $M_i$ is nontrivial.

Consider first the case where $L_i = D_5$. Then, (7.1) of [12] and (1.14) imply $\rho_i(L_A')$ acts irreducibly on the two fundamental spin representations of $L_i$. Hence, $\langle U_{\pm\beta_1}\rangle \le L_i$, else $Q_Y/K_{\beta_1}$ is a 16-dimensional irreducible $L_A'$ module containing a nontrivial image of $Q_A$. The same argument applied to $Q_Y/K_{\beta_6}$ implies $\langle U_{\pm\beta_7}\rangle \le L_Y'$. Thus, $L_Y' = L_i \times L_j$ where $L_j = \langle U_{\pm\beta_7}\rangle$ or $\langle U_{\pm\beta_7}, U_{\pm\beta_8}\rangle$. Also, (2.7) implies $M_j$ is trivial. Now, using (1.30) and (1.32), and recalling that $p>3$ in this configuration, we see that the bound on $\dim V_{\beta_6}(Q_Y)$ of (1.25) is exceeded. Thus, $L_i = D_4$.

Since $p\ne 2$, (7.1) of [12] and (1.14) imply $M_i|L_A'$ has high weight $(q_1+3q_2)\mu_\beta$, for $q_1$ and $q_2$ distinct p-powers. By considering $\rho_i(h_\beta(c))$, one checks that $\rho_i(L_A')$ acts irreducibly on two of the three restricted

8-dimensional irreducible $L_i$ modules. On the third $\rho_i(L_A')$ acts with composition factors of dimensions 3 and 5. Thus, $Q_A \leq K_{\beta_1}$, contradicting (2.3). This completes the consideration of Case I.

Case II: Suppose $M_i$ is trivial.

Note that there does not exist $\gamma \in \Pi(Y)-\Pi(L_Y)$ and $1\leq j\leq r$, $j\neq i$ such that $(\Sigma L_i,\gamma) \neq 0 \neq (\Sigma L_j,\gamma)$. For otherwise, by size restrictions, $M_j$ is nontrivial and (1.32) and (1.36) imply that the bound on $\dim V_\gamma(Q_Y)$ of (1.25) is exceeded in every possible such configuration. Thus, $Y = E_8$ and $L_Y' = L_i \times \langle U_{\pm\beta_8}\rangle$, where $L_i = \langle U_{\pm\beta_j} \mid 2\leq j\leq 5 \rangle$ or $\langle U_{\pm\beta_j} \mid 1\leq j\leq 5 \rangle$. Also, $\langle\lambda,\gamma\rangle = 0$ for all $\gamma \in \Pi(Y)-\Pi(L_Y)$ such that $(\gamma,\Sigma L_i) \neq 0$, else the bound on $\dim V_\gamma(Q_Y)$ is exceeded. Thus, (6.3) holds.□

(6.4). Let $\gamma \in \Pi(Y)-\Pi(L_Y)$ and $1\leq i\neq j\leq r$ such that $(\Sigma L_i,\gamma) \neq 0 \neq (\Sigma L_j,\gamma)$. Then $M_i$ or $M_j$ is trivial.

Proof: Suppose false; i.e., suppose $M_i$ and $M_j$ are both nontrivial. By Theorem (7.1) of [12] and (6.3), $L_i$ and $L_j$ have type $A_{k_i}$, $A_{k_j}$, respectively, for some $k_i, k_j \geq 1$. Thus, by (1.5), $Z_A \leq Z_Y$. Let $W_k$ denote the natural module for $L_k$, $k = i, j$. By (7.1) of [12], if rank $L_k > 1$, $M_k \cong W_k$ or $W_k^*$ for $k = i, j$.

Case I: $V_{L_\ell}(-\gamma) \cong W_\ell$ or $W_\ell^*$ for $\ell = i$ and $j$.

By (2.5) and (2.7) and the preceeding remarks, $\gamma = \beta_4$ and there exists $k \neq i, j$, such that $M_k$ is trivial and $(\Sigma L_k,\beta_4) \neq 0$. Then, (1.15) implies that $L_i \times L_j \times L_k$ has type $A_1 \times A_1 \times A_m$, $m = 1$ or 3, or $A_1 \times A_2 \times A_m$, $m = 2, 3$ or 4 (with a possible reordering of the triples). Then (1.36) implies that the bound on $\dim V_{\beta_4}(Q_Y)$ of (1.25) is exceeded unless the triples are $\{A_1,A_1,A_m\}$, $m = 1$ or 3, or $\{A_1,A_2,A_2\}$ with the $A_1$ component acting trivially on $V^1(Q_Y)$.

Suppose rank $L_m = 1$ for $m = i, j, k$. Let $\gamma_m \in \Pi(Y)$ such that $L_m = \langle U_{\pm\gamma_m}\rangle$ for $m = i, j, k$. Let $q_i$ (respectively, $q_j$) be the field twist on the embedding of $L_A'$ in $\langle U_{\pm\gamma_i}\rangle$ (respectively, $\langle U_{\pm\gamma_j}\rangle$), where $q_i \neq q_j$.

Then, by (2.7), we may assume that the field twist on the embedding of $L_A'$ in $\langle U_{\pm\gamma_k}\rangle$ is also $q_j$. Then the $L_A'$ composition factors of $Q_Y/K_{\beta_4}$ have high weights $(2q_j + q_i)\mu_\beta$ and $q_i\mu_\beta$. Since $V_{T_Y}(\lambda-\gamma_i-\beta_4) \neq 0$, some element from the group $U_{-\beta_4}\cdot U_{-\gamma_i-\beta_4}$ must appear in the factorization of an element in $Q_A-Q_A'$. But $-\beta_4$ (respectively, $-\gamma_i-\beta_4$) affords $T(L_A')$ weight $(2q_j + q_i)\mu_\beta$ (respectively, $(2q_j - q_i)\mu_\beta$). Neither of these weights occurs in $(Q_A{}^\alpha)^{q_i}$. Thus, rank $L_m > 1$ for $m = i$, $j$, or $k$.

We consider now the case where $L_i{\times}L_j{\times}L_k$, in some reordering, has type $A_1{\times}A_1{\times}A_3$, so $\Pi(L_Y) = \{\beta_2,\beta_3,\beta_5,\beta_6,\beta_7\}$. Let $\langle U_{\pm\beta_5},U_{\pm\beta_6},U_{\pm\beta_7}\rangle = L_0$. Let $q_1$ (respectively, $q_2$) be the field twist on the embedding of $L_A'$ in $\langle U_{\pm\beta_2}\rangle$ (respectively, $\langle U_{\pm\beta_3}\rangle$). Since $p>2$, (1.15) implies $V_{L_0}(-\beta_4)|L_A'$ is tensor indecomposable; in particular, $p>3$. Let the field twist on the embedding of $L_A'$ in $L_0$ be $q_3$. By (2.5), (2.6) and (2.7), $\{q_1,q_2,q_3\}$ consists of two distinct powers of p. It is then an easy check to see that there is no 2-dimensional $L_A'$ composition factor of $Q_Y/K_{\beta_4}$. Thus $Q_A \leq K_{\beta_4}$, contradicting (2.3).

We have, therefore, $\Pi(L_i{\times}L_j{\times}L_k) = \{\beta_1,\beta_3,\beta_2,\beta_5,\beta_6\}$ with $\langle\lambda,\beta_2\rangle = 0$. The bound on $\dim V_{\beta_4}(Q_Y)$ of (1.25) implies that $\langle\lambda,\beta_1\rangle = 1 = \langle\lambda,\beta_6\rangle$ and $\langle\lambda,\beta_m\rangle = 0$ for $2\leq m\leq 5$. Now, $f_{134}v^+$ affords an $L_Y'$ composition factor in $V_{\beta_4}(Q_Y)$ of dimension 14, if p=3, or 16 otherwise. Also, $\dim V_{T_Y}(\lambda-\beta_1-\beta_3-\beta_4-\beta_5-\beta_6) \geq 4$, by (1.34) and if p=3, a 1-space from this weight space occurs in the above mentioned composition factor, and otherwise, a 2-space from this weight space occurs. Hence, 3 (respectively, 2) distinct composition factors of $V_{\beta_4}(Q_Y)$ of dimension 2 are afforded by vectors in $V_{T_Y}(\lambda-\beta_1-\beta_3-\beta_4-\beta_5-\beta_6)$, if $p\neq 3$ (respectively, p=3). In either case, the bound on $\dim V_{\beta_4}(Q_Y)$ is exceeded. Thus, $L_i{\times}L_j{\times}L_k$ does not have type $A_2{\times}A_2{\times}A_1$ and this completes the consideration of Case I.

<u>Case II</u>: For $\ell = i$ or $j$, $V_{L_\ell}(-\gamma) \not\cong W_\ell$ or $W_\ell{}^*$.

Since $W_\ell$ is an irreducible $L_A'$ module for $\ell = i$, $j$, if rank $L_\ell > 1$, there does not exist $\delta \in \Pi(Y)-\Pi(L_Y)$ such that $Q_Y/K_\delta \cong W_\ell$ or $W_\ell^*$, else $Q_A \le K_\delta$. Applying (1.36) and the various techniques for obtaining lower bounds on dimensions of modules (e.g., (1.30), (1.32) and (1.35)), the bound on $\dim V_\gamma(Q_Y)$ of (1.25) restricts the situation still further. We find that $\Pi(L_i \times L_j) = \{\beta_1,\beta_2,\beta_4,\beta_5\}$ and p=3. In particular, $W_j|L_A'$ is tensor decomposable. Let $q_1$ be the field twist on the embedding of $L_A'$ in $\langle U_{\pm\beta_1}\rangle$. Suppose $W_j|L_A'$ has high weight $(q_2+q_3)\mu_\beta$, for $q_2$ and $q_3$ distinct p-powers, with $q_2 \neq q_1 \neq q_3$. Then, $Q_Y/K_{\beta_3}$ has $L_A'$ composition factors with high weights $(q_1+2q_2)\mu_\beta$ and $(q_1+2q_3)\mu_\beta$. In particular, there is no 2-dimensional $L_A'$ composition factor of $Q_Y/K_{\beta_3}$, contradicting (2.3).

This completes the consideration of Case II and the proof of (6.4).□

<u>(6.5).</u> Suppose there exist distinct $1 \le i,j,k \le r$ such that $(\Sigma L_m,\beta_4) \neq 0$ for m = i, j, k and suppose $M_m$ is nontrivial for m = i, j or k. Then one of the following holds:

(i) $A \neq G_2$, $Y = E_6$, $\lambda|T_Y = \lambda_1$ or $\lambda_6$ and $\Pi(L_Y) = \{\beta_i \mid i \neq 4\}$.

(ii) $A \neq A_2$, $Y = E_8$, $\lambda|T_Y = \lambda_1$ and $\Pi(L_i \times L_j \times L_k) = \{\beta_1,\beta_3,\beta_2,\beta_5,\beta_6\}$.

(iii) $Y = E_7$, $p\neq 3$, $L_i \times L_j \times L_k$ has type $A_1 \times A_1 \times A_3$ and $\lambda|T_Y = x\lambda_1 + \lambda_7$, $p > x \ge 0$.

(iv) $A \neq A_2$, rank $L_m = 1$ for m = i, j, k, and $\dim(M_i \otimes M_j \otimes M_k) = 2 = \dim V^1(Q_Y)$ and $\langle\lambda,\beta_4\rangle = 0$.

<u>Proof</u>: Since all components of $L_Y'$ are necessarily of classical type, (1.5) implies $Z_A \le Z_Y$. Let $W_m$ denote the natural module for $L_m$, m = i, j, k. Then Theorem (7.1) of [12] implies that if rank$L_m > 1$ and $M_m$ is nontrivial, $M_m \cong W_m$ or $W_m^*$, and (6.4) implies that only one of $M_i$, $M_j$ and $M_k$ is nontrivial. As well, (1.15) restricts the situation somewhat. We then use the bound on $\dim V_{\beta_4}(Q_Y)$ of (1.25) in conjunction with (1.36) to see that $\langle\lambda,\beta_4\rangle = 0$ and one of the following holds:

(a) rank $L_m = 1$ for m = i, j, k and $\dim(M_i \otimes M_j \otimes M_k) = 2$.

(b) $\Pi(L_Y) = \{\beta_2,\beta_3,\beta_5,\beta_6,\beta_7\}$, $\langle\lambda,\beta_m\rangle = 0$ for $2\le m\le 6$ and $\langle\lambda,\beta_7\rangle = 1$.

(c) $L_i \times L_j \times L_k$ has type $A_2\times A_1\times A_2$, the $A_1$ component acts trivially on $V^1(Q_Y)$ and if $M_\ell$ is nontrivial for $\ell\in\{i,j,k\}$, $M_\ell \cong (V_{L_\ell}(-\gamma))^*$.

(d) $L_i\times L_j\times L_k$ has type $A_2\times A_1\times A_3$, $\langle\lambda,\beta_m\rangle = 0$ for $1\le m\le 6$, $\langle\lambda,\beta_7\rangle = 1$ and $V^1(Q_A)$ is a tensor decomposable $L_A'$ module.

(Recall, $\lambda|T_Y \ne \lambda_8$ if $Y = E_8$.) In the configuration of (d), $Y = E_7$, else $Q_Y/K_{\beta_8}$ is a 4-dimensional irreducible $L_A'$-module containing a nontrivial image of $Q_A{}^\alpha$. But now, using (1.23), (1.32) and (1.35), we find that $\dim V|A \ne \dim V|Y$. Thus the configuration of (d) does not occur.

Consider now the configuration of (c). If $Y = E_6$, $\dim V|Y = 27$. Using the methods of (1.30), (1.32), (1.33) and (1.35) we see that if $A$ has type $G_2$, $\dim V|A > 27$. Thus, (i) holds. If $Y = E_7$, $Q_A \le K_{\beta_7}$ since $Q_Y/K_{\beta_7}$ is a 3-dimensional irreducible $L_A'$-module. But then, $\lambda|T_Y = \lambda_1$. Consider the case where $Y = E_8$. We must study the image of $Q_A$ in $Q_Y/K_{\beta_4}$. For the purposes of this argument we may assume $\langle\lambda,\beta_1\rangle = 1$, $\langle\lambda,\beta_m\rangle = 0$ for $2\le m\le 6$. Let $q_1$, $q_2$ and $q_3$ be the field twists on the embeddings of $L_A'$ in $\langle U_{\pm\beta_1},U_{\pm\beta_3}\rangle$, $\langle U_{\pm\beta_2}\rangle$ and $\langle U_{\pm\beta_5},U_{\pm\beta_6}\rangle$, respectively. By (2.7), $q_1$, $q_2$ and $q_3$ are not all distinct. We will show that, in fact, $q_1 = q_3$.

Suppose $q_1 = q_2 \ne q_3$. Then the $L_A'$ composition factors of $Q_Y/K_{\beta_4}$ have high weights $(3q_2+2q_3)\mu_\beta$ and $(q_1+2q_3)\mu_\beta$. (If $p=3$ and $3q_1 = q_3$, the weights are $3q_3\mu_\beta$, $q_3\mu_\beta$ and $(q_1+2q_3)\mu_\beta$.) Thus, (2.4) implies that $p=3$, $3q_1 = q_3$ and $\beta_4|Z_A = 9q_1\alpha$ or $3q_1\alpha$. Since $V_{T_Y}(\lambda-\beta_1-\beta_2-\beta_3-\beta_4) \ne 0$, a nonidentity element from the set $U_{-24}\cdot U_{-234}\cdot U_{-1234}$ must appear in the factorization of some element in $Q_A - Q_A'$. Also, $-\beta_2-\beta_4$ (respectively, $-\beta_2-\beta_3-\beta_4$, $-\beta_1-\beta_2-\beta_3-\beta_4$) affords $T(L_A')$ weight $7q_1\mu_\beta$ (respectively, $5q_1$, $3q_1$). So (2.4) implies $\beta_4|Z_A = 3q_1\alpha = q_3\alpha$. Then, examining the $T(L_A')$ weights of root group elements in $Q_Y/K_{\beta_4}$, we see that $x_{-\alpha}(t) = x_{-1234}(c_1t^{q_3})u_1$ and $x_{-\alpha-\beta}(t) = x_{-12345}(c_2t^{q_3})u_2$, where $c_i \in k^*$, $u_i \in K_{\beta_4}$. However, there is then a nontrivial contribution to the root group $U_{-1345}$ in the expression for $[x_\beta(t), x_{-\alpha-\beta}(t)]$, contradicting (6.2)(2) and the given factorization of $x_\alpha(t)$. Thus, we do not have $q_1 = q_2 \neq q_3$. A similar argument shows that the configuration $q_1 \neq q_2 = q_3$ cannot occur. In fact, $q_1 = q_3$, regardless of the labelling of $\lambda|T_Y$.

If $Q_A \leq K_{\beta_7}$, $\lambda|T_Y = \lambda_1$. Suppose $Q_A \nleq K_{\beta_7}$. Then $\langle U_{\pm\beta_8}\rangle \leq L_Y'$, else $Q_Y/K_{\beta_7}$ is a 3-dimensional irreducible $L_A'$ module containing a nontrivial image of $Q_A{}^\alpha$. Then (2.7) implies that the field twist on the embedding of $L_A'$ in $\langle U_{\pm\beta_8}\rangle$ is $q_1 = q_3$. Thus, (2.5) implies $\langle\lambda,\beta_8\rangle = 0$. Moreover, $\langle\lambda,\beta_7\rangle = 0$ else the bound on $\dim V_{\beta_7}(Q_Y)$ of (1.25) is exceeded. Thus, $\lambda|T_Y = \lambda_1$ or $\lambda_6$. In either case, $A \neq A_2$, as (1.23) implies $\dim V|A < \dim V|Y$. If $\lambda|T_Y = \lambda_6$, (7.1) of [12] implies $\langle\lambda,\alpha\rangle = 0, q, 2q, 3q, 4q, 5q, q+q_0$ or $q+2q_0$, for $q$ and $q_0$ distinct p-powers. Then (1.27) implies $\dim V|A \leq 2\cdot 3^3\cdot 7^2\cdot 11$. But by (1.38) $\dim V|Y > \dim V|A$. Hence, $\lambda|T_Y \neq \lambda_6$ and (ii) holds.

For $L_Y$ as in (b), $p\neq 3$, else $V^1(Q_Y)$ is tensor decomposable and (1.15) is contradicted. Also $Y = E_7$, else $Q_A/K_{\beta_7}$ is a 4-dimensional irreducible $L_A'$-module containing a nontrivial image of $Q_A$. Thus, (iii) holds.

Finally, we must consider the case where $\mathrm{rank}(L_m) = 1$ for $m = i, j, k$. Suppose $q_0$ is the field twist on the embedding of $L_A'$ in two of the components $L_i, L_j, L_k$ and $q_1$ is the twist on the embedding in the third. If $q_0 \neq q_1$, the $L_A'$ composition factors of $Q_A/K_{\beta_4}$ have high weights $(2q_0 + q_1)\mu_\beta$ and $q_1\mu_\beta$. Since $p>2$, (2.4) implies $\beta_4|Z_A = q_1\alpha$.

Temporarily label as follows: $\Pi(L_m) = \{\gamma_m\}$ for $m = i, j, k$, and let $\langle\lambda,\gamma_i\rangle \neq 0$. So $\langle\lambda,\gamma_j+\gamma_k\rangle = 0$. Let $q_m$ be the field twist on the embedding of $L_A'$ in $L_m$ for $m = i, j, k$. Suppose $q_j = q_k \neq q_i$. From above, $\beta_4|Z_A = q_i\alpha$. However, $V_{T_Y}(\lambda-\gamma_i-\beta_4) \neq 0$ implies that a nonidentity element from the set $U_{-\beta_4}\cdot U_{-\gamma_i-\beta_4}$ must appear in the factorization of some element in $Q_A - Q_A'$. But $-\beta_4$ (respectively, $-\gamma_i-\beta_4$) affords $T(L_A')$ weight $(2q_j + q_i)\mu_\beta$ (respectively, $(2q_j - q_i)\mu_\beta$) which is not a weight in $(Q_A{}^\alpha)^{q_i}$ if $q_i \neq q_j$. Thus, $q_i = q_j$ or $q_i = q_k$.

We are now able to show that there does not exist $1 \leq \ell \leq r$, $\ell \notin \{i,j,k\}$ such that $M_\ell$ is nontrivial. For, suppose there exists such an $\ell$. Then, $L_\ell = \langle U_{\pm\beta_7}\rangle$, $\langle U_{\pm\beta_8}\rangle$ or $\langle U_{\pm\beta_7}, U_{\pm\beta_8}\rangle$. If $L_\ell = \langle U_{\pm\beta_7}\rangle$, (2.17) and (2.4) imply $Q_A \leq K_{\beta_6}$, contradicting (2.3). If $L_\ell = \langle U_{\pm\beta_7}, U_{\pm\beta_8}\rangle$, (2.17) implies that the field twists on the embeddings of $L_A'$ in $\langle U_{\pm\beta_5}\rangle$ and in $L_\ell$ are equal. Also, if this twist is q, $\beta_6|Z_A = q\alpha$. The above work on field twists and (2.5) imply $\beta_4|Z_A = q\alpha$. However, $\dim V_{\beta_4}(Q_Y) = 4\dim M_\ell = \dim V^2(Q_A)_{\lambda - q\alpha}$, by (1.22). But there is a nontrivial contribution to $V^2(Q_A)_{\lambda - q\alpha}$ from $V_{\beta_6}(Q_Y)$. Contradiction. Thus, $L_\ell = \langle U_{\pm\beta_8}\rangle$. If $Q_A \not\leq K_{\beta_6}$, (2.8) implies that the field twists on the embedding of $L_A'$ in $\langle U_{\pm\beta_5}\rangle$ and in $L_\ell$ are equal. If $Q_A \leq K_{\beta_6}$, we consider the $L_A'$ composition factors of $Q_Y(\beta_7,\beta_6)$. Since $p>2$, (2.10) and (2.11) give the same result Now proceed as before to produce a contradiction. We have, therefore, $\dim V^1(Q_Y) = 2$. Moreover, (1.23) implies $\dim V|A < \dim V|Y$ if $A = A_2$. Thus, (iv) holds and the proof of (6.5) is complete.□

(6.6). Let $\gamma \in \Pi(Y)-\Pi(L_Y)$. Suppose there exists a unique pair $1\leq i,j\leq r$ such that $(\Sigma L_i,\gamma) \neq 0 \neq (\Sigma L_j,\gamma)$, $L_i$, $L_j$ are of type $A_{k_i}$, $A_{k_j}$ for some $k_i$, $k_j \geq 1$ and $\dim(M_i \otimes M_j) > 1$. Then, one of the following holds:

(i) $A = G_2$ (so $p>3$), $Y = E_8$, $\Pi(L_i \times L_j) = \{\beta_1,\beta_2,\beta_3,\beta_4,\beta_6\}$, $\lambda|T_Y = \lambda_1$, $\langle\lambda,\beta\rangle = 4q_1$ and $\langle\lambda,\alpha\rangle = q+q_1$, $2q_1+q$ or $3q_1+q$, for q and $q_1$ distinct p-powers

(ii) $A \neq A_2$, $Y = E_7$ or $E_8$, $\langle U_{\pm\beta_1}\rangle$ and $\langle U_{\pm\beta_2}, U_{\pm\beta_4}, U_{\pm\beta_5}\rangle$ are components of $L_Y'$, and $\langle U_{\pm\beta_7}\rangle \leq L_Y'$. Also, $\langle\lambda,\beta_2\rangle = 1$, $\langle\lambda, \beta_\ell\rangle = 0$ for $\ell = 1, 3, 4, 5, 6, 7$ and $\dim V^1(Q_Y) = 4$.

(iii) $L_i \times L_j$ has type $A_1 \times A_2$ and $\langle\lambda,\gamma\rangle = 0$.

(iv) $A = A_2$ or $B_2$, $p>3$, $Y = E_7$, $\lambda|T_Y = \lambda_7$, $\Pi(L_Y) = \{\beta_1,\beta_3,\beta_5,\beta_6,\beta_7\}$ and $V^1(Q_A)$ is a tensor indecomposable $L_A'$ module.

(v) $A = B_2$ or $G_2$, $Y = E_8$, $\lambda|T_Y = \lambda_2$ and $\Pi(L_Y) = \{\beta_1,\beta_2,\beta_4,\beta_6,\beta_7,\beta_8\}$. Moreover, $\beta_3|Z_A = q\alpha = \beta_5|Z_A$, where q is the field twist on the embedding of $L_A'$ in each component of $L_Y'$.

(vi) $A = B_2$ or $G_2$, $Y = E_7$ or $E_8$, $L_i$ (respectively, $L_j$) has type $A_1$ (respectively, $A_3$), and $\langle\lambda,\gamma\rangle = 0$. $V_{L_m}(-\gamma) \cong W_m$ or $W_m^*$, where $W_m$ is the natural module for $L_m$, for $m = i,j$. If $L_iv^+ \neq v^+$, then $L_jv^+ = v^+$ and $M_i \cong W_i$. If $L_jv^+ \neq v^+$, then $L_iv^+ = v^+$ and $M_j \cong (V_{L_j}(-\gamma))^*$. Moreover, in each case, $V_{L_j}(-\gamma)|L_A'$ is tensor decomposable and $\dim V^1(Q_Y) = \dim(M_i \otimes M_j)$.

In addition, if there exists $\delta \in \Pi(Y) - \Pi(L_Y)$ with $(\delta,\Sigma L_i) \neq 0$, $(\delta,\Sigma L_k) = 0$ for $k \neq i$ and $Q_Y/K_\delta \cong M_i$ or $M_i^*$, then either $k_i = 1$ or $V_\delta(Q_Y) = 0$.

Proof: Since all components of $L_Y'$ are necessarily of classical type, (1.5) implies $Z_A \leq Z_Y$. Let $W_k$ denote the natural module for $L_k$, $k = i, j$. Theorem (7.1) of [12] implies that if rank $L_k > 1$ and $M_k$ is nontrivial, then $M_k \cong W_k$ or $W_k^*$ for $k = i, j$. Also (6.4) implies that only one of $M_i$ and $M_j$ is nontrivial.

Case I: $V_{L_k}(-\gamma) \cong W_k \wedge W_k$ (or $W_k^* \wedge W_k^*$) for $k = i$ or $j$.

The bound on $\dim V_\gamma(Q_Y)$ of (1.25) and (1.15) restrict the situation considerably. We are left with the following possibilities for the type of $L_i \times L_j$: $A_1 \times A_k$, $k = 3, 4, 5$ with the $A_1$ component acting trivially on $V^1(Q_Y)$, or $A_3 \times A_3$. The cases $L_i \times L_j$ of type $A_3 \times A_3$ or $A_1 \times A_5$ are easily ruled out by standard arguments.

Consider now the case $L_i \times L_j$ of type $A_1 \times A_4$. The bound on $\dim V_\gamma(Q_Y)$ of (1.25) implies $\langle\lambda,\beta_2\rangle = 0 = \langle\lambda,\gamma\rangle$. Thus, if $Y = E_6$, $\lambda|T_Y = \lambda_1$ or $\lambda_6$ and $\dim V|Y = 27$. However, if we recall that $p \geq 5$ and apply (1.29), (1.32) and (1.23), it is not difficult to see that $\dim V|A > 27$ in every case. Thus, $Y = E_7$ or $E_8$. Moreover, standard arguments imply $\Pi(L_i \times L_j) = \{\beta_1,\beta_2,\beta_3,\beta_4,\beta_6\}$ and $\langle\lambda,\beta_1\rangle = 1$, $\langle\lambda,\beta_i\rangle = 0$ for $2 \leq i \leq 6$. We consider the image of $Q_A$ in $Q_Y/K_{\beta_5}$. Let $q_1$ be the field twist on the embedding of $L_A'$ in the $A_4$ component and let $q_2$ be the field twist on the embedding of $L_A'$ in $\langle U_{\pm\beta_6}\rangle$. Examining the $L_A'$ composition factors of

$Q_Y/K_{\beta_5}$ we see that if $q_1 \neq q_2$ we must have $p = 5$ and $5q_1 = q_2$. Moreover, the only composition factor isomorphic to a twist of $Q_A{}^\alpha$ has twist $q_1$, so $\beta_5|Z_A = q_1\alpha$. Since $V_{T_Y}(\lambda-\beta_1-\beta_3-\beta_4-\beta_5) \neq 0$, a nonidentity element from the set $U_{-5}\cdot U_{-45}\cdot U_{-345}\cdot U_{-1345}$ must appear in the factorization of some element in $Q_A - Q_A'$. However, $-\beta_5$ (respectively, $-\beta_4-\beta_5$, $-\beta_3-\beta_4-\beta_5$, $-\beta_1-\beta_3-\beta_4-\beta_5$) affords $T(L_A')$ weight $11q_1\mu_\beta$ (respectively, $9q_1\mu_\beta$, $7q_1\mu_\beta$, $5q_1\mu_\beta$), none of which occur in $(Q_A{}^\alpha)^{q_1}$. Thus, (2.4) implies $q_1 = q_2$.

We now claim that $\beta_5|Z_A = q_1\alpha$. Examining the $L_A'$ composition factors of $Q_Y/K_{\beta_5}$ we see that this is clear if $p>7$. If $p=5$, possibly $\beta_5|Z_A = 5q_1\alpha$. But, in this case, examining the $T(L_A')$ weight vectors in $Q_Y/K_{\beta_5}$, we have $x_{-\alpha}(t) = x_{-45}(c_1t^{5q_1})x_{-56}(c_2t^{5q_1})w_1$ and $x_{-\alpha-\beta}(t) = x_{-(1,1,1,2,1,1)}(d_1t^{5q_1})\cdot x_{-(1,1,2,2,1,0)}(d_2t^{5q_1})w_2$, where $c_i$, $d_i \in k$, $c_1$ or $c_2$ nonzero, $d_1$ or $d_2$ nonzero, and $w_i \in K_{\beta_5}$. However, since a nonidentity element from the group $U_{\beta_4}$ appears in the factorization of $x_\beta(t)$, there is a nontrivial contribution to the root group $U_{-123456}$ in the expression for $[x_{-\alpha-\beta}(t), x_\beta(t)]$. This contradicts the given factorization of $x_{-\alpha}(t)$. Thus, $\beta_5|Z_A \neq 5q_1\alpha$. If $p=7$, the $L_A'$ composition factors of $Q_Y/K_{\beta_5}$ have high weights $7q_1\mu_\beta$, $5q_1\mu_\beta$, $3q_1\mu_\beta$, and $q_1\mu_\beta$ and $-\beta_5$ affords $T(L_A')$ weight $7q_1\mu_\beta$. By (2.16) $\beta_5|Z_A \neq 7q_1\mu_\beta$. So $\beta_5|Z_A = q_1\alpha$, as claimed.

Suppose $Y = E_7$; so $\lambda|T_Y = \lambda_1 + x\lambda_7$ where $p>x>0$. Thus, $Q_A \not\leq K_{\beta_7}$ and since $Q_Y/K_{\beta_7}$ is isomorphic to $(Q_A{}^\alpha)^{q_1}$, $\beta_7|T_A = q_1\alpha$. However, $f_{\beta_7}v^+$ and $f_{1345}v^+$ afford 2 distinct $L_Y'$ composition factors in the $Z_A$ weight space $V^2(Q_A)_{\lambda-q_1\alpha}$, exceeding the dimension bound of (1.22). So $Y \neq E_7$.

We now have $Y = E_8$. If $\langle U_{\pm\beta_8}\rangle$ is a component of $L_Y'$, (2.17) implies $Q_A \leq K_{\beta_7}$, so $\lambda|T_Y = \lambda_1$. If $\langle U_{\pm\beta_8}\rangle$ is not a component of $L_Y'$, $\langle\lambda,\beta_8\rangle = 0$ by (2.3) and $\langle\lambda,\beta_7\rangle = 0$ by (2.13). Again $\lambda|T_Y = \lambda_1$. Also, $f_{1345}v^+$ affords an $L_Y'$ composition factor in $V^2(Q_A)_{\lambda-q_1\alpha}$ of dimension 10, which is the upper bound on this dimension, by (1.22). So $\dim V^2(Q_A)_{\lambda-q_1\alpha} = 10$. This can occur only if $q_1$ has nonzero coefficient in the p-adic expansion of

$\langle\lambda,\alpha\rangle$. (See (1.22) for a description of $V^2(Q_A)_{\lambda-q_1\alpha}$.) Suppose now that $V|A$ is tensor indecomposable. We first claim that $Q_A \nleq K_{\beta_7}$. Otherwise, we have $P_A \leq P_Y\hat{} \geq B_Y^-$, a parabolic subgroup with Levi factor $L_Y\hat{} = \langle L_Y, U_{\pm\beta_7}\rangle$ and $Q_A \leq R_u(P_Y\hat{}) = Q_Y\hat{}$. But $\dim V^2(Q_Y\hat{})$ exceeds the bound on $\dim V^2(Q_A)$. So $Q_A \nleq K_{\beta_7}$, as claimed. This implies that $L_Y' = L_i \times L_j$. Moreover, $\beta_7|Z_A = q_1\alpha$ as in $E_7$. Also, $Q_A \leq K_{\beta_8}$, so if $P_2 \geq B_Y^-$ is the parabolic subgroup of $Y$ with Levi factor $L_2 = \langle L_Y, U_{\pm\beta_8}\rangle$, then $P_A \leq P_2$ and $Q_A \leq R_u(P_2) = Q_2$. Now $[V,Q_A{}^2] = [V,Q_2{}^2]$, since $V_{\beta_5}(Q_2) = V_{\beta_5}(Q_Y) = V^2(Q_A)_{\lambda-q_1\alpha} = V^2(Q_A)$. By (1.20) $\dim V^3(Q_A) \leq 20$. But $f_{(1,1,1,2,2,1,0,0)}v^+$ and $f_{134567}v^+$ afford distinct $L_2$ composition factors in $V^3(Q_2)$ of dimensions 15 and 10, respectively. Contradiction. Hence, $V|A$ is tensor decomposable.

By induction and the above remarks, $\langle\lambda,\alpha\rangle = q_1+q$, $q_1+2q$, $2q_1+q$, $q+q_1+q_0$, $q_1+3q$ or $3q_1+q$, for $q$ and $q_0$ distinct $p$-powers, each distinct from $q_1$. So (1.23) implies that $A$ does not have type $A_2$. By [8], $\dim V|Y = 3875$. Now, (1.27) implies $\dim V|A < \dim V|Y$ if $A = B_2$. If $A = G_2$, [8] and the methods of (1.30) and (1.32) imply $\dim V|A > 3875$ if $\langle\lambda,\alpha\rangle = q+q_0+q_1$, $q_1+3q$ or $q_1+2q$. Thus, (6.6)(i) holds.

Consider now $L_i \times L_j$ of type $A_1 \times A_3$. Since the $A_3$ component acts nontrivially on $V^1(Q_Y)$, $\Pi(L_i \times L_j) = \{\beta_1,\beta_2,\beta_4,\beta_5\}$. Otherwise, $Q_Y/K_{\beta_1}$ is a 4-dimensional irreducible $L_A'$-module with proper submodule $Q_A K_{\beta_1}/K_{\beta_1}$. A similar argument and (6.4) imply $Y = E_7$ or $E_8$ and $\langle U_{\pm\beta_7}\rangle$ is contained in a component of $L_Y'$ which acts trivially on $V^1(Q_Y)$. The bound of (1.25) on $\dim V_{\beta_3}(Q_Y)$ and $\dim V_{\beta_6}(Q_Y)$ implies $\langle\lambda,\beta_2\rangle = 1$ and $\langle\lambda,\beta_k\rangle = 0$ for $k = 1$, 3, 4, 5, 6. Moreover, $A$ does not have type $A_2$, else (1.23), (1.26) and (1.32) imply $\dim V|A < \dim V|Y$. Thus, if $L_i \times L_j$ has type $A_1 \times A_3$, (6.6)(ii) holds.

This completes the consideration of Case I.

Case II: $V_{L_k}(-\gamma) \cong W_k$ or $W_k{}^*$ for $k = i, j$.

We first note that since $p>2$, (1.15) allows us to reduce to the following pairs $L_i \times L_j$: $A_1 \times A_i$, $i = 2, 3, 4$, $A_2 \times A_3$ and $A_3 \times A_3$. Moreover, the bound on $\dim V_{\gamma}(Q_Y)$ and (1.36) imply $\langle\lambda,\gamma\rangle = 0$ in every case. Again, standard arguments show that $L_i \times L_j$ does not have type $A_3 \times A_3$.

Consider the configuration where $(L_i,L_j)$ has type $(A_2,A_3)$. Then $p>2$ and (1.15) imply that $W_j|L_A'$ is tensor indecomposable and so $p>3$. Let q be the field twist on the embedding of $L_A'$ in $L_i$ and in $L_j$. (There is only one twist by (2.7).) Temporarily label as follows: $\Pi(L_i) = \{\gamma_1,\gamma_2\}$, $\Pi(L_j) = \{\gamma_3,\gamma_4,\gamma_5\}$ with $(\gamma_2,\gamma) \neq 0 \neq (\gamma_3,\gamma)$ and $(\gamma_3,\gamma_4) \neq 0 \neq (\gamma_4,\gamma_5)$. Then there does not exist $\delta \in \Pi(Y)-\Pi(L_Y)$ such that $(\delta,\gamma_5) \neq 0$. For otherwise, $(\delta,\Sigma L_k) = 0$ for all $k \neq i$ so $Q_Y/K_\delta \cong W_i$ or $W_i^*$. Thus, $Q_A \leq K_\delta$ and $M_i$ is trivial. But since $p>2$, (2.10) and (2.11) imply that there is a nontrivial image of $Q_A{}^{\alpha}$ in $Q_Y(\gamma,\delta)$, a 3-dimensional irreducible $L_A'$ module. Contradiction. Similarly, there does not exist $\tau \in \Pi(Y)-\Pi(L_Y)$, $\tau \neq \gamma$ such that $(\tau,\gamma_3) \neq 0$. Otherwise, as above, $Q_A \leq K_\tau$ and (2.11) applies. However, $Q_Y(\gamma,\tau)$ has all even $T(L_A')$ weights and there can be no nontrivial image of $Q_A{}^{\alpha}$ in this module. (See (1.15).)

The considerations of the preceeding paragraph imply that either $Y = E_7$ with $\Pi(L_i \times L_j) = \{\beta_1,\beta_3,\beta_5,\beta_6,\beta_7\}$ or $Y = E_8$ with $\Pi(L_i \times L_j) = \{\beta_2,\beta_4,\beta_6,\beta_7,\beta_8\}$, $\{\beta_2,\beta_4,\beta_5,\beta_7,\beta_8\}$ or $\{\beta_3,\beta_4,\beta_6,\beta_7,\beta_8\}$. If $\Pi(L_i \times L_j) = \{\beta_3,\beta_4,\beta_6,\beta_7,\beta_8\}$, $Q_Y/K_{\beta_k}$ is a 3-dimensional irreducible $L_A'$ module, for $k = 1$ and 2, so $Q_A \leq K_{\beta_k}$. Hence $\langle\lambda,\beta_\ell\rangle = 0$ for $1 \leq \ell \leq 4$. But then the bound on $\dim V_{\beta_5}(Q_Y)$ implies $\lambda|T_Y = \lambda_8$. If $\Pi(L_i \times L_j) = \{\beta_2,\beta_4,\beta_5,\beta_7,\beta_8\}$, standard arguments allow us to reduce to a special case of (6.6)(ii). If $\Pi(L_i \times L_j) = \{\beta_1,\beta_3,\beta_5, \beta_6,\beta_7\}$, the bound on $\dim V_{\beta_4}(Q_Y)$ and (2.3) imply $\lambda|T_Y = \lambda_1$ or $\lambda_7$. Thus, $\lambda|T_Y = \lambda_7$ and $\dim V|Y = 56$. Using (1.30) and (1.32), it is not difficult to see that $\dim V|A > 56$ if $A = G_2$. Thus, (6.6)(iv) holds.

Finally, suppose $\Pi(L_i \times L_j) = \{\beta_2,\beta_4,\beta_6,\beta_7,\beta_8\}$. Consider $Q_A K_{\beta_5}/K_{\beta_5} \leq Q_Y/K_{\beta_5}$. If we examine the $L_A'$ composition factors of $Q_Y/K_{\beta_5}$ and recall that $p>3$, we see that $\beta_5|Z_A = q\alpha$, if $p>5$. If $p=5$, $-\beta_5$ affords $T(L_A')$ weight $5q\mu_\beta$ so $\beta_5|Z_A \neq 5q\alpha$, by (2.16). So again

$\beta_5|Z_A = q\alpha$. Note that $\langle U_{\pm\beta_1}\rangle$ is a component of $L_Y'$. Otherwise, $\langle\lambda,\beta_1\rangle = 0$ by (2.3) and $\langle\lambda,\beta_i\rangle = 0$ for $i = 2, 3, 4$ as $Q_A \leq K_{\beta_3}$; but then the bound on $\dim V_{\beta_5}(Q_Y)$ implies $\lambda|T_Y = \lambda_8$. Notice that if $Q_A \not\leq K_{\beta_3}$, (2.17) implies that the field twist on the embedding of $L_A'$ in $\langle U_{\pm\beta_1}\rangle$ is also q, and that $\beta_3|Z_A = q\alpha$. Thus, $\langle\lambda,\beta_1\rangle = 0$, by (2.5). As well, the bound on $\dim V^2(Q_A)_{\lambda - q\alpha}$ of (1.22) implies $\lambda|T_Y = \lambda_2$. Also, if we apply (1.23) when $A = A_2$, we see that $\dim V|A < \dim V|Y$. Thus (6.6)(v) holds.

Now, consider $L_iXL_j$ of type $A_1XA_4$. Temporarily label as follows: $\Pi(L_iXL_j) = \{\gamma_0,\gamma_1,\gamma_2,\gamma_3,\gamma_4\}$ with $(\gamma_0,\gamma) \neq 0 \neq (\gamma,\gamma_1)$ and $(\gamma_i,\gamma_{i+1}) \neq 0$ for $i = 1, 2, 3$. We have already $\langle\lambda,\gamma\rangle = 0$. Actually, the bound on $\dim V_\gamma(Q_Y)$ implies $\langle\lambda,\gamma_k\rangle = 0$ for $k = 0, 1, 2, 3$ and $\langle\lambda,\gamma_4\rangle = 1$. Let W be the natural module for the $A_4$ component. Then $L_A'$ acts irreducibly on W and $W^*$. Moreover, $L_A'$ acting on $W\wedge W$ (or $W^*\wedge W^*$) has all even weights. Thus, since $p>2$, there does not exist $\tau \in \Pi(Y)-\Pi(L_Y)$ such that $(\tau,\gamma_i) \neq 0$ for some $1\leq i\leq 4$ and $(\tau,\Sigma L_k) = 0$ for all $k \neq i, j$. These considerations imply that $Y = E_8$ and either (a) $\Pi(L_Y) = \{\beta_i \mid i = 3, 5\leq i\leq 8\}$, with $\langle\lambda,\beta_8\rangle = 1$ or (b) $\Pi(L_iXL_j) = \{\beta_i \mid i = 2, 5\leq i\leq 8\}$, with $\langle\lambda,\beta_8\rangle = 1$ or (c) $\Pi(L_iXL_j) = \{\beta_2,\beta_4,\beta_5, \beta_6,\beta_8\}$, with $\langle\lambda,\beta_2\rangle = 1$. As well, (2.17) implies that there exists a p-power, q, such that q is the field twist on the embedding of $L_A'$ in $L_i$ and in $L_j$ and such that $\gamma|Z_A = q\alpha$.

In the configuration of (a), suppose that $Q_A \not\leq K_{\beta_1}$. Then $Q_Y/K_{\beta_1}$ is isomorphic to $(Q_A{}^\alpha)^q$ as $L_A'$-module,so (2.4) implies $\beta_1|Z_A = q\alpha$. But $\beta_4|Z_A = q\alpha$; so the bound on $\dim V^2(Q_A)_{\lambda-q\alpha}$ of (1.22) implies $\langle\lambda,\beta_1\rangle = 0$. Also, if $Q_A \leq K_{\beta_1}$, $\langle\lambda,\beta_1\rangle = 0$. But then in either case $\lambda|T_Y = \lambda_8$.

If $L_iXL_j$ is as in (b), $\langle U_{\pm\beta_1}\rangle \leq L_Y'$, else (2.3) and the preceeding remarks imply $\lambda|T_Y = \lambda_8$. If $Q_A \not\leq K_{\beta_3}$, (2.8) implies $\beta_3|Z_A = q\alpha$. Since $Q_Y/K_{\beta_3}$ is a 2-dimensional $L_A'$ irreducible, the field twist on the embedding of $L_A'$ in $\langle U_{\pm\beta_1}\rangle$ is q and therefore $\langle\lambda,\beta_1\rangle = 0$, by (2.5). Moreover, $\langle\lambda,\beta_3\rangle = 0$, else the bound on $\dim V^2(Q_A)_{\lambda-q\alpha}$ is exceeded. Also, if $Q_A \leq K_{\beta_3}$, $\langle\lambda,\beta_1+\beta_3\rangle = 0$. But in either case $\lambda|T_Y = \lambda_8$.

If $L_i \times L_j$ is as in (c), previous remarks imply $\langle U_{\pm\beta_1}\rangle \leq L_Y'$. However, now we have a configuration of Case I which was ruled out. This completes the consideration of $L_i \times L_j$ of type $A_1 \times A_4$.

We now consider $(L_i, L_j)$ of type $(A_1, A_3)$. We have $\langle\lambda,\gamma\rangle = 0$. Moreover, the bound on $\dim V_\gamma(Q_Y)$ of (1.25) implies that if $M_i$ is nontrivial, $M_i \cong W_i$ and if $M_j$ is nontrivial $M_j \cong (V_{L_j}(-\gamma))^*$ ($\cong W_j$ or $W_j^*$ by the first paragraph of the proof). As well, (1.15) implies $W_j|L_A'$ is tensor decomposable. Temporarily label as follows: $\Pi(L_i) = \{\gamma_0\}$, $\Pi(L_j) = \{\gamma_1,\gamma_2,\gamma_3\}$, $(\gamma_1,\gamma) \neq 0$, $(\gamma_i,\gamma_{i+1}) \neq 0$, $i = 1, 2$. Let $q$ be the field twist on the embedding of $L_A'$ in $L_i$. Say $V_{L_j}(-\gamma)|L_A'$ has high weight $(q + q_0)\mu_\beta$. (The power $q$ appears by (2.7).) Then the $L_A'$ composition factors of $Q_Y/K_\gamma$ have high weights $(2q + q_0)\mu_\beta$ and $q_0\mu_\beta$. Since $p>2$, $\gamma|Z_A = q_0\alpha$, by (2.4).

<u>Claim:</u> If $\dim M_k > 1$ for some $k$, then $k = i$ or $j$.

Reason: Suppose false; i.e., suppose there exists $k \neq i, j$ such that $\dim M_k > 1$. If $L_k$ is separated from $L_i$ or $L_j$ by exactly one node of the Dynkin diagram, size restrictions, (6.4) and the work of this result thus far imply $Y = E_8$ and $\Pi(L_Y) = \{\beta_1,\beta_4,\beta_5,\beta_6,\beta_8\}$. Also, $\langle\lambda,\beta_1\rangle = 1 = \langle\lambda,\beta_8\rangle$ and $\langle\lambda,\beta_i\rangle = 0$, $3\leq i\leq 7$. Let $q_1$, $q_2$ be the distinct field twists on the embeddings of $L_A'$ in $\langle U_{\pm\beta_1}\rangle$, $\langle U_{\pm\beta_8}\rangle$, respectively. (They must be distinct by (2.5).) Then, by the preceeding paragraph, $Q_Y/K_{\beta_2}$ is an irreducible $L_A'$-module with high weight $(q_1 + q_2)\mu_\beta$ and $\beta_3|Z_A = q_2\alpha$ and $\beta_7|Z_A = q_1\alpha$. For the purposes of this argument, we may assume $h_\beta(c) = h_{\beta_1}(c^{q_1})h_{\beta_4}(c^{q_1+q_2})h_{\beta_5}(c^{2q_2})h_{\beta_6}(c^{q_1+q_2})h_{\beta_8}(c^{q_2})$. Examining the $T(L_A')$ weight vectors in $Q_Y/K_{\beta_3}$ and $Q_Y/K_{\beta_7}$, we find that $x_{-\alpha}(t) = x_{-13}(c_1t^{q_2})\cdot x_{-34}(c_2t^{q_2})x_{-78}(d_1t^{q_1})x_{-567}(d_2t^{q_1})w$, where $c_i, d_i \in k$, $c_1$ or $c_2$ nonzero, $d_1$ or $d_2$ nonzero, and $w \in K_{\beta_3} \cap K_{\beta_7}$. Suppose $c_2 = 0$. Since $V_{T_Y}(\lambda-\beta_1-\beta_3-\beta_4) \neq 0$, a nonidentity element from the group $U_{-134}$ must occur in the factorization of some element from $Q_A - Q_A'$. However, $-\beta_1-\beta_3-\beta_4$ affords $T(L_A')$ weight $(q_2-2q_1)\mu_\beta$, which does not occur in

$(Q_A{}^{\alpha})^{q_2}$, contradicting (2.4). Thus, $c_2 \neq 0$. A similar argument shows that $d_2 \neq 0$. But $c_2 d_2 \neq 0$ contradicts (2.8).

Thus, $L_k$ is separated from $L_i$ and $L_j$ by more than one node of the Dynkin diagram. Size restrictions imply $Y = E_7$ or $E_8$ and $\Pi(L_Y) = \{\beta_1,\beta_2,\beta_5,\beta_6,\beta_7\}$. Since $Q_A \nleq K_{\beta_3}$ and $Q_Y/K_{\beta_3}$ is a 2-dimensional irreducible $L_A'$-module, $\beta_3|T_A = q_0\alpha$ where $q_0$ is the field twist on the embedding of $L_A'$ in $\langle U_{\pm\beta_1}\rangle$. Then (2.8) implies $\beta_4|Z_A = q_0\alpha$. Thus, if $L_j = \langle U_{\pm\beta_5}, U_{\pm\beta_6}, U_{\pm\beta_7}\rangle$, $V_{L_j}(-\beta_4)|L_A'$ has high weight $(q+q_0)\mu_\beta$ and the field twist on the embedding of $L_A'$ in $\langle U_{\pm\beta_2}\rangle$ is q. (See the previous general work on $Q_AK_\gamma/K_\gamma \leq Q_Y/K_\gamma$, in this configuration.) Thus, (2.5) implies that $M_j$ is trivial and so $\langle\lambda,\beta_2\rangle = 1$, $\langle\lambda,\beta_i\rangle = 0$ for $4 \leq i \leq 7$. Now, using (1.36) if $\langle\lambda,\beta_3\rangle \neq 0$, we see that the bound on $\dim V^2(Q_A)_{\lambda-q_0\alpha}$ of (1.22) is exceeded. This completes the proof of the Claim.

Now apply (1.23) when $A = A_2$ to find that $\dim V|A < \dim V|Y$, unless $Y = E_7$ and $\lambda|T_Y = \lambda_7$. But then $\dim V|Y = 56$ and $\dim V|A = 8$ or 64 (7 or 49 if p=3). (See (1.36).) Thus, A does not have type $A_2$. Moreover, (1.23) also implies Y does not have type $E_6$. Thus, (6.6)(vi) holds. This completes the consideration of Case II.

The final statement of (6.6) follows from (2.3).□

<u>(6.7).</u> Suppose there exists $1 \leq i \leq r$ such that $L_i$ is separated from all other components of $L_Y'$ by more than one node of the Dynkin diagram and such that $M_i$ is nontrivial. Then rank $L_i = 1$ and $\dim V^1(Q_Y) = \dim M_i$.

<u>Proof</u>: Since all components of $L_Y'$ are necessarily of classical type, (1.5) implies $Z_A \leq Z_Y$. Moreover, by (6.3), $L_i$ has type $A_k$ for some k. Let W denote the natural module for $L_i$. Theorem (7.1) of [12] implies $M_i \cong W$ or $W^*$ if rank$L_i > 1$. Thus, if rank $L_i > 1$, there does not exist $\gamma \in \Pi(Y)-\Pi(L_Y)$ such that $(\gamma,\Sigma L_i) \neq 0$ and $V_{L_i}(-\gamma) \cong W$ or $W^*$, as $Q_Y/K_\gamma$ would be a (rank$(L_i)$+1)-dimensional irreducible $L_A'$-module containing a nontrivial image of $Q_A$. Thus, if rank $L_i > 1$, $\Pi(L_i) = \{\beta_1,\beta_2,\beta_3,\beta_4\}$.

However, (1.15) implies that $Q_A \le K_{\beta_5}$, contradicting (2.3). So rank $L_i = 1$. Let q be the field twist on the embedding of $L_A'$ in $L_i$. We will suppose $\dim V^1(Q_Y) > \dim M_i$. In particular, there exists $1 \le m \le r$, $m \ne i$, such that $M_m$ is nontrivial. Thus, by (6.3) each component of $L_Y'$ has type $A_{m_k}$, for some $m_k \ge 1$.

Now, (6.5), (6.6) and the above remarks imply that either

(a) there exists $k \ne i,m$ such that $L_m \times L_k$ has type $A_1 \times A_2$ (or$A_2 \times A_1$), $\Pi L_m$ and $\Pi L_k$ are separated by exactly one node of the Dynkin diagram, corresponding to a simple root $\gamma$, and $(\Sigma L_j, \gamma) = 0$ for $j \ne k,m$, or

(b) rank$L_m = 1$ and $\Pi(L_m)$ is separated by more than one node of the Dynkin diagram from all other components of $L_Y'$.

<u>Claim.</u> There does not exist $\delta, \tau \in \Pi(Y) - \Pi(L_Y)$ such that $(\delta,\tau) < 0$, $(\delta, \Sigma L_i) \ne 0$ and $(\tau, \Sigma L_Y) = 0$.

Reason: Suppose false. Then by (2.13) and (2.3), $\langle \lambda, \delta \rangle = 0 = \langle \lambda, \tau \rangle$, $\delta | T_A = q\alpha$ and $\tau | T_A = 0$. Also, by (2.15), $\Pi L_i$ corresponds to an end node of the Dynkin diagram and $\Pi L_i \ne \{\beta_1\}$. If $\Pi L_i = \{\beta_2\}$, (2.15) implies $\tau = \beta_5$ and $\Pi(L_Y) = \{\beta_2, \beta_1, \beta_7\}$. If $\Pi L_i = \{\beta_7\}$, we argue similarly that $\Pi(L_Y) = \{\beta_7, \beta_1, \beta_2\}$ or $\{\beta_7, \beta_1, \beta_3, \beta_2\}$. Finally, if $\Pi L_i = \{\beta_8\}$, $\Pi(L_Y) = \{\beta_8, \beta_1, \beta_2, \beta_4\}$.

Consider the case where $\Pi(L_Y) = \{\beta_1, \beta_2, \beta_7\}$. Then $Q_A \nleq K_{\beta_3}$ and $Q_A \nleq K_{\beta_4}$; otherwise, $p > 2$, (2.10) and (2.11) imply that there is a nontrivial image of $Q_A{}^\alpha$ in $Q_Y(\beta_3, \beta_4)$, contradicting (1.15). Now (2.8) implies that the field twists on the embeddings of $L_A'$ in $\langle U_{\pm\beta_1} \rangle$ and in $\langle U_{\pm\beta_2} \rangle$ are equal; call this twist q. (In particular, only one of $\langle \lambda, \beta_1 \rangle$ and $\langle \lambda, \beta_2 \rangle$ is nonzero.) Let $q_0 \ne q$ be the field twist on the embedding of $L_A'$ in $\langle U_{\pm\beta_7} \rangle$. Then $\beta_3 | T_A = q\alpha = \beta_4 | T_A$, $\beta_5 | T_A = 0$ and $\beta_6 | T_A = q_0\alpha$. Moreover, $\langle \lambda, \beta_4 + \beta_5 + \beta_6 \rangle = 0$. If $\langle \lambda, \beta_2 \rangle \ne 0$, $\langle \lambda, \beta_3 \rangle = 0$, else $(V_{\beta_3}(Q_Y) \oplus V_{\beta_4}(Q_Y)) \le V^2(Q_A)_{\lambda - q\alpha}$ and the bound on $\dim V^2(Q_A)_{\lambda - q\alpha}$ is exceeded. But now we see that there is no vector in $V|Y$ of weight $\lambda - q_0\alpha$, contradicting (2.14). Hence, $\langle \lambda, \beta_1 \rangle \ne 0$ and $\langle \lambda, \beta_2 \rangle = 0$. If $\langle \lambda, \beta_3 \rangle \ne 0$, $f_{34}v^+$ and $f_{345}v^+$ are 2 linearly independent vectors in $V_{T_A}(\lambda - 2q\alpha)$, contradicting (1.31). So $\langle\lambda,\beta_3\rangle = 0$ and again (2.14) is contradicted. Hence, $\Pi(L_Y) \neq \{\beta_1,\beta_2,\beta_7\}$.

If $\Pi(L_Y) = \{\beta_1,\beta_3,\beta_2,\beta_7\}$, let $P_Y\hat{}$ be the parabolic of (2.12), so $P_Y\hat{}$ has Levi factor $L_Y\hat{} = \langle L_Y, U_{\pm\beta_5}\rangle$. Then the bound on $\dim V_{\beta_4}(R_u(P_Y\hat{}))$ implies that $\langle\lambda,\beta_2+\beta_3+\beta_4\rangle = 0$ and $\langle\lambda,\beta_1\rangle = 1$. By (2.17), there is a p-power $q_0$ which is the field twist on the embedding of $L_A'$ in $\langle U_{\pm\beta_1},U_{\pm\beta_3}\rangle$ and $\langle U_{\pm\beta_2}\rangle$ and such that $\beta_4|Z_A = q_0\alpha$. Examining the $T(L_A)$ weight vectors in $Q_Y/K_{\beta_4}$, we have $x_{-\alpha}(t) = x_{-\beta_3-\beta_4}(c_1t^{q_0})\cdot x_{-\beta_2-\beta_4}(c_2t^{q_0})u_0$, for $c_i \in k$, $c_1$ or $c_2$ nonzero and $u_0 \in K_{\beta_4}$. Hence $(\beta_3+\beta_4)|T_A = q_0\alpha$ and $\beta_4|T_A = q_0(\alpha-\beta)$. But now one can check that there is no vector in $V|Y$ with weight $\lambda-q\alpha$, where q is the field twist on the embedding of $L_A'$ in $L_i$, contradicting (2.14).

Finally, consider the case where $\Pi(L_Y) = \{\beta_1,\beta_2,\beta_4,\beta_8\}$. Let $q_0$ be the field twist on the embedding of $L_A'$ in $\langle U_{\pm\beta_1}\rangle$ and $\langle U_{\pm\beta_2},U_{\pm\beta_4}\rangle$. Then $Q_A \leq K_{\beta_5}$, so $\langle\lambda,\beta_2+\beta_4+\beta_5+\beta_6+\beta_7\rangle = 0$. Also, $-\beta_5$ is not involved in $L_A'$, else $\beta_5|Z_A = 0$ and the bound on $\dim V^2(Q_A)_{\lambda-q_0\alpha}$ is exceeded. Let $P_Y\hat{}$ be the parabolic of (2.12); so $P_Y\hat{}$ has Levi factor $L_Y\hat{} = \langle L_Y, U_{\pm\beta_6}\rangle$. Then, $Q_A \leq K_{\beta_5} \leq R_u(P_Y\hat{}) = Q_Y\hat{}$ and (2.11) implies that there is a nontrivial image of $Q_A{}^\alpha$ in $Q_Y\hat{}(\beta_7,\beta_5)$. But this implies that the field twist on the embedding of $L_A'$ in $\langle U_{\pm\beta_8}\rangle$ is also $q_0$, contradicting (2.5).

This completes the proof of the claim.

Now consider the configuration of (a). Note that by size retrictions and the above claim $\Pi(L_i)$ is separated by exactly two nodes of the Dynkin diagram from $\Pi(L_k{\times}L_m)$. In fact, $\Pi(L_k{\times}L_m) = \{\beta_5,\beta_7,\beta_8\}$ or $\{\beta_5,\beta_6,\beta_8\}$, $\Pi(L_i) = \{\beta_1\}$ and in each case $\langle U_{\pm\beta_2}\rangle$ is a component of $L_Y'$. For in the other possible configurations (2.8), (2.17), and (2.18) would force the field twist on the embedding of $L_A'$ in $L_m$ to be q, contradicting (2.5) and (2.6). In fact, $\Pi(L_k{\times}L_m) \neq \{\beta_5,\beta_6,\beta_8\}$. For otherwise, $Q_A \not\leq K_{\beta_4}$ and we argue that $L_m = \langle U_{\pm\beta_8}\rangle$ and the field twist on the embedding of $L_A'$ in $L_m$ is q. This again produces a contradiction.

Consider now $\Pi(L_Y) = \{\beta_1,\beta_2,\beta_5,\beta_7,\beta_8\}$, with $\langle\lambda,\beta_1\rangle \neq 0$. Since $p>2$, (1.15) implies $Q_A \leq K_{\beta_4}$. However, $-\beta_4$ is not involved in $L_A'$; for otherwise, $\beta_4|Z_A = 0$ and using the parabolic $P_Y\hat{}$ of (2.11), we see that the bound on $\dim V_{\beta_3}(Q_Y\hat{})$ is exceeded. If $q$ is the field twist on the embedding of $L_A'$ in $\langle U_{\pm\beta_1}\rangle$, then $q \neq q_0$, where $q_0$ is the field twist on the embedding of $L_A'$ in $\langle U_{\pm\beta_5}\rangle$ and $\langle U_{\pm\beta_7},U_{\pm\beta_8}\rangle$. By (2.11), there is a nontrivial image of $Q_A{}^\alpha$ in $Q_Y(\beta_6,\beta_4)$; so the field twist on the embedding of $L_A'$ in $\langle U_{\pm\beta_2}\rangle$ is also $q_0$. Now there is a nontrivial image of $Q_A{}^\alpha$ in the $L_Y'$ module $Q_Y(\beta_3,\beta_4)$, which has $L_A'$ composition factors of high weights $(2q_0+q)\mu_\beta$ and $q\mu_\beta$. Hence, $(\beta_3+\beta_4)|Z_A = q\alpha$ and $\beta_4|Z_A = 0$. But as above, the bound on $\dim V_{\beta_3}(Q_Y\hat{})$ of (1.25) is exceeded. Thus, the configuration of (a) cannot occur.

It remains to consider the case where $\mathrm{rank}(L_m) = 1$ and $L_m$ is separated from all other components of the Dynkin diagram by more than one node of the Dynkin diagram. Note that the above claim implies that if $\Pi L_i$ or $\Pi L_m = \{\beta_\ell\}$, then $\ell \notin \{3,4,5\}$ and if $Y = E_8$, $\ell \neq 6$. Also, (2.8) implies that $\Pi(L_i)$ and $\Pi(L_m)$ are not separated by exactly two nodes of the Dynkin diagram. Finally, there does not exist $\tau \in \Pi(L_Y)$ such that $L_Y' = L_i \times L_m \times \langle U_{\pm\tau}\rangle$ and $\Pi(L_i)$ and $\Pi(L_m)$ are separated from $\tau$ by exactly two nodes of the Dynkin diagram. For otherwise, $p>2$, (2.18) and (2.8) imply that the field twists on the embeddings of $L_A'$ in $L_i$, $\langle U_{\pm\tau}\rangle$ and $L_m$ are all equal. These remarks and the above claim allow us to reduce to $\Pi(L_Y) = \{\beta_1,\beta_2,\beta_5,\beta_8\}$. Then since $p>2$, $Q_A \leq K_{\beta_4}$, so $\Pi(L_i \times L_m) = \{\beta_1,\beta_8\}$. Let $q$ (respectively, $q_0$) be the field twist on the embedding of $L_A'$ in $\langle U_{\pm\beta_1}\rangle$ (respectively, $\langle U_{\pm\beta_8}\rangle$). Then (2.18) and (2.8) imply that $q_0$ is also the field twist on the embedding of $L_A'$ in $\langle U_{\pm\beta_5}\rangle$. As in the previous case, (2.11) implies that there is a nontrivial image of $Q_A{}^\alpha$ in $Q_Y(\beta_3,\beta_4)$. Thus, the field twist on the embedding of $L_A'$ in $\langle U_{\pm\beta_2}\rangle$ is either $q$ or $q_0$.

Since $p>2$, examining the $L_A'$ composition factors of $Q_Y(\beta_3,\beta_4)$ gives that $(\beta_3+\beta_4)|Z_A = q\alpha$ or $q_0\alpha$ (depending on the field twist on the embedding of $L_A'$ in $\langle U_{\pm\beta_2}\rangle$). But if $(\beta_3+\beta_4)|Z_A = q\alpha$, then $\beta_4|Z_A = 0$, as $\beta_3|Z_A = q\alpha$. Using the parabolic $P_Y\hat{}$ of (2.11), we see that the bound on $\dim V_{\beta_3}(Q_Y\hat{})$ is exceeded. Thus, $(\beta_3+\beta_4)|Z_A = q_0\alpha$. However, using (1.36) if $\langle\lambda,\beta_1\rangle$ or $\langle\lambda,\beta_8\rangle = p-1$, we see that the contribution of $V_{\beta_7}(Q_Y)$ and the $L_Y'$ composition factor(s) afforded by $f_{\beta_3+\beta_4}v^+$ and/or $f_{\beta_1+\beta_3+\beta_4}v^+$ exceed the bound on $\dim V^2(Q_A)_{\lambda-q_0\alpha}$.

This completes the proof of (6.7).□

<u>(6.8).</u> Let $\gamma\in\Pi(Y)-\Pi(L_Y)$ such that there exists a unique pair $1\le i,j\le r$ with $(\Sigma L_i,\gamma)\ne 0\ne(\Sigma L_j,\gamma)$ and $L_i$ has type $A_1$, $L_j$ has type $A_2$ and $\dim(M_i\otimes M_j) > 1$. Then $\dim V^1(Q_Y)\le 3$ and if $A = A_2$, then $Y = E_6$.

<u>Proof</u>: Suppose there exists $k\ne i, j$ such that $M_k$ is nontrivial. By size restrictions, $L_m$ is necessarily of classical type for all $1\le m\le r$, so by (1.5), $Z_A\le Z_Y$. Also, by size restrictions, $L_k$ has type $A_{n_k}$ for some $n_k$. So (6.5), (6.6) and (6.7) imply $L_k$ has type $A_1$ or $A_2$ and is separated by exactly one node of the Dynkin diagram, corresponding to a root $\delta\in\Pi(Y)-\Pi(L_Y)$, from a component of $L_Y'$ of type $A_2$ or $A_1$, respectively. Moreover, there are exactly two components of $L_Y'$ whose root systems are not orthogonal to $\delta$. By size restrictions, $(\delta,\Sigma(L_i\times L_j))\ne 0$. However, (2.17) implies that the field twists on the embeddings of $L_A'$ in $L_i$, $L_j$ and $L_k$ are equal, contradicting (2.5) and (2.6). Hence, $M_\ell$ is trivial for all $\ell\ne i, j$.

Now (6.6) implies $\langle\lambda,\gamma\rangle = 0$. Moreover, if $L_i = \langle U_{\pm\gamma_i}\rangle$ for $\gamma_i\in\Pi(L_Y)$, and if $\langle\lambda,\gamma_i\rangle\ne 0$, the bound on $\dim V_\gamma(Q_Y)$ implies $\langle\lambda,\gamma_i\rangle\le 2$. Thus, $\dim V^1(Q_Y) = \max\{\dim M_i, \dim M_j\}\le 3$. Then, if $Y = E_7$ or $E_8$, A does not have type $A_2$, as (1.23) shows $\dim V|A\le 27 < \dim V|Y$. This completes the proof of (6.8).□

<u>Proof of (6.0)</u>: Let $\alpha_1$, $\alpha_2$, $\mu_1$ and $\mu_2$ be as in the statement of (6.0). Without loss of generality, we may assume $\langle\lambda,\alpha_1\rangle\ne 0$. Let

$L_A = \langle U_{\pm\alpha_1}\rangle T_A$ and fix notation as before. By (6.1), $L_Y'$ is not a simple algebraic group.

Suppose there exists $1\le i\le r$ such that $L_i$ has type $D_k$ for some k. Then by (6.3), (1.23) and (1.10), $Y = E_8$, $\lambda|T_Y = x\lambda_7+c\lambda_8$, $\lambda|T_A = c\mu_1+c\mu_2$ and $\Pi(L_Y) = \{\beta_2,\beta_3,\beta_4,\beta_5,\beta_8\}$ or $\{\beta_1,\beta_2,\beta_3,\beta_4,\beta_5,\beta_8\}$. Also, (2.9) implies that $Q_A \le K_{\beta_6}$. Then using the parabolic $P_Y\hat{}$ of (2.11) and applying (1.36), we see that the bound on $\dim V^2(Q_A)$ implied by (1.22) is exceeded. Thus, there does not exist i such that $L_i$ has type $D_k$.

Suppose there exist distinct $1\le i,j,k\le r$ such that $(\Sigma L_m,\beta_4) \ne 0$ for $m = i, j, k$ and such that $M_m$ is nontrivial for $m = i$, j or k. Then (6.5)(i) or (iii) holds. If (6.5)(i) holds, it is established in the proof that the field twists on the embeddings of $L_A'$ in $\langle U_{\pm\beta_1},U_{\pm\beta_3}\rangle$ and in $\langle U_{\pm\beta_5},U_{\pm\beta_6}\rangle$ are equal. Call this twist q. Then by (1.23) and (1.10), $\lambda|T_A = 2\mu_1 + 2\mu_2$. Now, $\dim V|Y = 27$; but (1.36) implies $\dim V|A < 27$ if $p=5$. Hence, (6.0)(a) holds.

Now, consider the configuration described in (6.5)(iii). Then, by (1.23), $\lambda|T_A = (q_1+q_2)(\mu_1 + \mu_2)$ for $q_1$ and $q_2$ distinct p-powers, or $\lambda|T_A = q(3\mu_1 + 3\mu_2)$, for some p-power q, $p>3$. If $x\ne 0$, (1.32) and (1.26) imply $\dim V|A < \dim V|Y$. Thus, $x=0$ and $\dim V|Y = 56$. If $\lambda|T_A = (q_1+q_2)(\mu_1 + \mu_2)$, $\dim V|A = 64$ or $p=3$ and $\dim V|A = 49$. (Use (1.35).) Hence, $\lambda|T_A = q(3\mu_1 + 3\mu_2)$ and $p>3$. So, by (1.10), $q=1$. The Weyl module for A with high weight $3\mu_1 + 3\mu_2$ has dimension 64 by (1.27). Using (1.33) and the fact that $p>3$, we see that $\dim V|A < 64$ if and only if $p=7$ and $\dim V_{T_A}(\lambda-\alpha_1-\alpha_2) = 1$ or $p=5$ and $\dim V_{T_A}(\lambda-3\alpha_1-3\alpha_2) < 4$. However, if $p=7$ and $\dim V_{T_A}(\lambda-\alpha_1-\alpha_2) = 1$, (1.33) implies $\dim V|A \le 64 - \dim V(2\mu_1 + 2\mu_2) < 56$. And if $p=5$, $\dim V|A \ge 64 - 4 = 60$. Hence $\dim V|A \ne 56$. Thus, the configuration of (6.5)(iii) does not occur.

Suppose there exists $\gamma \in \Pi(Y)-\Pi(L_Y)$ such that there exists a unique pair $1\le i,j\le r$ with $(\Sigma L_i,\gamma) \ne 0 \ne (\Sigma L_j,\gamma)$ and $M_i$ or $M_j$ is nontrivial. Then $L_i$ and $L_j$ have type $A_{k_i}$, respectively $A_{k_j}$ for some $k_i, k_j \ge 1$ and

(6.6)(iii) or (iv) holds. If (6.6)(iv) holds, p>3, (1.23) and (1.10) imply $\lambda|T_A = 3\mu_1 + 3\mu_2$. Then the argument of the preceeding paragraph shows dim $V|A \neq$ dim $V|Y$.

Suppose (6.6)(iii) holds. Then by (6.8), $Y = E_6$ and $\dim V^1(Q_Y) \leq 3$. If $\Pi(L_i \times L_j) = \{\beta_1,\beta_4,\beta_5\}$, (6.6) and (6.8) imply $\langle\lambda,\beta_k\rangle = 0$ for $2\leq k\leq 6$ and $\langle\lambda,\beta_1\rangle = c \leq 2$. By (1.23), $\lambda|T_A = q(c\mu_1)$. But then dim $V|A <$ dim $V|Y$. Suppose $\Pi(L_i \times L_j) = \{\beta_3,\beta_5,\beta_6\}$. Recall that the field twists on the embeddings of $L_A'$ in $L_i$ and in $L_j$ are equal. Call this twist q. Then, (2.17) implies $\beta_4|Z_A = q\alpha$. Moreover, $Q_A \nleq K_{\beta_1}$, else p > 2, (2.10) and (2.11) imply that there is a nontrivial image of $Q_A{}^\alpha$ in $Q_Y(\beta_4,\beta_1)$, a 3-dimensional irreducible $L_A'$-module. Hence, by (2.4), $\beta_1|Z_A = q\alpha$, also. This implies $\langle\lambda,\beta_1\rangle = 0$ and $\langle\lambda,\beta_3\rangle \leq 1$, else the bound on $\dim V^2(Q_A)_{\lambda-q\alpha}$ of (1.22) is exceeded. Hence by (1.23) and (2.3), $\lambda|T_Y = \lambda_3$ or $\lambda_5$ and $\lambda|T_A = q(2\mu_1 + \mu_2)$ or $\lambda|T_Y = \lambda_6$ and $\lambda|T_A = q(2\mu_1)$. However, in each case, dim $V|A <$ dim $V|Y$. Consider next the configuration $\Pi(L_i \times L_j) = \{\beta_1,\beta_2,\beta_4\}$. If $L_Y' = L_i \times L_j$, $Q_A \leq K_{\beta_5}$, $\lambda|T_Y = c\lambda_1$ and by (1.23) $\lambda|T_A = q(c\mu_1)$, for c = 1 or 2. However, dim $V|A <$ dim $V|Y$. Thus, $\langle U_{\pm\beta_6}\rangle \leq L_Y'$. If $\langle\lambda,\beta_2+\beta_4\rangle \neq 0$, $Q_A \nleq K_{\beta_5}$, so (2.17) implies that the field twist on the embedding of $L_A'$ in $\langle U_{\pm\beta_6}\rangle$ is also q and that $\beta_5|Z_A = q\alpha$. Thus, we see that $\lambda|T_Y = \lambda_2$, else the bound on $\dim V^2(Q_A)_{\lambda-q\alpha}$, of (1.22), is exceeded. But this is a contradiction. Hence, $\langle\lambda,\beta_1\rangle \neq 0$. Applying (6.8) and (1.23), we see that dim $V|A <$ dim $V|Y$.

So if (6.6)(iii) holds, we may assume by symmetry that $\Pi(L_i \times L_j) = \{\beta_1,\beta_3,\beta_2\}$. If $L_Y' = L_i \times L_j$, $\langle\lambda,\beta_5+\beta_6\rangle = 0$. But (1.23) then implies $\dim V|A < \dim V|Y$. Thus, $L_Y' = L_i \times L_j \times \langle U_{\pm\beta_6}\rangle$ and by (6.8), $\langle\lambda,\beta_6\rangle = 0$. In fact, $\langle\lambda,\beta_5\rangle = 0$, for otherwise, $Q_A \nleq K_{\beta_5}$, and (2.8) implies $\beta_5|Z_A = q\alpha$ which means the bound on $\dim V^2(Q_A)_{\lambda - q\alpha}$, of (1.22), is exceeded. But now (1.23) implies dim $V|A <$ dim $V|Y$. Hence, the hypothesis of (6.6) cannot be satisfied.

Consider now the possibility that there exists $1\leq i\leq r$ such that $L_i$ is separated from all other components of $L_Y'$ by more than one node of the Dynkin diagram. Then all components of $L_Y'$ are necessarily of classical type, so by (1.5), $Z_A \leq Z_Y$. By (6.7), rank $L_i = 1$ and $\dim V^1(Q_Y) = \dim M_i = c+1$, for some $p>c>0$. Let q be the field twist on the embedding of $L_A'$ in $L_i$. Consider first the case where $Y = E_7$ or $E_8$. Then (1.23) and (1.10) imply $\lambda|T_A = c\mu_1 + c\mu_2$. Let $\gamma$, $\delta \in \Pi(Y)-\Pi(L_Y)$ such that $(\Sigma L_i,\gamma) \neq 0$, $(\gamma,\delta) < 0$. Then, by (2.4), $\gamma|Z_A = \alpha$ and by (2.3) and (2.9), $Q_A \leq K_\delta$. If $-\delta$ is not involved in $L_A'$, (2.11) implies $(\gamma+\delta)|Z_A = \alpha$. Thus, $\delta|Z_A = 0$. If $-\delta$ is involved in $L_A'$, (2.10) implies $\delta|Z_A = 0$. Hence, we may use the parabolic $P_Y\hat{}$ of (2.11) to see that the bound on $\dim V_\gamma(Q_Y)$ and (1.36) imply $\langle\lambda,\gamma\rangle = 0$. Moreover, $(\Sigma L_Y,\delta) = 0$, else $c\leq 2$ and $\dim V|A \leq 27 < \dim V|Y$. Now (2.15) implies $L_i = \langle U_{\pm\beta_2}\rangle$, $\gamma = \beta_4$ and $\delta = \beta_5$ or $L_i = \langle U_{\pm\beta_n}\rangle$ if $Y = E_n$. Now, if $\langle\lambda,\beta_j\rangle \neq 0$ for some $\beta_j \in \Pi(Y) - \Pi(L_i)$, then $\beta_j \in \Pi(Y) - \Pi(L_Y)$ and the bound on $\dim V^2(Q_A)$ is exceeded. So, $\langle\lambda,\beta_j\rangle = 0$ for all $\beta_j \notin \Pi(L_i)$. In each case, there is a parabolic of Y with Levi factor, L, of type $A_5$ such that $v^+$ affords an L composition factor with dimension $(c+1)(c+2)(c+3)\cdot(c+4)(c+5)/5!$ by (1.12). Since $\dim V|A \leq (c+1)^3$, by (1.27), $c\leq 3$. Using the methods of (1.30) and (1.32), it is easy to check that $\dim V|A < \dim V|Y$. Hence, $Y = E_6$.

Since $L_Y'$ is not simple, we may take $L_i = \langle U_{\pm\beta_j}\rangle$ for j = 1, 2, 3. In fact, by (2.15), we may exclude j=3. If $L_i = \langle U_{\pm\beta_2}\rangle$, (2.15) implies $\langle U_{\pm\beta_1}\rangle \leq L_Y'$. By (2.9), $Q_A \leq K_{\beta_3}$. However, since $p > 2$, this contradicts (2.18). Hence, $L_i = \langle U_{\pm\beta_1}\rangle$. Then (2.15) implies $(\Sigma L_Y,\beta_4) \neq 0$, so by (2.9), $Q_A \leq K_{\beta_4}$ and $V_{\beta_4}(Q_Y) = 0$. The considerations of the case where $L_i = \langle U_{\pm\beta_2}\rangle$ and previous general remarks imply that $\Pi(L_Y) = \{\beta_1,\beta_2,\beta_5\}$ or $\{\beta_1,\beta_5,\beta_6\}$ or $\{\beta_1,\beta_2,\beta_5,\beta_6\}$. In the second and third cases, (1.23) and Theorem (7.1) of [12] imply $\lambda|T_Y = \lambda_1$ and $\lambda|T_A = q\mu_1 + 2q_0\mu_2$ for some p-power $q_0$. However, $\dim V|A \leq 18 < \dim V|Y$. In the first case, by considering the action of A on $V^*$, we see that $\langle\lambda,\beta_3\rangle = 0$. Then (1.23) and (1.10) imply $\lambda|T_A = c\mu_1 + y\mu_2$ and $\lambda|T_Y = c\lambda_1 + y\lambda_6$. Hence, if $y\neq 0$,

(2.4) implies $\beta_6|Z_A = \alpha$ and the bound on $\dim V^2(Q_A)$ is exceeded. So $y=0$. But then clearly $\dim V|A < \dim V|Y$. Therefore, the conditions of (6.7) cannot be satisfied.

It remains to consider the possibility that there exists $1 \le i \le r$ such that $L_i$ has exceptional type. Then $Y = E_8$, $\Pi(L_Y) = \{\beta_1, \beta_2, \beta_3, \beta_4, \beta_5, \beta_6, \beta_8\}$ and by (7.1) of [12], $\lambda|T_Y = x\lambda_7 + c\lambda_8$ for $p>c>0$. Then (1.23) and (1.10) imply $\lambda|T_A = c\mu_1 + c\mu_2$. However, this contradicts Remark (6.2).

This completes the proof of (6.0).□

(6.9): Summary of Results in Chapter 6

| Description of V\|Y | $\langle\lambda,\beta\rangle$ | Reference |
|---|---|---|
| 0 0 0 0 0 0<br>1 | $3q, q+q_1$ | (6.6)(ii) |
| 0 0 0 0 0 0 a   $a \geq 0$<br>1 | $3q, q+q_1$ | (6.6)(ii) |
| 0 0 0 0 0 b c   $c>0, b \geq 0$<br>0 | $cq$ | (6.2) or (6.3) |
| 1 0 0 0 0<br>0 | $2q$ | (6.5)(i) |
| 1 0 0 0 0 0 0<br>0 | $2q, 4q$ | (6.5)(ii) or (6.6)(i) |
| a 0 0 0 0 1   $a \geq 0$<br>0 | $3q, q+q_1$ | (6.5)(iii) |
| 0 0 0 0 0 1<br>0 | $3q$ | (6.6)(iv) |
| 0 0 0 0 0 0 0<br>1 | $2q$ | (6.6)(v) |
| <u>Partial description of V\|Y</u> | | |
| 0 0 0 <u>or</u> 1 0 0 <u>or</u> 0 0 1<br>1   0   0 | $q$ | (6.5)(iv) |
| ···1 0 0 0··· <u>or</u> ···0 1 0 0··· | $2q$ | (6.8) |
| ···c 0 0 0···   $c \leq 2$ | $cq$ | (6.8) |
| ···1 0 0 0 0··· | $q, q+q_1$ | (6.6)(vi) |
| ···c···   $c \geq 0$ | $cq$ | (6.7) |

# CHAPTER 7: $A = B_2$

In this chapter, we will prove that there are no examples in the solution of the main problem with $A$ of type $B_2$, $Y$ of type $E_n$ and $p \neq 2$. We adopt Notation and Hypothesis (2.0), with the following additions and/or modifications. With $\Pi(A) = \{\alpha_1, \alpha_2\}$, with $\alpha_1$ long, we will take $L_A = \langle U_{\pm\alpha_2} \rangle T_A$. Note that since $p \neq 2$, $Q_A{}^{\alpha_1}$ is a 3-dimensional irreducible $L_A'$ module.

<u>Remark (7.1).</u> Note that $h_{\alpha_2}(-1) \in Z(A) \leq Z(Y)$. Since $Z(Y) \cong \mathbb{Z}_3$ (respectively, $\mathbb{Z}_2$, 1) if $Y = E_6$ (respectively, $E_7$, $E_8$), $h_{\alpha_2}(-1) \neq 1$ will imply $Y$ has type $E_7$. In particular, if $L_i$ has type $A_1$ for some $1 \leq i \leq r$, then $Y = E_7$.

<u>(7.2).</u> Let $q$ be a $p$-power.

(1) If $p > 3$ and $\langle \lambda, \alpha_2 \rangle = 3q$, then $\dim V|A \neq 56$.

(2) If $p \geq 3$ and $\langle \lambda, \alpha_2 \rangle = q$, then $\dim V|A = 56$ if and only if $\langle \lambda, \alpha_1 \rangle = 2q_0$ for some $p$-power $q_0 \neq q$.

(3) If $\lambda | T_Y = \lambda_j$ for some $j$, then $\langle \lambda, \alpha_2 \rangle \neq 2q$.

<u>Proof</u>: Suppose $p > 3$, $\langle \lambda, \alpha_2 \rangle = 3q$ and $\dim V|A = 56$. Then (1.27) implies that $\langle \lambda, \alpha_1 \rangle \neq 0$. The methods of (1.30) and (1.32) imply that $V|A$ is tensor indecomposable, else $\dim V|A > 56$. So $\lambda | T_A = xq\mu_1 + 3q\mu_2$, for some $x < p$. If $x > 1$, the methods of (1.30), (1.32) and (1.35) imply that $\dim V|A > 56$. So we may assume $\lambda | T_A = \mu_1 + 3\mu_2$. Now, $p \neq 7$, else (1.27) and (1.35) imply that $\dim V|A \leq 64 - \dim V(3\mu_2) < 56$. Also, $p \neq 5$, else the last proposition of [4] implies $\dim V|A \leq 64 - \dim V(\mu_1 + \mu_2) < 56$. But now, $p > 7$ and (1.33) imply $\dim V|A = 64$. This completes the proof of (1).

Now suppose $p>2$, $\langle\lambda,\alpha_2\rangle = q$ and $\dim V|A = 56$. We first claim that $V|A$ is tensor decomposable. For if $\lambda|T_A = xq\mu_1+q\mu_2$ for some $0\le x<p$, (1.27) implies $x\ge 3$. In fact $x=3$, else the methods of (1.30) and (1.32) imply $\dim V|A > 56$. By (1.33), $\dim V(3\mu_1+\mu_2) < 80$ only if $p=13$ or $p=7$. If $p=7$, $\dim V|A \ge 80 - 4(\dim V_{T_A}(\lambda-3\alpha_1-3\alpha_2)) \ge 80 - 16$. (Use (1.29) to find a spanning set for the indicated weight space.) If $p=13$, $\dim V|A < 80$ only if $\dim V_{T_A}(\lambda-2\alpha_1-\alpha_2) < 2$. However, $\dim V_{T_A}(\lambda-2\alpha_1-\alpha_2) \ge 1$, so $\dim V|A \ge 80 - \dim V(3\mu_2) \ge 60$. ($\lambda-2\alpha_1-2\alpha_2 = 3\mu_2$.) Thus, $V|A$ is tensor decomposable as claimed. It is now easy to see that $\langle\lambda,\alpha_1\rangle = 2q_0$ for some p-power $q_0 \ne q$. This completes the proof of (2).

Suppose $\langle\lambda,\alpha_2\rangle = 2q$ and $\lambda|T_Y = \lambda_j$ for some $1 \le j \le \operatorname{rank} Y$. Then $\langle\lambda,\alpha_1\rangle \ne 0$ else $\dim V|A < \dim V|Y$. If $\langle\lambda,\alpha_1\rangle = 1\cdot q_0$ for some p-power $q_0$, $\dim V|A \le 50$; so (1.32) implies that $Y = E_6$ and $\dim V|A = 27$. So $q=q_0$. But now applying the last proposition of [4], we see that $\dim V(\mu+2\mu_2) \ne 27$. Consider now the possibility that $\langle\lambda,\alpha_1\rangle = 2q_0$, for some p-power $q_0$. If $q \ne q_0$, using (1.27), (1.30) and (1.32), we find that $117 \le \dim V|A \le 140$. However, using (1.32) and [8], we see that $\dim V|Y > 140$ or $\dim V|A < 117$. Thus, $q=q_0$, and by (1.27), (1.30) and (1.32), $36 \le \dim V|A \le 81$. Thus, [8] implies that $Y = E_7$, $\lambda|T_Y = \lambda_7$ and $\dim V|A = 56$. Now, $\dim V|A < 81$ and (1.33) imply $p=5$ or $p=7$ (recall $p \ne 2$). If $p=7$, (1.33) and the last proposition of [4] imply $\dim V|A = 81 - t$, where $t$ is the dimension of the irreducible kA-module with high weight $2\mu_2$. Using (1.33) again for this module when $p=7$, we see that $t = 10$ and $\dim V|A = 71 \ne 56$. So $p=5$. Then (1.33) and [4] imply that $\dim V|A \ge (81-14) - x$, where $x$ is the multiplicity of the weight $\lambda-3\alpha_1-4\alpha_2$ in the Weyl module $W(\lambda)$. But by (1.29), a spanning set for this weight space has size 8. Thus $x \le 8$ and $\dim V|A > 56$. So, $\dim V|A \ne 56$, as claimed. Hence, $\langle\lambda,\alpha_1\rangle \ne 2q_0$, for $q_0$ a p-power.

For the remaining possibilities, we refer to (6.9) and the configurations in which $\lambda|T_Y$ is more explicitly described. In each case, the methods of (1.30), (1.32) and (1.35), (1.27), [8] and the work of the preceeding paragraphs show that $\dim V|A \neq \dim V|Y$. This completes the proof of (3) and of (7.2).□

(7.3). If $\dim V^1(Q_Y) > 1$, $L_Y'$ is not a simple algebraic group.

Proof: Suppose false. Then Theorem (7.1) of [12] implies $L_Y'$ is of classical type, so by (1.5), $Z_A \leq Z_Y$. Consider first the case where $L_Y'$ has type $D_k$ for some $k \geq 4$. Arguing as in the proof of (6.1), we reduce to $L_Y'$ of type $D_6$, with $V^1(Q_Y)$ isomorphic to the natural module for $L_Y$. Moreover, $Y = E_7$, else $Q_Y/K_{\beta_8}$ is a 12-dimensional irreducible $L_A'$ module containing a nontrivial image of $Q_A{}^{\alpha_1}$. The bound on $\dim V_{\beta_1}(Q_Y)$, (1.32) and (1.34) imply $\langle\lambda,\beta_1\rangle = 0$. Therefore, $\dim V|Y = 56$ and $\langle\lambda,\alpha_2\rangle = q_1+5q_2$ or $q_1+q_2+2q_3$, for $q_1$, $q_2$ and $q_3$ distinct p-powers. Using the methods of (1.30) and (1.32), we see that $\dim V|A > 56$ in either case. Hence, $L_Y'$ has type $A_k$, for some k.

Assume for now that rank $L_Y' > 2$. Then (7.1) of [12] implies $V^1(Q_Y) \cong W$ or $W^*$, where W is the natural module for $L_Y'$. Since $L_Y'$ acts irreducibly on W (and $W^*$), there does not exist $\gamma \in \Pi(Y) - \Pi(L_Y)$ such that $V_{L_Y'}(-\gamma) \cong W$ or $W^*$. If $L_Y'$ has type $A_{n-1}$ when Y has type $E_n$, the bound on $\dim V_{\beta_2}(Q_Y)$ of (1.25) implies $\lambda|T_Y = \lambda_n$. In particular, $Y \neq E_8$. But now induction, (1.30) and (1.32) imply that $\dim V|A > \dim V|Y$ in each case.

Now, consider $L_Y' = \langle U_{\pm\beta_j} \mid 1 \leq j \leq 4\rangle$. The bound on $\dim V_{\beta_5}(Q_Y)$ and (2.3) imply $\langle\lambda,\beta_j\rangle = 0$ for $j \geq 5$. Also, $\langle\lambda,\alpha_2\rangle = 4q$, for some p-power q. If $Y = E_6$ (respectively, $E_7$), $\lambda|T_Y = \lambda_1$ (respectively, $\lambda_2$). Thus, $Y \neq E_6$ as $\dim V|A > 27 = \dim V|Y$. So $Y = E_7$ and $\lambda|T_Y = \lambda_2$, or $Y = E_8$ and $\lambda|T_Y = \lambda_1$ or $\lambda_2$. The $L_A'$ composition factors of $Q_Y/K_{\beta_5}$ have high weights $6q\mu_2$ and $2q\mu_2$; if $p = 5$, the high weights are $(q+5q)\mu_2$ and $2q\mu_2$. Thus, (2.12) applies to give $\beta_j|Z_A = 0$ for $j \geq 6$. Then, using the parabolic $P_Y\hat{}$ of (2.12), we see that the bound on $\dim V_{\beta_5}(Q_Y\hat{})$, of (1.25), is exceeded. So, $L_Y' \neq \langle U_{\pm\beta_j} \mid 1 \leq j \leq 4\rangle$. But if $L_Y' = \langle U_{\pm\beta_j} \mid j \neq 1,3\rangle$ in Y of type $E_8$, the bound on $\dim V_{\beta_3}(Q_Y)$, of (1.25), implies $\lambda|T_Y = \lambda_8$. Also,

$L_Y' \neq \langle U_{\pm\beta_j} \mid j = 2,4,5,6,7\rangle$ as there is no 3-dimensional composition factor of $Q_Y/K_{\beta_3}$, contradicting (2.3). Thus, we have shown that $\mathrm{rank}(L_Y') \leq 2$.

Note that rank $L_Y' \neq 1$, else $\dim(Q_Y/K_\gamma) \leq 2$ for all $\gamma \in \Pi(Y) - \Pi(L_Y)$ and $Q_A \leq K_\gamma$ for all such $\gamma$, contradicting (2.3). Hence, $L_Y'$ has type $A_2$. Let q be the field twist on the embedding of $L_A'$ in $L_Y'$. Then $\beta|T_A = q\alpha_2$ for $\beta \in \Pi(L_Y)$. Since $Q_AK_\gamma/K_\gamma = Q_Y/K_\gamma$ for all $\gamma \in \Pi(Y) - \Pi(L_Y)$ such that $(\gamma,\Sigma L_Y) \neq 0$, $\gamma|T_A = q\alpha_1$, for all such $\gamma$. By (2.12), $\delta|T_A = 0$ for all $\delta \in \Pi(Y) - \Pi(L_Y)$ such that $(\delta,\Sigma L_Y) = 0$. Also, $\langle\lambda,\alpha_1\rangle \neq 0$, else $\dim V|A < \dim V|Y$. So there exists $\gamma \in \Pi(Y) - \Pi(L_Y)$ with $(\gamma,\Sigma L_Y) \neq 0$ and $\langle\lambda,\gamma\rangle \neq 0$, else $V_{T_A}(\lambda-q\alpha_1) = 0$. Moreover, there exists a unique such $\gamma$, else $\dim V_{T_A}(\lambda-q\alpha_1) \geq 2$, contradicting (1.31). Also, (2.13) implies $\gamma$ corresponds to an end node of the Dynkin diagram. Finally, we need to note that $V|A$ is a conjugate of a restricted module as there are no nontrivial $T_Y$ weights in $V|Y$ restricting to $\lambda - q_0\alpha_1$, for $q_0 \neq q$. So by (1.10), $\lambda|T_A = x\mu_1 + 2\mu_2$ for $p > x > 0$. Then by (1.29), $\dim V_{T_A}(\lambda-\alpha_1-2\alpha_2) \leq 3$. However, it is easy to check that if the above conditions are satisfied, there exist 4 linearly independent vectors in $V|Y$ which lie in $V_{T_A}(\lambda-\alpha_1-2\alpha_2)$. This completes the proof of (7.3).

<u>(7.4)</u>. If $\dim V^1(Q_Y) > 1$, each $L_i$ has type $A_{k_i}$, for some $k_i \geq 1$.

<u>Proof</u>: Suppose false. Then $L_i$ has type $D_k$ for some i and k. Otherwise, by (7.1) of [12], $Y = E_8$ and $L_Y'$ has type $E_6 \times A_1$. But this contradicts Remark (7.1). Since $p > 2$, (7.1) of [12], (7.3) and size restrictions imply $L_i$ has type $D_4$ or $D_5$ and $Y = E_7$ or $E_8$. Since all components of $L_Y'$ are necessarily of classical type, (1.5) implies $Z_A \leq Z_Y$.

Arguing as in the proof of (6.3) and applying (7.1) and (7.3), we reduce to $L_Y'$ of type $D_4$. Note that $M_i$ is nontrivial. For otherwise, (7.1), (7.3) and the bound on $\dim V_{\beta_6}(Q_Y)$ and on $\dim V_{\beta_1}(Q_Y)$ imply $Y = E_8$ and

$\lambda|T_Y = \lambda_8$. Then, $M_i$ nontrivial, $p>2$, (7.1) of [12], and (1.14) imply $L_A'$ acts on $M_i$ ($\cong$ the "natural" module for $L_i$) with high weight $(q_1 + 3q_2)\mu_2$, for $q_1$ and $q_2$ distinct p-powers. Since $Q_Y/K_{\beta_1}$ and $Q_Y/K_{\beta_6}$ must each have a 3-dimensional $L_A'$ composition factor, the remarks in the proof of (6.3) imply that $\langle\lambda,\beta_3\rangle = 0$, $V_{L_i}(-\beta_6)|L_A' \cong M_i|L_A'$ and $\langle U_{\pm\beta_7}\rangle \le L_Y'$. Moreover, (1.15) implies $Y = E_8$ and $\langle U_{\pm\beta_7}, U_{\pm\beta_8}\rangle \le L_Y'$. Then by (2.7), $\langle\lambda,\beta_7\rangle = 0 = \langle\lambda,\beta_8\rangle$ and the bound on $\dim V_{\beta_1}(Q_Y)$ and on $\dim V_{\beta_6}(Q_Y)$ implies $\lambda|T_Y = \lambda_2$. By (6.9), $\langle\lambda,\alpha_1\rangle = 0$, $3q$, $q_0 + q$, $2q$, or $q$, for $q$ and $q_0$ distinct p-powers. But (1.27) and (1.38) imply that, in every case, $\dim V|A < \dim V|Y$. Contradiction.

This completes the proof of (7.4).□

<u>(7.5).</u> If $\Pi(L_Y) = \{\beta_1,\beta_2,\beta_3,\beta_5,\beta_7\}$, then $\dim V^1(Q_Y) = 1$.

<u>Proof</u>: Suppose false. By (1.5), $Z_A \le Z_Y$. Let $q_1$, $q_2$, $q_3$ and $q_4$ be the field twists on the embeddings of $L_A'$ in $\langle U_{\pm\beta_1}, U_{\pm\beta_3}\rangle$, $\langle U_{\pm\beta_2}\rangle$, $\langle U_{\pm\beta_5}\rangle$, $\langle U_{\pm\beta_7}\rangle$, respectively. Note that (7.1) implies that $Y = E_7$.

<u>Claim 1.</u> If $\langle\lambda,\beta_1+\beta_2+\beta_3+\beta_5\rangle > 0$, exactly one of $\langle\lambda,\beta_1+\beta_3\rangle$, $\langle\lambda,\beta_2\rangle$ and $\langle\lambda,\beta_5\rangle$ is nonzero.

Reason: Suppose false. Then (2.7), (2.5) and (2.6) imply that $\{q_1,q_2,q_3\}$ consists of exactly two distinct p-powers. If $q_2 = q_3$, the $L_A'$ composition factors of $Q_Y/K_{\beta_4}$ have high weights $(2q_1 + 2q_2)\mu_2$ and $2q_1\mu_2$. Hence, $\beta_4|Z_A = q_1\alpha_1$. Since $\langle\lambda,\beta_1\rangle$ or $\langle\lambda,\beta_3\rangle$ is nonzero, a nonidentity element from the set $U_{-4}\cdot U_{-34}\cdot U_{-134}$ must occur in the factorization of some element in $Q_A - Q_A'$. However, $-\beta_4$ (respectively, $-\beta_3-\beta_4$, $-\beta_1-\beta_3-\beta_4$) affords $T(L_A')$ weight $(2q_1 + 2q_2)\mu_2$ (respectively, $2q_2\mu_2$, $(-2q_1 + 2q_2)\mu_2$). And since $p > 2$ and $q_1 \ne q_2$, none of these weights occurs in $(Q_A{}^{\alpha_1})^{q_1}$. Hence, we may assume $q_1 = q_2$. Then, the $L_A'$ composition factors of $Q_Y/K_{\beta_4}$ have high weights $(3q_1+ q_3)\mu_2$ and $(q_1+ q_3)\mu_2$. If $p = 3$ and $3q_1 = q_3$, the weights are $2q_3\mu_2$, $(q_1+ q_3)\mu_2$ and 0. Thus, $p = 3$, $3q_1=q_3$ and $\beta_4|Z_A = q_3\alpha_1$. Then, we find that

$x_{-\alpha_1}(t) = x_{-\beta_4}(c_1t^{q_3})u_1$, $x_{-\alpha_1-\alpha_2}(t) = x_{-1234}(c_2t^{q_3})x_{-45}(c_3t^{q_3})u_2$, and $x_{-\alpha_1-2\alpha_2}(t) = x_{-12345}(c_4t^{q_3})\ u_3$, where $c_1, c_4 \in K^*$, $c_2, c_3 \in K$, $c_2$ or $c_3$ nonzero, and $u_i \in K_{\beta_4}$. Then, there is a nontrivial contribution to the root group $U_{-1345}$ in the expression for $[x_{\alpha_2}(t), x_{-\alpha_1-2\alpha_2}(t)]$, contradicting the given factorizations of $x_{-\alpha_1}(t)$ and $x_{-\alpha_1-\alpha_2}(t)$. This completes the proof of Claim 1.

Claim 2. $\langle\lambda,\beta_1+\beta_2+\beta_3+\beta_5\rangle = 0$.

Reason: Suppose false; then Claim 1 implies that exactly one of $\langle\lambda,\beta_1+\beta_3\rangle$, $\langle\lambda,\beta_2\rangle$, $\langle\lambda,\beta_5\rangle$ is nonzero. By (2.7), $\{q_1,q_2,q_3\}$ consists of at most two distinct p-powers. Arguing as in the proof of Claim 1, we find that $q_2=q_3$ and $\beta_4|Z_A = q_1\alpha_1$.

Suppose $\langle\lambda,\beta_1+\beta_3\rangle \neq 0$. Then, the proof of Claim 1 shows that $q_1=q_2$. Also, the bound on $\dim V_{\beta_4}(Q_Y)$ implies $\langle\lambda,\beta_4+\beta_3\rangle = 0$. Now, if $Q_A \nleq K_{\beta_6}$, (2.17) implies that the field twist on the embedding of $L_A'$ in $\langle U_{\pm\beta_7}\rangle$ is also $q_1$. Thus, by (2.5) and (2.6), $\langle\lambda,\beta_7\rangle = 0$. Also, $\langle\lambda,\beta_6\rangle = 0$, else the bound on $\dim V_{\beta_6}(Q_Y)$ is exceeded. But now we have $\lambda|T_Y = \lambda_1$. So $\langle\lambda,\beta_1+\beta_3\rangle = 0 \neq \langle\lambda,\beta_2+\beta_5\rangle$.

Suppose $\langle\lambda,\beta_2\rangle \neq 0$, so $\langle\lambda,\beta_\ell\rangle = 0$ for $\ell = 1, 3, 5$. If $Q_A \nleq K_{\beta_6}$, (2.17) implies that the field twist on the embedding of $L_A'$ in $\langle U_{\pm\beta_7}\rangle$ is $q_3 = q_2$. Then (2.5) and (2.6) imply $\langle\lambda,\beta_7\rangle = 0$. Moreover, $\langle\lambda,\beta_6\rangle = 0$, else the bound on $\dim V_{\beta_6}(Q_Y)$ is exceeded. Finally, using (1.36) and the bound on $\dim V_{\beta_4}(Q_Y)$, we see that $\lambda|T_Y = \lambda_2$, and $\langle\lambda,\alpha_2\rangle = 1\cdot q_2$. By (6.9), $\langle\lambda,\alpha_1\rangle = 0, q, q + q_0, 3q$ or $2q$, for $q$ and $q_0$ distinct p-powers. In every case, (1.27) and [8] imply $\dim V|A < \dim V|Y$. Thus, $\langle\lambda,\beta_1+\beta_2+\beta_3\rangle = 0$ and $\langle\lambda,\beta_5\rangle \neq 0$.

The arguments of the preceeding paragraph imply $\lambda|T_Y = \lambda_5 + x\lambda_6$, for $x \geq 0$ and by (2.17), $\beta_6|T_A = q_2\alpha_1$. Moreover, $q_1 \neq q_2$, else the bound on $\dim V^2(Q_A)_{\lambda-q_2\alpha_1}$, of (1.22), is exceeded. So $V^2(Q_A)_{\lambda-q_1\alpha_1} \neq 0 \neq V^2(Q_A)_{\lambda-q_2\alpha_1}$ and $V|A$ is tensor decomposable. In particular, $\langle\lambda,\alpha_1\rangle \neq 0$. By (6.9), $\langle\lambda,\alpha_1\rangle = q, 2q$ or $xq$, for some p-power $q \neq q_2$. Then by (1.27) and (1.32), $\dim V|A < \dim V|Y$ unless $\langle\lambda,\alpha_1\rangle = xq$ with $x \neq 0$. However, then $\dim V^2(Q_A)_{\lambda-q_2\alpha_1} = 2$ while $f_{\beta_6}v^+$ affords an $L_Y'$ composition factor in $V^2(Q_A)_{\lambda-q_2\alpha_1}$ of dimension 6. Contradiction. This completes the proof of Claim 2.

Claim 2 implies that $\langle\lambda,\beta_7\rangle \neq 0$; so by (2.7), $q_3 = q_4$.

<u>Claim 3.</u> $Q_A \leq K_{\beta_4}$.

Reason: Suppose $Q_A \nleq K_{\beta_4}$. We examine the image of $Q_A{}^{\alpha_1}$ in $Q_Y/K_{\beta_4}$. Arguing as in the proof of Claim 1, we see that $q_2 = q_3$ and $\beta_4|Z_A = q_1\alpha_1$. Suppose $q_1 \neq q_3$. Examining the $T(L_A')$ weights of $Q_Y/K_{\beta_4}$, we see that $x_{-\alpha_1}(t) = x_{-45}(c_1t^{q_1})x_{-24}(c_2t^{q_1})x_{-\beta_6}(c_3t^{q_3})u_0$, where $c_i \in k$, $c_1$ or $c_2$ nonzero, $c_3 \neq 0$, and $u_0 \in K_{\beta_4} \cap K_{\beta_6}$. In fact, $c_1c_2 \neq 0$, else there is a nontrivial contribution to the root group $U_{-\beta_4}$ in the expression for $[x_{\alpha_2}(t),x_{-\alpha_1}(t)]$. We also find that $x_{-\alpha_1-2\alpha_2}(t) = x_{-1345}(d_1t^{q_1})\cdot x_{-1234}(d_2t^{q_1})x_{-567}(d_3t^{q_3})u_1$, where $d_i \in k$, $d_3 \neq 0$, $d_1$ or $d_2$ nonzero and $u_1 \in K_{\beta_4} \cap K_{\beta_6}$. Thus, in the expression for $[x_{-\alpha_1}(t),x_{-\alpha_1-2\alpha_2}(t)]$, there is a nontrivial contribution to the root group $U_{-24567}$. Contradiction.

So $q_1 = q_3$ and again examining the $T(L_A')$ weight vectors in $Q_Y/K_{\beta_4}$, we find that $x_{-\alpha_1}(t) = x_{-34}(a_1t^{q_3})x_{-24}(a_2t^{q_3})x_{-45}(a_3t^{q_3})x_{-\beta_6}(a_4t^{q_3})w_0$ and $x_{-\alpha_1-2\alpha_2}(t) = x_{-1345}(b_1t^{q_3})x_{-1234}(b_2t^{q_3})x_{-2345}(b_3t^{q_3})x_{-567}(b_4t^{q_3})w_1$, where $a_i, b_i \in k$, $a_4b_4 \neq 0$, $w_0, w_1 \in K_{\beta_4} \cap K_{\beta_6}$ and $a_ib_j \neq 0$ for some $1 \leq i, j \leq 3$. In fact, at least two of $a_1, a_2, a_3$ are nonzero, else there is a nontrivial contribution to the root group $U_{-\beta_4}$ in the expression for $[x_{-\alpha_1}(t),x_{\alpha_2}(t)]$. So $a_1$ or $a_2$ is nonzero. But then there is a nontrivial contribution to the group $U_{-24567}\cdot U_{-34567}$ in the expression for $[x_{-\alpha_1}(t),x_{-\alpha_1-2\alpha_2}(t)]$. Contradiction. This completes the proof of Claim 3.

Now, $Q_A \leq K_{\beta_4}$ implies that $\lambda|T_Y = x\lambda_6 + c\lambda_7$, for $x \geq 0$, $c>0$ and $\langle\lambda,\alpha_2\rangle = cq_4$. Referring to (6.9), we see that if $x \neq 0$ or if $c>1$, $\langle\lambda,\alpha_1\rangle = 0$, $q$, $2q$, $xq$ or $cq$. So $\dim V^2(Q_A) \leq 5c+3$, by (1.22). Let $w = f_{456}v^+$ if $\langle\lambda,\beta_6\rangle \neq 0$, or $w = f_{4567}v^+$ if $\langle\lambda,\beta_6\rangle = 0$. Then $w \notin [V, K_{\beta_4}{}^2]$, so $w \notin [V,Q_A{}^2]$; hence, $w$ affords an $L_Y'$ composition factor in $V^2(Q_A)$.

Adding the dimension of the $L_Y'$ composition factor afforded by $f_{\beta_6}v^+$ (or $f_{67}v^+$), we find that if $x \neq 0$ or $c>1$, $\dim V^2(Q_A) > 5c+3$. (Use (1.36).) Hence, $\lambda|T_Y = \lambda_7$ and (7.2) implies that $\langle\lambda,\alpha_1\rangle = 2q$, for some p-power $q \neq q_4$.

One checks that in the action of $L_A'$ on the kA module $V(2q\mu_1+q_4\mu_2)$, if $q_4 \neq 3q$, there is an 8- or 10-dimensional $L_A'$ composition factor. However, the given embedding of $L_A'$ in $P_Y$ affords no such $L_A'$ composition factor on $V(\lambda_7)$. So p=3 and $q_4 = 3q$. Then, there are exactly six 6-dimensional $L_A'$ composition factors of $V(2q\mu_1 + q_4\mu_2)$, two of which have distinct high weights. However, one checks that the given embedding of $L_A'$ in $P_Y$ does not afford such an $L_A'$ composition series of $V(\lambda_7)$.

This completes the proof of (7.5).□

(7.6). Let $\gamma \in \Pi(Y)-\Pi(L_Y)$ and $1 \leq i,j \leq r$ such that $(\Sigma L_i,\gamma) \neq 0 \neq (\Sigma L_j,\gamma)$.

(i) Then $M_i$ or $M_j$ is trivial.

(ii) If in addition there exists $k \neq i,j$ such that $(\Sigma L_k,\gamma) \neq 0$, then $\dim(M_\ell) = 1$ for $\ell = i,j,k$.

Proof: By (7.4), all components of $L_Y'$ have classical type, so by (1.5), $Z_A \leq Z_Y$. Let $W_m$ denoted the natural module for $L_m$. By (7.1) of [12], if $\dim M_m > 1$ and $\operatorname{rank} L_m > 1$, $M_m \cong W_m$ or $W_m^*$. Consider first the case where there exists k as in (ii), so $\gamma = \beta_4$. Then (2.5) and (2.7) imply that at most two of $M_i$, $M_j$, and $M_k$ are nontrivial. Since $\operatorname{rank}(L_m) = 1$ for m = i, j or k, (7.1) implies that $Y = E_7$ and $\langle U_{\pm\beta_3}\rangle$ is not a component of $L_Y'$. This observation, together with (1.15), implies that $L_i \times L_j \times L_k$ has type $A_1\times A_2\times A_3$ or $\Pi(L_i\times L_j\times L_k) = \{\beta_1,\beta_3,\beta_2,\beta_5\}$. In the second case, $L_Y' = L_i \times L_j\times L_k\times \langle U_{\pm\beta_7}\rangle$, else $h_{\alpha_2}(-1)$ does not centralize $U_{\beta_6}$. But this configuration does not occur, by (7.5). If $L_i\times L_j\times L_k$ has type $A_1\times A_2\times A_3$, p>2, the bound on $\dim V_{\beta_4}(Q_Y)$ and (1.36), imply $\lambda|T_Y = \lambda_1$ or $\lambda_7$. But

$\lambda|T_Y \neq \lambda_1$, so $\lambda|T_Y = \lambda_7$. Also, considering $Q_Y/K_{\beta_4}$ in view of (1.15), we see that $V^1(Q_A)$ is tensor indecomposable; so $\langle\lambda,\alpha_2\rangle = 3q$, for some p-power, q, and $p > 3$. But this contradicts (7.2). Hence, (ii) holds.

Now suppose $M_i$ and $M_j$ are both nontrivial. We may assume $V_{L_i}(-\gamma) \not\cong W_i$ or $W_i^*$. For otherwise, (2.5), (2.6) and (2.7) imply that there exists $k \neq i,j$ with $(\Sigma L_k,\gamma) \neq 0$. But then (ii) implies $\dim(M_i) = 1 = \dim(M_j)$. Hence, $\text{rank}(L_i) > 2$ and $V_{L_i}(-\gamma) \cong W_i \wedge W_i$ or $W_i^* \wedge W_i^*$. Moreover, since $L_A'$ acts irreducibly on $W_i$, there does not exist $\delta \in \Pi(Y) - \Pi(L_Y)$ such that $Q_Y/K_\delta \cong W_i$ or $W_i^*$, else $Q_A \leq K_\delta$, contradicting (2.3). Since $(W_i \wedge W_i)|L_A'$ has all even $T(L_A')$ weights, (1.15) implies that $L_j$ has type $A_2$ or $L_j$ has type $A_3$ and $M_j|L_A'$ is tensor decomposable. Thus $L_i$ has type $A_4$ and $p > 3$. Let $q_1$ be the field twist on the embedding of $L_A'$ in $L_i$. Then $(W_i \wedge W_i)|L_A'$ has composition factors with high weights $6q_1\mu_2$ and $2q_1\mu_2$ (or $(5q_1+q_1)\mu_2$ and $2q_1\mu_2$, if p=5). Using (2.5) and (2.6), and the above remarks, it is a check to see that there is no composition factor of $Q_Y/K_\gamma$ isomorphic to a twist of $Q_A^{\alpha_1}$. Thus, $Q_A \leq K_\gamma$, contradicting (2.3). Hence, (i) holds.

This completes the proof of (7.6).

<u>(7.7)</u>. Let $\gamma \in \Pi(Y) - \Pi(L_Y)$ and $1 \leq i, j \leq r$, such that $(\Sigma L_i,\gamma) \neq 0 \neq (\Sigma L_j,\gamma)$. Then $\dim(M_i \otimes M_j) = 1$.

<u>Proof</u>. Suppose $\dim(M_i \otimes M_j) > 1$. By (7.4), each component of $L_Y'$ has type $A_{k_i}$ for some $k_i \geq 1$; so (1.5) implies $Z_A \leq Z_Y$. Let $W_m$ denote the natural module for $L_m$, m = i, j. By (7.1) of [12], if $M_m$ is nontrivial and $\text{rank}(L_m) > 1$, $M_m \cong W_m$ or $W_m^*$. Also, (7.6) implies that only one of $M_i$ and $M_j$ is nontrivial.

<u>Case I</u>: Suppose $V_{L_i}(-\gamma) \not\cong W_i$ or $W_i^*$.

Then $\text{rank}(L_i) > 2$ and $V_{L_i}(-\gamma) \cong W_i \wedge W_i$ or $W_i^* \wedge W_i^*$. Since $(W_i \wedge W_i)|L_A'$ has all even weights, p>2 and (1.15) implies $L_j$ has type $A_2$ or $A_3$, so $Y = E_7$ or $E_8$. Note that if $M_m$ is nontrivial and $\text{rank}(L_m) > 2$, there does not exist $\delta \in \Pi(Y) - \Pi(L_Y)$ such that $Q_A \nleq K_\delta$, $(\delta, \Sigma L_m) \neq 0$ and $Q_Y/K_\delta \cong W_m$ or $W_m{}^*$. These remarks and the bound on $\dim V_\gamma(Q_Y)$ imply that $\gamma = \beta_5$ and either $Y = E_8$, with $\Pi(L_Y) = \{\beta_i \mid i \neq 5,8\}$ and $\lambda | T_Y = \lambda_1 + x\lambda_8$, or $Y = E_7$ (respectively, $E_8$), with $\Pi(L_Y) = \{\beta_2, \beta_3, \beta_4, \beta_6, \beta_7\}$ and $\lambda | T_Y = \lambda_7$ (respectively, $\lambda_7 + x\lambda_8$, for $x \geq 0$).

In the first case, the argument in the second paragraph of the proof of (7.6) implies that the field twists on the embeddings of $L_A'$ in the two components of $L_Y'$ are equal. Call this twist q. Then, as $p>3$, the only $L_A'$ composition factors of $Q_Y/K_{\beta_5}$ isomorphic to a twist of $Q_A{}^{\alpha_1}$ have high weight $2q\mu_2$. Thus, $\beta_5 | Z_A = q\alpha_1$. If $x \neq 0$, $Q_A \nleq K_{\beta_8}$ and $Q_Y/K_{\beta_8}$ is the irreducible $L_A'$ module with high weight $2q\mu_2$, so $\beta_8 | Z_A = q\alpha_1$ also. However, the bound on $\dim V^2(Q_A)_{\lambda - q\alpha_1}$, of (1.22), is exceeded. Thus, $x = 0$. By (6.9), $\langle \lambda, \alpha_1 \rangle = 0$, $2q_0$ or $q_0$, for some p-power $q_0$. But (1.27) and [8] imply $\dim V|A < \dim V|Y$. Thus, the first configuration does not occur.

In the second case, (7.2) implies that $Y \neq E_7$. If $x \neq 0$, one checks that $V_{T_Y}(\lambda - \beta_8) \leq V_{T_A}(\lambda - q\alpha_1)$, where $\langle \lambda, \alpha_2 \rangle = 2q$, for some p-power q. By (6.9), $\langle \lambda, \alpha_1 \rangle = 0$, q, 2q or xq. But then $9 \geq \dim V^2(Q_A) \geq V^2(Q_Y) \geq \dim(V_{\beta_8}(Q_Y) + V_{\beta_5}(Q_Y)) \geq 12$. Hence $x = 0$, contradicting (7.2).

This completes the consideration of Case I.

<u>Case II</u>: Suppose $V_{L_m}(-\gamma) \cong W_m$ or $W_m{}^*$ for $m = i, j$.

By (7.6), we have $(\Sigma L_k, \gamma) = 0$ for $k \neq i, j$. Also, (1.15) and (7.1) imply $L_i \times L_j$ has type $A_1 \times A_\ell$, $\ell = 1$ or 3 and $Y = E_7$, or $L_i \times L_j$ has type $A_2 \times A_\ell$, $\ell = 2$, 3 or 4, or $A_3 \times A_3$. Actually, $L_i \times L_j$ cannot have type $A_3 \times A_3$, else $Y = E_8$ and $Q_A \leq K_{\beta_2}$; so $\langle \lambda, \beta_\ell \rangle = 0$ for $1 \leq \ell \leq 4$. But then the bound on $\dim V_\gamma(Q_Y)$ of (1.25) implies $\lambda | T_Y = \lambda_8$.

Now consider $L_i \times L_j$ of type $A_2 \times A_4$. Using (2.3) and the bound on $\dim V_\gamma(Q_Y)$ of (1.25), we restrict the possibilities for $\lambda$. We are left with $Y = E_8$, $\lambda | T_Y = \lambda_1$ and $\langle \lambda, \alpha_2 \rangle = 2q$ or $4q$, for some p-power q. But (7.2) and a dimension argument from Case I imply that $\dim V|A \neq \dim V|Y$.

Suppose $L_i \times L_j$ has type $A_2 \times A_3$. Temporarily label as follows:

$L_i = \langle U_{\pm\gamma_1}, U_{\pm\gamma_2}\rangle$, $L_j = \langle U_{\pm\gamma_k} \mid k = 3,4,5\rangle$, $(\gamma_2,\gamma) \neq 0 \neq (\gamma,\gamma_3)$, $(\gamma_k,\gamma_{k+1}) < 0$, $k = 3, 4$. Note that (1.15) implies $V_{L_j}(-\gamma) \mid L_A'$ is tensor decomposable. Let $V_{L_j}(-\gamma)$ have high weight $(q_1 + q_2)\mu_2$ as $L_A'$ module, where $q_1$ and $q_2$ are distinct p-powers. Then, by (2.7) and (2.6) we may assume that the field twist on the embedding of $L_A'$ in $L_i$ is $q_1$. Then $Q_Y/K_\gamma$ has $L_A'$ composition factors with high weights $(3q_1 + q_2)\mu_2$ and $(q_1+ q_2)\mu_2$. Thus, $p = 3$ and $3q_1 = q_2$, else there is no 3-dimensional composition factor. In this case, $\gamma \mid Z_A = q_2\alpha_1$, and we find that

$$x_{-\alpha_1}(t) \in U_{-\gamma}K_\gamma,$$

$$x_{-\alpha_1-\alpha_2}(t) \in (U_{-\gamma-\gamma_3} \cdot U_{-\gamma-\gamma_3-\gamma_4} \cdot U_{-\gamma_1-\gamma_2-\gamma-\gamma_3} \cdot U_{-\gamma_1-\gamma_2-\gamma-\gamma_3-\gamma_4})K_\gamma, \text{ and}$$

$$x_{-\alpha_1-2\alpha_2}(t) = x_{-\gamma_1-\gamma_2-\gamma-\gamma_3-\gamma_4-\gamma_5}(ct^{q_2})u_0,$$

where $c \in k^*$, $u_0 \in K_\gamma$. Then, there is a nontrivial contribution to the root group $U_{-\gamma_2-\gamma-\gamma_3-\gamma_4-\gamma_5}$ in the expression for $[x_{\alpha_2}(t), x_{-\alpha_1-2\alpha_2}(t)]$, contradicting the given information about $x_{-\alpha_1}(t)$ and $x_{-\alpha_1-\alpha_2}(t)$. Thus, $L_i \times L_j$ does not have type $A_2 \times A_3$.

Consider now the pair $A_1 \times A_3$ in Y of type $E_7$. Temporarily label as follows: $L_i = \langle U_{\pm\gamma_0}\rangle$, $L_j = \langle U_{\pm\gamma_k} \mid 1 \leq k \leq 3\rangle$, $(\gamma,\gamma_1) \neq 0$ and $(\gamma_k,\gamma_{k+1}) < 0$ for $k = 1, 2$. By (1.15), $W_j \mid L_A'$ is tensor indecomposable, so $p > 3$. Note that there does not exist $\delta \in \Pi(Y) - \Pi(L_Y)$, with $(\delta,\gamma_0) \neq 0$ and $(\delta,\Sigma L_k) = 0$ for $k \neq i$. For otherwise, since $Q_Y/K_\delta$ is a 2-dimensional irreducible $L_A'$ module and $Q_Y(\gamma,\delta)$ is a 4-dimensional irreducible $L_A'$ module, (2.11) implies $-\delta$ is involved in $L_A'$. But this cannot occur as $p \neq 2$. (See (2.10).) Arguing similarly, one shows that there does not exist $\delta \in \Pi(Y) - \Pi(L_Y)$ with $(\delta,\gamma_3) \neq 0$ and $(\delta,\Sigma L_k) = 0$ for all $k \neq j$. These remarks, and the bound on $\dim V_\gamma(Q_Y)$, together with (1.36), imply that $\langle\lambda,\gamma\rangle = 0$ and either (a) $\Pi(L_i \times L_j) = \{\beta_2,\beta_5,\beta_6,\beta_7\}$ or (b) $\Pi(L_i \times L_j) = \{\beta_2,\beta_4,\beta_5,\beta_7\}$. By (2.17), there exists a p-power q, which is the field twist on the embeddings of $L_A'$ in $L_i$ and in $L_j$ and such that $\gamma \mid Z_A = q\alpha_1$.

In each case, $L_Y' = L_i \times L_j$, else $h_{\alpha_2}(-1) \notin Z(Y)$. In case (a), the bound on $\dim V_{\beta_5}(Q_Y)$ and (2.3) imply $\lambda|T_Y = c\lambda_2$ for $1 \le c \le 3$ or $\lambda|T_Y = \lambda_5$ or $\lambda_7$. If $\lambda|T_Y = c\lambda_2$ for c = 2 or 3, then $\langle\lambda,\alpha_1\rangle = 0$ or $cq_1$ for some p-power $q_1$. But (1.27) and (1.32) imply $\dim V|A < \dim V|Y$. In the remaining cases, (6.9) implies $\langle\lambda,\alpha_1\rangle = 0, q_1, 2q_1, 3q_1$ or $q_1+q_2$, for $q_1$ and $q_2$ distinct p-powers. Now (1.27) and [8] imply $\dim V|A \neq \dim V|Y$ unless $\lambda|T_Y = \lambda_7$. In this case $\dim V|Y \neq \dim V|A$ by (7.2). Thus the configuration of (a) does not occur. If $L_Y' = L_i \times L_j$ as in (b), $Q_A \le K_{\beta_3}$ as $Q_Y/K_{\beta_3}$ has $L_A'$ composition factors of dimensions 5 and 1. Hence $\lambda|T_Y = c\lambda_7$, for $1 \le c \le 3$. If $\lambda|T_Y \neq \lambda_7$, we may argue as above to see that $\dim V|A \neq \dim V|Y$. If $\lambda|T_Y = \lambda_7$, (7.2) implies $\lambda|T_A = 2q_0\mu_1+q\mu_2$, for q and $q_0$ distinct p-powers. However, the $Z_A$ weight space $V^2(Q_A)_{\lambda-q\alpha_1}$ has dimension 2, while $0 \neq w \in V_{T_Y}(\lambda-\beta_7-\beta_6)$ affords an $L_Y'$ composition factor of $V^2(Q_A)_{\lambda-q\alpha_1}$ of dimension 4. Thus, $L_i \times L_j$ does not have type $A_1 \times A_3$.

Consider now the case where $L_i \times L_j$ has type $A_1 \times A_1$ in Y of type $E_7$. Temporarily label as follows: let $\gamma_i, \gamma_j \in \Pi(L_Y)$, with $L_k = \langle U_{\pm\gamma_k}\rangle$, k = i, j and let q be the field twist on the embeddings of $L_A'$ in $L_i$ and in $L_j$. (See (2.17).) Then, it is easy to check that $\gamma|T_A = q\alpha_1$. As in the previous case, there does not exist $\delta \in \Pi(Y) - \Pi(L_Y)$ such that $(\delta,\gamma_i) \neq 0$, $(\delta,\Sigma L_k) = 0$ for all $k \neq i$. Similarly, there does not exist $\delta \in \Pi(Y) - \Pi(L_Y)$ such that $(\delta,\gamma_j) \neq 0$, $(\delta,\Sigma L_k) = 0$ for all $k \neq j$. These remarks imply that $\{\gamma_i,\gamma_j\} = \{\beta_2,\beta_5\}$ or $\{\beta_5,\beta_7\}$.

Consider the case where $\{\gamma_i,\gamma_j\} = \{\beta_2,\beta_5\}$. Then $\langle U_{\pm\beta_1}\rangle$ is not a component of $L_A'$, else $h_{\alpha_2}(-1) \notin Z(Y)$, contradicting (7.1). Thus, (2.12) implies $\beta_1|T_A = 0 = \beta_3|T_A$. This forces $\langle\lambda,\beta_4\rangle = 0$, else $f_{34}v^+$ and $f_{\beta_4}v^+$ are linearly independent vectors in $V_{T_Y}(\lambda-q\alpha_1)$, contradicting (1.31). Suppose $\langle\lambda,\beta_2\rangle \neq 0$. Then $f_{24}v^+$, $f_{234}v^+$ and $f_{1234}v^+$ are 3 linearly independent vectors in $V_{T_Y}(\lambda-q\alpha_1-q\alpha_2)$, contradicting (1.37). Thus, $\langle\lambda,\beta_2\rangle = 0$. A similar argument shows that $\langle\lambda,\beta_5\rangle = 0$. So $\{\gamma_i,\gamma_j\} \neq \{\beta_2,\beta_5\}$.

Consider now the case where $\{\gamma_i,\gamma_j\} = \{\beta_5,\beta_7\}$. The previous remarks imply that there exists another component of $L_Y'$, say $L_k$, with $(\Sigma L_k,\beta_4) \neq 0$. Note that $\langle U_{\pm\beta_3}\rangle$ is not a component of $L_Y'$, else $h_{\alpha_2}(-1)$ does not centralize $U_{\beta_1}$. So if $U_{\pm\beta_3} \leq L_Y'$, $\langle U_{\pm\beta_1}, U_{\pm\beta_3}\rangle$ is a component of $L_Y'$. In fact, if $\langle U_{\pm\beta_1}, U_{\pm\beta_3}\rangle$ is a component of $L_Y'$ then $\langle U_{\pm\beta_2}\rangle$ is also, else $h_{\alpha_2}(-1)$ does not centralize $U_{\beta_4}$. Thus, (7.5) implies $L_Y' = L_i \times L_j \times \langle U_{\pm\beta_2}\rangle$.

By the previous case, $\langle\lambda,\beta_2+\beta_5\rangle = 0$ and $\langle\lambda,\beta_7\rangle \neq 0$. We claim that $Q_A \leq K_{\beta_4}$. Otherwise, (2.17) implies that the field twist on the embedding of $L_A'$ in $\langle U_{\pm\beta_2}\rangle$ is also q and we find that $x_{-\alpha_1}(t) = x_{-\beta_4}(c_1t^q)\cdot x_{-\beta_6}(c_2t^q)u_1$, $x_{-\alpha_1-2\alpha_2}(t) = x_{-245}(c_3t^q)x_{-567}(c_4t^q)u_2$, where $c_i \in k^*$, $u_i \in K_{\beta_4} \cap K_{\beta_6}$. But then there is a nontrivial contribution to the group $U_{-2456}\cdot U_{-4567}$ in the expression for $[x_{-\alpha_1}(t), x_{-\alpha_1-2\alpha_2}(t)]$. So $Q_A \leq K_{\beta_4}$ and (2.3) implies $\langle\lambda,\beta_k\rangle = 0$ for $1 \leq k \leq 5$.

If $-\beta_4$ is involved in $L_A'$, $\beta_4|Z_A = 0$ by (2.10). Otherwise, (2.11) implies that there is a nontrivial image of $Q_A{}^{\alpha_1}$ in $Q_Y(\beta_6,\beta_4)$. So the field twist on the embedding of $L_A'$ in $\langle U_{\pm\beta_2}\rangle$ is q, $(\beta_4+\beta_6)|Z_A = q\alpha_1$ and again $\beta_4|Z_A = 0$. Using the parabolic $P_Y{}^\wedge$ of (2.11), we see that the bound on $\dim V_{\beta_6}(Q_Y{}^\wedge)$ is exceeded unless $\langle\lambda,\beta_6\rangle = 0$ and $\langle\lambda,\beta_7\rangle \leq 3$. (Refer to (1.36) in case $\langle\lambda,\beta_7\rangle = p-1$.) So $\lambda|T_Y = c\lambda_7$, $c \leq 3$ and $\langle\lambda,\alpha_2\rangle = cq$. By (6.9), $\langle\lambda,\alpha_1\rangle = 0$ or $cq_0$ if $c > 1$, or $\langle\lambda,\alpha_1\rangle = 0, 3q_0, q_1 + q_0, q_0$ or $2q_0$, for $q_0$ and $q_1$ distinct p-powers, if $c = 1$. But (1.27) and (1.38) imply $\dim V|A < \dim V|Y$ if $c>1$. Thus, $c = 1$, $\dim V|Y = 56$, and by (7.2), $\lambda|T_A = 2q_0\mu_1 + q\mu_2$ where $q_0 \neq q$. Note that the $Z_A$ weight space $V^2(Q_A)_{\lambda-q\alpha_1}$ has dimension 2. But $V_{T_Y}(\lambda-\beta_7-\beta_6)$, $V_{T_Y}(\lambda-\beta_7-\beta_6-\beta_5)$ and $V_{T_Y}(\lambda-\beta_7-\beta_6-\beta_5-\beta_4)$ are 3 nonzero weight spaces lying in $V^2(Q_A)_{\lambda-q\alpha_1}$. Contradiction.

It remains to consider the case where $L_i \times L_k$ has type $A_2 \times A_2$. We first claim that there does not exist a third component of $L_Y'$. For if $L_Y'$ has 3 components, size restrictions and the fact that $(\gamma,\Sigma L_k) = 0$ for $k \neq i, j$, imply that the third component has type $A_1$. Then by (7.1), $Y = E_7$ and $\Pi(L_Y) = \{\beta_1,\beta_2,\beta_4,\beta_6,\beta_7\}$. Now, (1.15) implies $Q_A \leq K_{\beta_3}$, so

$\langle\lambda,\beta_\ell\rangle = 0$ for $1 \le \ell \le 4$ and $\langle\lambda,\beta_6\rangle$ or $\langle\lambda,\beta_7\rangle$ is nonzero. Since $p>2$ and all $T(L_A')$ weights in $Q_Y/K_{\beta_3}$ are odd, (2.10) implies that $-\beta_3$ is not involved in $L_A'$. Thus (2.11) implies that there is a nontrivial image of $Q_A{}^{\alpha_1}$ in $Q_Y(\beta_5,\beta_3)$, an $L_A'$ module with no 3-dimensional composition factor. Thus, $L_Y' = L_i \times L_j$, as claimed.

Now, the bound on $\dim V_\gamma(Q_Y)$, of (1.25), implies that $\langle\lambda,\gamma\rangle = 0$. Hence, there exists $\delta \in \Pi(Y) - \Pi(L_Y)$ with $\langle\lambda,\delta\rangle \neq 0$. For otherwise, $\lambda|T_Y = \lambda_\ell$ for some $\ell$ and $\langle\lambda,\alpha_2\rangle = 2q$, contradicting (7.2). Then $\delta \neq \gamma$ and by (2.3), $(\delta,\Sigma L_Y) \neq 0$. Say $(\delta,\Sigma L_i) \neq 0$. Let q be the field twist on the embedding of $L_A'$ in $L_i$ and in $L_j$. (Use (2.7) to get equal twists.) Then the $L_A'$ composition factors of $Q_Y/K_\gamma$ have high weights $4q\mu_2$, $2q\mu_2$ and 0. Thus, $\gamma|Z_A = q\alpha_1$. Moreover, $Q_AK_\delta/K_\delta = Q_Y/K_\delta$ and by (2.13), $\delta|T_A = q\alpha_1$. Then, the bound on $\dim V^2(Q_A)_{\lambda-q\alpha_1}$, of (1.22), implies that $M_i$ is nontrivial. Also, by (2.13), there does not exist $\delta_1 \in \Pi(Y) - \Pi(L_Y)$ with $(\delta_1,\delta) < 0$. So $\delta$ corresponds to an end node of the Dynkin diagram. We now claim that there does not exist $\gamma_1 \in \Pi(Y) - \Pi(L_Y)$ with $(\gamma,\gamma_1) < 0$. For if there exists such a $\gamma_1$, $Q_A \le K_{\gamma_1}$ and (2.12) and the above remarks imply that $\gamma_1|Z_A = 0$. Then using the parabolic $P_Y\hat{}$ of (2.11), we find that $\dim V_\gamma(Q_Y\hat{}) + \dim V_\delta(Q_Y\hat{})$ exceeds the bound on $\dim V^2(Q_A)_{\lambda-q\alpha_1}$, of (1.22). Finally, we note that there does not exist $\delta_1 \in \Pi(Y) - \Pi(L_Y)$ with $\gamma \neq \delta_1 \neq \delta$ and $(\delta_1,\Sigma L_i) \neq 0$. For, as with $\delta$, $\delta_1|T_A = q\alpha_1$ and the bound on $\dim V^2(Q_A)_{\lambda-q\alpha_1}$ is exceeded. These remarks imply that $Y = E_8$ and $\delta = \beta_8$. The bound on $\dim V^2(Q_A)_{\lambda-q\alpha_1}$ implies, even more explicitly, that $\lambda|T_Y = \lambda_7 + x\lambda_8$ for $p > x > 0$.

By (6.9), $\langle\lambda,\alpha_1\rangle = 0$, $xq_0$, $q_0$ or $2q_0$, for some p-power $q_0$. Then (1.27) and (1.32) imply $\langle\lambda,\alpha_1\rangle = xq_0$, else $\dim V|A < \dim V|Y$. In fact, since $\beta_8|T_A = q\alpha_1$, $f_{\beta_8}v^+$ is a nonzero vector in $V_{T_A}(\lambda-q\alpha_1)$; so $q_0 = q$ and by (1.10), $\lambda|T_A = x\mu_1 + 2\mu_2$. Now, let $P_0 \ge B_Y{}^-$ be the parabolic subgroup of Y with Levi factor $L_0 = \langle U_{\pm\beta_\ell} \mid 5 \le \ell \le 8\rangle T_Y$. Then, $L_0$ has a natural subgroup, B, of type $B_2$. Moreover, $V^1(R_u(P_0))|B$ has a composition factor with the same high weight, as a $B_2$-module, as that of $V|A$. Thus, $\dim V|A < \dim V|Y$. Contradiction.

This completes the proof of (7.7).

<u>(7.8).</u> Suppose there exists $1 \le i \le r$ such that $L_i$ is separated from all other components of $L_Y'$ by more than one node of the Dynkin diagram. Then $M_i$ is trivial.

<u>Proof:</u> Suppose false; i.e., suppose $L_i$ is as described and $M_i$ is nontrivial. By (7.4), each component of $L_Y'$ is of classical type, so (1.5) implies $Z_A \le Z_Y$. Let $W$ be the natural module for $L_i$ (of type $A_k$). Then by (7.1) of [12], if $\mathrm{rank}(L_i) > 1$, $M_i \cong W$ or $W^*$.

<u>Case I:</u> Suppose $\mathrm{rank}(L_i)>2$.

Arguing as in the proof of (6.7) and applying (7.1) and (7.3), we see that $L_i = \langle U_{\pm\beta_j} \mid 1 \le j \le 4\rangle$ and $L_Y' = L_i \times \langle U_{\pm\beta_7}, U_{\pm\beta_8}\rangle$. Let $q_1$ (respectively, $q_2$) be the field twist on the embedding of $L_A'$ in $L_i$ (respectively, $\langle U_{\pm\beta_7}, U_{\pm\beta_8}\rangle$). Then the $L_A'$ composition factors of $Q_Y/K_{\beta_5}$ have high weights $6q_1\mu_2$ and $2q_1\mu_2$. So $\beta_5|Z_A = q_1\alpha_1$. Since $Q_Y/K_{\beta_6} \cong (Q_A{}^{\alpha_1})^{q_2}$ as $L_A'$ modules, if $Q_A \not\le K_{\beta_6}$, $\beta_6|Z_A = q_2\alpha_1$. Then (2.8) implies that $q_1 = q_2$. Thus, either $Q_A \le K_{\beta_6}$ and $\langle\lambda,\beta_k\rangle = 0$ for $k = 6, 7, 8$, or $q_1 = q_2$ and (2.5) and (2.6) imply $\langle\lambda,\beta_7+\beta_8\rangle = 0$. In fact, even in the second case, $\langle\lambda,\beta_6\rangle = 0$, else the bound on $\dim V^2(Q_A)_{\lambda-q_1\alpha_1}$ of (1.22) is exceeded. Also, $\langle\lambda,\beta_5\rangle = 0$, else the bound on $\dim V_{\beta_5}(Q_Y)$ of (1.25) is exceeded. So $\lambda|T_Y = \lambda_1$ or $\lambda_2$ and $\langle\lambda,\alpha_2\rangle = 4q_1$. Referring to (6.9), we find that if $\lambda|T_Y = \lambda_1$, then $\langle\lambda,\alpha_1\rangle = 0, 2q, q$, or $q+q_0$, for $q$ and $q_0$ distinct $p$-powers. However in each case, by (1.27) and [8], $\dim V|A < \dim V|Y$. Thus, $\lambda|T_Y = \lambda_2$. Now, by (6.9), $\langle\lambda,\alpha_1\rangle = 0, 3q, 2q, q$, or $q+q_0$. However, (1.27) and (1.38) imply $\dim V|A < \dim V|Y$. This completes the consideration of Case I.

<u>Case II:</u> Suppose $\mathrm{rank}(L_i) \le 2$.

Then, in fact, $\mathrm{rank}(L_i) = 2$, else there exists a 2-dimensional $L_A'$

irreducible, $Q_Y/K_\delta$, containing a nontrivial image of $Q_A{}^{\alpha_1}$. Suppose there exists $1 \le k \le r$, $k \ne i$ with $M_k$ nontrivial. Then (7.7) implies that $L_k$ is separated from all other components of $L_Y'$ by more than one node of the Dynkin diagram. Then previous remarks of this result imply that $L_k$ is of type $A_2$. Let $q_i$ (respectively, $q_k$) be the field twist on the embedding of $L_A'$ in $L_i$ ($L_k$). So $q_i \ne q_k$. If $\delta \in \Pi(Y) - \Pi(L_Y)$ with $(\delta, \Sigma L_j) \ne 0$, for $j = i$ or $j = k$, then $Q_Y/K_\delta \cong (Q_A{}^{\alpha_1})^{q_j}$ and $\delta|T_A = q_j\alpha_1$. (See (2.4).) By (2.8), $L_i$ and $L_k$ are separated by more than two nodes of the Dynkin diagram. So $Y = E_8$ and $\Pi(L_Y) = \{\beta_1,\beta_3,\beta_7,\beta_8\}$. Moreover, by (2.13) and (2.3), $\langle\lambda,\beta_j\rangle = 0$ for $j = 2,4,5,6$. If we take $L_i = \langle U_{\pm\beta_1}, U_{\pm\beta_3}\rangle$ and $L_k = \langle U_{\pm\beta_7}, U_{\pm\beta_8}\rangle$, the above remarks imply $\beta_4|T_A = q_i\alpha_1$ and $\beta_6|T_A = q_k\alpha_1$. Also, (2.13) implies $\beta_2|T_A = 0 = \beta_5|T_A$. We also have $\beta_1|T_A = q_i\alpha_2 = \beta_3|T_A$ and $\beta_7|T_A = q_k\alpha_2 = \beta_8|T_A$. In particular, $V_{T_A}(\lambda - q\alpha_1) = 0$ for all p-powers q. Thus, $\langle\lambda,\alpha_1\rangle = 0$ and $\lambda|T_A = (2q_i + 2q_k)\mu_2$; so $\dim V|A \le 100 < \dim V|Y$ by (1.27) and (1.32). Hence, there does not exist $1 \le k \le r$, $k \ne i$ with $M_k$ nontrivial.

Now (7.2) implies that there exists $\delta \in \Pi(Y) - \Pi(L_Y)$ with $\langle\lambda,\delta\rangle \ne 0$. By (2.3), $(\delta,\Sigma L_Y) \ne 0$. We claim that $(\delta,\Sigma L_i) \ne 0$. Otherwise, the bound on $\dim V_\delta(Q_Y)$ implies that there exists a unique $1 \le k \le r$, $k \ne i$ with $\mathrm{rank}(L_k) \le 2$ and with $(\Sigma L_k,\delta) \ne 0$. Actually, $L_k$ has type $A_2$, else $Q_Y/K_\delta$ is a 2-dimensional irreducible $L_A'$ module containing a nontrivial image of $Q_A{}^{\alpha_1}$. If $L_i$ and $L_k$ are separated by more than two nodes of the Dynkin diagram, $L_i \times L_k$ is as in the above paragraph. Then $\langle\lambda,\delta\rangle \ne 0$ contradicts (2.13). Thus, $L_i$ and $L_k$ are separated by exactly two nodes of the Dynkin diagram. Let $\gamma_i,\gamma_k \in \Pi(Y) - \Pi(L_Y)$ with $(\gamma_i,\gamma_k) < 0$, $(\gamma_j,\Sigma L_j) \ne 0$ for $j = i,k$. Let q be the field twist on the embedding of $L_A'$ in $L_i$. Note that if $\gamma_k|Z_A = 0$, (so $\gamma_k \ne \delta$) then $0 \ne w \in V_{T_Y}(\lambda-\delta)$ affords an $L_Y\hat{}$ composition factor of $V_\delta(R_u(P_Y\hat{}))$ which exceeds the bound of (1.25), where $P_Y\hat{} \ge B_Y^-$ is the parabolic of Y with Levi factor $L_Y\hat{} = \langle L_Y, U_{\pm\gamma_k}\rangle$. Hence, if $Q_A \le K_{\gamma_k}$, so $\delta \ne \gamma_k$, (2.10) implies that $-\gamma_k$ is not involved in $L_A'$ and by (2.11), there is a nontrivial image of $Q_A{}^{\alpha_1}$ in $Q_Y(\gamma_i,\gamma_k)$. Hence, the field twist on the embedding of $L_A'$ in $L_k$ is q. If $Q_A \nsubseteq K_{\gamma_k}$, (2.8) implies that $\gamma_k|Z_A = q\alpha_1$, which in turn implies, by (2.6) that the field twist on the embedding of $L_A'$ in $L_k$ is again q. Thus, $Q_Y/K_\delta \cong (Q_A{}^{\alpha_1})^q$ and $\delta|T_A = q\alpha_1$. (See (2.4).) But then the bound on $\dim V^2(Q_A)_{\lambda-q\alpha_1}$ of (1.22) is exceeded. Thus, if $\delta \in \Pi(Y) - \Pi(L_Y)$ with $\langle\lambda,\delta\rangle \neq 0$, then $(\delta,\Sigma L_i) \neq 0$ as claimed.

Now, there exists a unique such $\delta$. For otherwise, $\dim V_{T_A}(\lambda-q\alpha_1) > 1$, contradicting (1.31). Hence, $\lambda|T_Y = \lambda_\ell + x\lambda_m$ for some $\ell$,m. Moreover, the nodes of the Dynkin diagram corresponding to $\beta_\ell$ and $\beta_m$ are separated by at most one node. By (6.9), $\langle\lambda,\alpha_1\rangle = 0, q_0, 2q_0, 3q_0, q_0+q_1$ or $xq_0$ for distinct p-powers $q_0$ and $q_1$. Using (1.32) and (1.27), we see that $\dim V|A < \dim V|Y$ unless $\langle\lambda,\alpha_1\rangle = xq_0$. Moreover, since $f_\delta v^+ \in V_{T_A}(\lambda-q\alpha_1)$, $q_0 = q$ and by (1.10), V|A is restricted.

Temporarily label as follows: $\Pi L_i = \{\gamma_1,\gamma_2\}$ and $(\delta,\gamma_1) < 0$. We claim that $\gamma_2$ must correspond to an end node of the Dynkin diagram. For otherwise, if $\delta_0 \in \Pi(Y) - \Pi(L_Y)$ with $(\delta_0,\gamma_2) < 0$, arguing as above, $\delta_0|T_A = \alpha_1$. The bound on $\dim V^2(Q_A)$ implies $\langle\lambda,\gamma_1\rangle = 1$ and $\langle\lambda,\gamma_2\rangle = 0$. Consider the subgroup $L_0 = \langle U_{\pm\delta}, U_{\pm\gamma_1}, U_{\pm\gamma_2}, U_{\pm\delta_0}\rangle$. Then $L_0$ has a natural subgroup, B, of type $B_2$. Moreover, $v^+$ affords an $L_0$ composition factor of V which restricted to B produces a B composition factor with the same high weight as V|A, as $B_2$ module. But $L_0$ lies in a proper parabolic of Y and so acts reducibly on V. Hence, $\dim V|A < \dim V|Y$. Thus, $\gamma_2$ corresponds to an end node, as claimed. Also, (2.13) implies that $L_i \neq \langle U_{\pm\beta_1}, U_{\pm\beta_3}\rangle$ and if $L_i = \langle U_{\pm\beta_2}, U_{\pm\beta_4}\rangle$, $\delta = \beta_5$ and $Y = E_7$ or $E_8$. In fact, $L_i \neq \langle U_{\pm\beta_2}, U_{\pm\beta_4}\rangle$. For otherwise, (2.12) implies that $\beta_1|T_A = 0$ and using the parabolic $P_Y\hat{}$ of (2.12), we see that the bound on $\dim V^2(Q_A)$ is exceeded. Hence, either $Y = E_7$ with $\lambda|T_Y = x\lambda_5+\lambda_6$ or $x\lambda_5+\lambda_7$ or $Y = E_8$, with $\lambda|T_Y = x\lambda_6+\lambda_7$ or $x\lambda_6+\lambda_8$.

In Y of type $E_7$, let $L_1 = \langle U_{\pm\beta_1}, U_{\pm\beta_3}, U_{\pm\beta_4}, U_{\pm\beta_5}\rangle$, a group of type $A_4$, which has a natural subgroup, B, of type $B_2$. Then $f_{245}f_6v^+$ affords an $L_1$ composition factor of V which restricts to B to produce a composition factor having the same high weight, as $B_2$ module, as $V|A$. Similarly, in Y of type $E_8$, let $L_2 = \langle U_{\pm\beta_3}, U_{\pm\beta_4}, U_{\pm\beta_5}, U_{\pm\beta_6}\rangle$. Here, the vector $f_{123456}f_7v^+$ serves the same purpose as above. In each case, $L_i$ lies in a proper parabolic of Y, so acts reducibly on V. Hence, $\dim V|A < \dim V|Y$ and the result of (7.8) holds.□

<u>(7.9).</u> There are no examples in the Main Theorem with Y of type $E_n$ and A of type $B_2$, when $p>2$.

<u>Proof:</u> Suppose false; i.e., suppose $V|Y$ is a nontrivial kY-module. Then, (7.3), (7.7), and (7.8) imply $\langle\lambda,\alpha_2\rangle = 0$. So $\langle\lambda,\alpha_1\rangle \neq 0$. Let $P \geq B_A^-$ be the parabolic subgroup of A with Levi factor $L = \langle U_{\pm\alpha_1}\rangle T_A$. Let R be a parabolic subgroup of Y with $P \leq R$ and $Q = R_u(P) \leq Q_0 = R_u(R)$. Let R be minimal with these properties. Let $L_0$ be a Levi complement of $Q_0$ in R such that $T_0 \leq L_0$, for some maximal torus of R, with $T_A \leq T_0$. Fix a base $\Pi_0(Y)$ of the root system, $\Sigma_0^+(Y)$, of Y such that $L \cap U_A \leq Q_0(L_0 \cap U_0)$, where $U_0$ is the product of $T_0$ root subgroups corresponding to roots in $\Sigma_0^+(Y)$ and $Q_0$ is the product of $T_0$ root subgroups corresponding to roots in $\Sigma_0^-(Y) - \Sigma(L_0)$. Let $\Pi_0(Y) = \{\gamma_1,\ldots,\gamma_n\}$, with Dynkin diagrams labelled as throughout. Let $\langle w^+\rangle$ be the unique 1-space fixed by $U_0$; let $\lambda$ be the $T_0$ weight of $w^+$. Then by (6.9), $\dim V|A < \dim V|Y$ unless $L_0' = L_1 \times L_2$, with $L_1$ a simple algebraic group of type $A_1$ and $L_2$ a semisimple algebraic group acting trivially on $V^1(Q_0)$. So if $L_1 = \langle U_{\pm\beta}\rangle$ for some $\beta \in \Pi_0(Y)$, then $\langle\lambda,\beta\rangle = c = \langle\lambda,\alpha_1\rangle$, for some $p>c>0$. (Use (1.10).) It is easy to check that $\dim V^2(Q) = c$, in this case. Thus, if $\gamma \in \Pi_0(Y) - \Pi(L_Y)$, $\langle\lambda,\gamma\rangle = 0$. (Use (1.36) if $(\gamma,\beta) \neq 0$.) Also, $\beta \neq \gamma_4$, else $f_{\gamma_3+\gamma_4}w^+$, $f_{\gamma_2+\gamma_4}w^+$ and $f_{\gamma_5+\gamma_4}w^+$ afford distinct $L_0'$ composition factors of $V^2(Q_0)$, exceeding $\dim V^2(Q)$. Hence we may choose $\gamma_j,\gamma_k,\gamma_\ell \in \Pi_0(Y)$ with $(\beta,\gamma_j) < 0$, $(\gamma_j,\gamma_k) < 0$ and $(\gamma_k,\gamma_\ell) < 0$. The subgroup $N = \langle U_{\pm\gamma_j}, U_{\pm\gamma_k}, U_{\pm\gamma_\ell}, U_{\pm\beta}\rangle \leq Y$ has type $A_4$, and therefore has a natural subgroup of type $B_2$, say $A_0$. Also, the N-composition factor of V afforded by $v^+$ is not all of $V|Y$, as N

is contained in the Levi factor of a proper parabolic of Y. But the $A_0$ composition factor of V|Y has the same high weight, as $B_2$ module, as does V|A. Thus, dimV|A < dimV|Y. Contradiction.□

# CHAPTER 8: $A = G_2$

Let $A < Y$ be simple algebraic groups, with $Y$ simply connected, having a root system of type $E_n$. Let $V = V(\lambda)$ be a restricted irreducible $kY$-module. In this chapter, we consider the main problem in case $A$ has type $G_2$. Let $T_A$ (respectively, $T$) be a fixed maximal torus of $A$ (respectively, $Y$) with $T_A \leq T$. Let $\Pi(A) = \{\alpha_1, \alpha_2\}$ be a base of the root system $\Sigma(A)$ and $\Pi(Y)$ a base of $\Sigma(Y)$. Label the Dynkin diagrams of $\Sigma(A)$ and $\Sigma(Y)$ as throughout. Let $\{\mu_1, \mu_2\}$ (respectively, $\{\lambda_1, \lambda_2, \ldots, \lambda_n\}$) denote the fundamental dominant weights corresponding to the given ordered bases. The result is the following:

<u>Theorem (8.0).</u> (a) If $V|A$ is irreducible, then $p \neq 2,7$, $Y = E_6$, $\lambda|T = \lambda_1$ (or $\lambda_6$), $\lambda|T_A = 2\mu_1$.

(b) If $p \neq 2,7$ and $Y = E_6$, then there exists a closed subgroup $B < Y$, $B$ of type $G_2$, such that $V(\lambda_1)|B$ is irreducible.

<u>Remark:</u> The proof of (8.0)(b) is given in [16]. We prove (8.0)(a) in this chapter in case $p > 3$. The case where $p = 2$ or $3$ is handled in Chapter 9.

We adopt Notation and Hypothesis (2.0) throughout this chapter, with the additional conditions: Assume <u>$p > 3$</u> and $L_A = \langle U_{\pm\alpha_1}\rangle T_A$, so $Q_A{}^{\alpha} = Q_A{}^{\alpha_2}$ is a 4-dimensional, tensor indecomposable $L_A'$ module.

The following technical lemma which will be used in many of the successive results.

<u>(8.1).</u> (i) For $p \geq 5$, $\dim V(4\mu_2) > 156$.

(ii) $\dim V = 27$ if and only if $\lambda | T_A = 2q\mu_1$, for some p-power q, and $p \neq 7$.

(iii) There does not exist an irreducible kA-module of dimension 56; so if Y has type $E_7$, $\lambda | T_Y \neq \lambda_7$.

<u>Proof</u>: By applying the methods of (1.30) and (1.33), repeatedly, and recalling that the Weyl group of A has order 12, we obtain (i).

Now, suppose $\dim V = 27$. Then (1.32) implies V is tensor indecomposable, so we may assume V is restricted. Suppose $\langle\lambda,\alpha_1\rangle = a \neq 0 \neq b = \langle\lambda,\alpha_2\rangle$. Then, by [8], $a>1$ or $b>1$. Breaking the argument up into separate cases for $a=2$ or $a>2$, and $b=2$ or $b>2$, the methods of (1.30), (1.32) and (1.35) show that $\dim V > 27$. Thus, $a=0$ or $b=0$. Moreover, by (1.27), if $a \neq 0$, $a>1$ and if $b \neq 0$, $b>1$. Since $p>3$, [8] implies $b\neq 2$. Then, the methods of (1.30) and (1.32) imply that $\dim V > 27$ if $b\neq 0$. So $a\neq 0$ and $b=0$. If $a=2$, [8] implies the result. By [8], $a\neq 3$. But if $a>3$, the methods of (1.30) and (1.32) imply $\dim V > 27$. Thus, (ii) holds. Arguing similarly, we obtain (iii).□

<u>(8.2).</u> If $\dim V^1(Q_Y) > 1$, $L_Y'$ is not a simple algebraic group.

<u>Proof</u>: Suppose false. Then Theorem (7.1) of [12] implies that $L_Y'$ is of classical type, so by (1.5), $Z_A \leq Z_Y$. Consider first the case where $L_Y'$ has type $D_k$ for some $k \geq 4$. We may argue as in the proof of (6.1), to obtain: $L_Y' = D_6$, $Y = E_7$ and $V^1(Q_Y) \cong W$, the natural module for $L_Y'$. Also, $\langle\lambda,\beta_1\rangle = 0$, else the bound on $\dim V_{\beta_1}(Q_Y)$, of (1.25), is exceeded. But now $\lambda | T_Y = \lambda_7$, contradicting (8.1). Thus, $L_Y'$ does not have type $D_k$ for $k \geq 4$.

If $L_Y'$ has type $A_k$ for $k>3$, we may argue as in the proof of (6.1) to reduce to $L_Y'$ of type $A_{n-1}$ in Y of type $E_n$, with $\lambda | T_Y = \lambda_n$, contradicting (8.1) and previous general remarks.

We have, therefore, $L_Y'$ of type $A_k$ for $k \leq 3$. Actually, $k=3$, else there exists $\gamma \in \Pi(Y)-\Pi(L_Y)$, with $Q_Y/K_\gamma$ an irreducible $L_A'$-module of dimension 2 or 3, containing a nontrivial image of $Q_A{}^{\alpha_2}$. Note also that there does not exist $\delta \in \Pi(Y)-\Pi(L_Y)$ with $V_{L_Y'}(-\gamma) \cong W\wedge W$. For $(W\wedge W)|L_A'$ has composition factors of dimensions 5 and 1 or two factors of dimension 3. Thus, for each $\gamma \in \Pi(Y)-\Pi(L_Y)$ with $(\gamma,\Sigma L_Y) \neq 0$, $Q_Y/K_\gamma \cong W$ or $W^*$. Since $p > 2$, $Q_A{}^{\alpha_2}$ is tensor indecomposable, so $W|L_A'$ is tensor indecomposable. Suppose $W|L_A'$ has high weight $3q\mu_2$ for some p-power q. Then comparing high weight vectors in $Q_Y/K_\gamma$ and $Q_A{}^{\alpha_2}$, we see that $\gamma|T_A = q\alpha_2$ for all $\gamma \in \Pi(Y)-\Pi(L_Y)$ with $(\gamma,\Sigma L_Y) \neq 0$. Also, (2.12) implies $\tau|T_A = 0$ for all $\tau \in \Pi(Y)-\Pi(L_Y)$ with $(\tau,\Sigma L_Y) = 0$. As in the proof of (2.16), $\beta|T_A = q\alpha_1$, for each $\beta \in \Pi(L_Y)$.

Now $\langle\lambda,\alpha_2\rangle \neq 0$, else $\dim V|A \neq \dim V|Y$. (Use [8], (1.30) and (1.32).) Hence, there exists $\gamma \in \Pi(Y)-\Pi(L_Y)$ with $(\gamma,\Sigma L_Y) \neq 0$ and $\langle\lambda,\gamma\rangle \neq 0$. For otherwise, there is no vector in $V|Y$ with $T_A$ weight $\lambda - q_0\alpha_2$, for any p-power $q_0$. By (2.13), $\gamma$ must correspond to an end node of the Dynkin diagram. Applying this restriction and the bound on $\dim(V^2(Q_A)_{\lambda-q\alpha_2})$, we reduce to the following:

(a) $Y = E_6$, $L_Y' = \langle U_{\pm\beta_i} \mid i = 1,3,4\rangle$, $\lambda|T_Y = \lambda_1 + x\lambda_2$ $x>0$.

(b) $Y = E_6$, $L_Y' = \langle U_{\pm\beta_i} \mid 4\leq i\leq 6\rangle$, $\lambda|T_Y = \lambda_6 + x\lambda_2$, $x>0$.

(c) $Y$ $E_7$, $L_Y' = \langle U_{\pm\beta_i} \mid 4\leq i\leq 6\rangle$, $\lambda|T_Y = \lambda_6 + x\lambda_7$, $x>0$.

(d) $Y = E_8$, $L_Y' = \langle U_{\pm\beta_i} \mid 5\leq i\leq 7\rangle$, $\lambda|T_Y = \lambda_7 + x\lambda_8$, $x>0$.

Actually, the configurations of (a) and (b) can be ruled out by (1.23).

Now, $V|A$ is a conjugate of a basic module since there is no vector in $V|Y$ with $T_A$ weight $\lambda-q_0\alpha_2$ for $q_0 \neq q$. So by (1.10), q=1 and $\langle\lambda,\alpha_1\rangle = 3$. By (1.29), $\dim V_{T_A}(\lambda-3\alpha_1-\alpha_2) \leq 4$. Thus, $Y = E_7$; for otherwise, $f_{4567}v^+$, $f_{34567}v^+$, $f_{134567}v^+$, $f_{24567}v^+$ and $f_{5678}v^+$ are five linearly independent vectors in $V_{T_A}(\lambda-3\alpha_1-\alpha_2)$. Now, one checks that in the action of $L_A'$ on the 56-dimensional irreducible kY-module $V(\lambda_7)$, there are no 2- or 3-dimensional composition factors, and all composition factors are tensor indecomposable. But there is no 56-dimensional kA-module which affords such an $L_A'$ composition series. Hence, $L_Y'$ does not have type $A_3$.

This completes the proof of (8.2).□

(8.3). If dim $V^1(Q_Y) > 1$, each $L_i$ has type $A_{k_i}$ for some $k_i \geq 1$.

Proof: We first claim that each $L_i$ has classical type. For otherwise, $Y = E_8$ and $L_Y'$ has type $E_6 \times A_1$. By Theorem (7.1) of [12], $\lambda|T_Y = x\lambda_7 + c\lambda_8$ and $\langle\lambda,\alpha_1\rangle = c \cdot q$, for $c>0$ and some p-power q. By (6.9) and (1.22), $\dim V^2(Q_A) \leq 11c + 8$. But $f_{\beta_7}v^+$ and/or $f_{\beta_7+\beta_8}v^+$ afford(s) $L_Y'$ composition factors in $V^2(Q_A)$, forcing $\dim V^2(Q_A) \geq 27c$. (Use (1.36) if $x \neq 0$ and $c = p-1$.) Thus, each component of $L_Y'$ has classical type, so (1.5) implies $Z_A \leq Z_Y$.

Suppose $L_i$ has type $D_k$ for some k. Arguing as in the proof of (6.3), we see that $M_i$ is trivial. Now $\langle U_{\pm\beta_7}, U_{\pm\beta_8}\rangle$ is not a component of $L_Y'$, else the bounds on $\dim V_{\beta_6}(Q_Y)$ and $\dim V_{\beta_1}(Q_Y)$ imply $\lambda|T_Y = \lambda_8$. Hence, $L_Y' = L_i \times L_j$ with $L_j$ of type $A_1$ and $M_j$ nontrivial. Moreover, $(\Pi L_i, \Pi L_j) \neq 0$, else there exists $\delta \in \Pi(Y)-\Pi(L_Y)$ with $(\delta,\Sigma L_j) \neq 0$ and $Q_Y/K_\delta$ a 2-dimensional irreducible $L_A'$-module containing a nontrivial image of $Q_A{}^{\alpha_2}$. Let $\gamma \in \Pi(Y) - \Pi(L_Y)$ with $(\gamma,\Sigma L_i) \neq 0 \neq (\gamma,\Sigma L_j)$. Now (1.36) and the bound on $\dim V_\gamma(Q_Y)$ of (1.25) imply $L_i = D_4$. So $\Pi(L_Y) = \{ \beta_7, \beta_m \mid 2 \leq m \leq 5 \}$, with $\langle\lambda,\beta_m\rangle = 0$ for $2 \leq m \leq 5$ and $\langle\lambda,\beta_7\rangle > 0$. The previous remarks imply that $Y = E_7$. But the bound on $\dim V_{\beta_1}(Q_Y)$ and on $\dim V_{\beta_6}(Q_Y)$ (in conjunction with (1.36)) implies that $\lambda|T_Y = \lambda_7$, contradicting (8.1). This completes the proof of (8.3).□

(8.4). Suppose there exists $\gamma \in \Pi(Y)-\Pi(L_Y)$ and $1 \leq i,j \leq r$ such that $(\gamma,\Sigma L_i) \neq 0 \neq (\gamma,\Sigma L_j)$. Then $M_i$ or $M_j$ is trivial.

Proof: Suppose false; i.e., suppose $M_i$ and $M_j$ are both nontrivial. By (8.3), each component, $L_k$, of $L_Y'$ has type $A_{m_k}$ for some $m_k \geq 1$; so (1.5) implies $Z_A \leq Z_Y$. Let $W_m$ denote the natural module for $L_m$, m = i, j. By (7.1) of [12], if $\mathrm{rank}(L_m) > 1$, $M_m \cong W_m$ or $W_m{}^*$.

Case I: Suppose $V_{L_i}(-\gamma) \not\cong W_i$ or $W_i^*$.

Then $\mathrm{rank}(L_i) \geq 3$ and $V_{L_i}(-\gamma) \cong W_i \wedge W_i$ or $W_i^* \wedge W_i^*$. Now (1.15) implies that $L_j$ cannot have type $A_2$. Also, since $L_A'$ acts irreducibly on $W_m$, for $m = i, j$, if $\mathrm{rank} L_m \neq 3$, there does not exist $\delta \in \Pi(Y) - \Pi(L_Y)$ with $(\delta, \Sigma L_m) \neq 0$ and $Q_Y/K_\delta \cong W_m$ or $W_m^*$ for $m = i$ or $j$. Finally, the bound on $\dim V_\gamma(Q_Y)$, restricts the situation still further. (Use (1.34), (1.36) and $p>3$.) These remarks imply that $L_i \times L_j$ has type $A_3 \times A_1$, $A_3 \times A_3$ or $A_4 \times A_1$.

If $L_i \times L_j$ has type $A_4 \times A_1$ with $q_1 \neq q_2$ the field twists on the embeddings of $L_A'$ in $L_i$ and $L_j$ respectively, then one checks that there is no 4-dimensional $L_A'$ composition factor of $Q_Y/K_\gamma$. But this contradicts (2.3); so $L_i \times L_j$ does not have type $A_4 \times A_1$. If $\Pi(L_i \times L_j) = \{\beta_k \mid k \neq 1,5\}$ in $E_8$, the bound on $\dim V_{\beta_5}(Q_Y)$ implies $\lambda | T_Y = x\lambda_1 + \lambda_\ell + \lambda_8$ where $\ell = 2$ or 3. However, $f_{\beta_\ell + \beta_4 + \beta_5} v^+$ and $f_{\beta_5 + \beta_6 + \beta_7 + \beta_8} v^+$ afford distinct $L_Y'$ composition factors of $V_{\beta_5}(Q_Y)$ of dimensions 60 and 20, respectively, exceeding the bound of (1.25). Hence, $L_i \times L_j$ does not have type $A_3 \times A_3$.

Finally, consider the case where $L_i \times L_j$ has type $A_3 \times A_1$. We first note that $W_i | L_A'$ is tensor indecomposable. For otherwise, if $W_i | L_A'$ has high weight $(q_1 + q_2)\mu_1$ and if $q_3$ is the field twist on the embedding of $L_A'$ in $L_j$, for $q_1, q_2, q_3$ distinct powers of p, the $L_A'$ composition factors of $Q_Y/K_\gamma$ are 6 dimensional. But this implies $Q_A \leq K_\gamma$, contradicting (2.3). So $W_i | L_A'$ has high weight $3q\mu_1$ for some $q \neq q_3$. However, now the $L_A'$ composition factors of $Q_Y/K_\gamma$ have dimensions 10 and 2 so again $Q_A \leq K_\gamma$, contradicting (2.3). This completes the consideration of Case I.

Case II: $V_{L_m}(-\gamma) \cong W_m$ or $W_m^*$ for $m = i, j$.

Then (2.7) implies that there exists $k \neq i, j$ with $(\Sigma L_k, \gamma) \neq 0$ and $\dim M_k = 1$. Thus, $\gamma = \beta_4$. We first claim that $\langle U_{\pm\beta_3} \rangle$ is not a component of $L_Y'$. For otherwise, $Q_A \leq K_{\beta_1}$ since $Q_Y/K_{\beta_1}$ is a 2-dimensional irreducible $L_A'$-module. Also $p > 2$ and (2.10) imply that $-\beta_1$ is not involved in $L_A$. Hence, by (2.11), there is a nontrivial image of $Q_A^{\alpha_2}$ in $Q_Y(\beta_4, \beta_1)$. But $Q_Y(\beta_4, \beta_1)$ is an irreducible, tensor decomposable

$L_A'$-module. Thus, $\langle U_{\pm\beta_1}, U_{\pm\beta_3}\rangle$ and $\langle U_{\pm\beta_2}\rangle$ are components of $L_Y'$. Then (1.15) and $p>3$ imply that the third component adjacent to $\beta_4$, say $L_0$, has type $A_{k_0}$ for $k_0 = 2, 3$ or $4$. If $L_0$ has type $A_3$, again by (1.15), $V_{L_0}(-\beta_4)|L_A'$ is tensor decomposable. The bound on $\dim V_{\beta_4}(Q_Y)$ of (1.25), together with (1.36), implies $\langle\lambda,\beta_2\rangle = 0$ in case $L_0$ has type $A_3$ or $A_4$. Previous remarks then imply that if $L_0$ has type $A_2$ or $A_3$, $Q_Y/K_\gamma$ has no $L_A'$ composition factor isomorphic to a twist of $Q_A{}^{\alpha_2}$. Hence, $\Pi(L_Y) = \{\beta_m \mid m \neq 4\}$ in $E_8$ and $\lambda|T_Y = \lambda_1 + \lambda_8$. (The labelling of $\lambda$ is given by the bound on $\dim V_{\beta_4}(Q_Y)$.) But now, $f_{134}v^+$ and $f_{45678}v^+$ afford distinct $L_Y'$ composition factors of $V_{\beta_4}(Q_Y)$ of dimensions at least 46 and 16, respectively, exceeding the bound of (1.25).

This completes the proof of (8.4).□

<u>(8.5).</u> Suppose there exist distinct $1 \leq i, j, k \leq r$ such that $(\Sigma L_\ell,\beta_4) \neq 0$ for $\ell = i, j, k$ and $\dim(M_i \otimes M_j \otimes M_k) > 1$. Then $Y = E_6$, $\lambda|T_Y = \lambda_1$ (or $\lambda_6$) and $\lambda|T_A = 2\mu_1$. Moreover, $p \neq 7$.

<u>Proof</u>: Since each component of $L_Y'$ is necessarily of classical type, (1.5) implies $Z_A \leq Z_Y$. Let $W_m$ denote the natural module for $L_m$, $m = i,j,k$. By (7.1) of [12], if $\mathrm{rank}(L_m) > 1$ and $M_m$ is nontrivial, $M_m \cong W_m$ or $W_m{}^*$, for $m = i,j,k$. By (8.4), only one of $M_i$, $M_j$ and $M_k$ is nontrivial.

Since $p > 2$, (1.15) implies $L_i \times L_j \times L_k$ has type $A_1 \times A_1 \times A_\ell$, $\ell = 1$ or 3, or $A_2 \times A_1 \times A_\ell$, $\ell = 2, 3$, or 4. If $L_i \times L_j \times L_k$ has type $A_1 \times A_1 \times A_\ell$, $Q_A \leq K_{\beta_1}$, as $Q_Y/K_{\beta_1}$ is a 2-dimensional irreducible $L_A'$-module. Moreover, (2.10) implies that $-\beta_1$ is not involved in $L_A'$. Hence, by (2.11) and (1.15), applied to $Q_Y(\beta_4,\beta_1)$, $\ell \neq 1$. And in the case where $L_i \times L_j \times L_k$ has type $A_1 \times A_1 \times A_3$, (1.15) applied to $Q_Y/K_{\beta_4}$ implies $V_{L_k}(-\beta_4)|L_A'$ is tensor indecomposable. But (2.11) and (1.15) (applied to $Q_Y(\beta_4,\beta_1)$) produce a contradiction. Thus, $L_i \times L_j \times L_k$ has type $A_2 \times A_1 \times A_\ell$, $\ell = 2, 3, 4$.

Consider the case where $L_i \times L_j \times L_k$ has type $A_2 \times A_1 \times A_4$. The bound on $\dim V_{\beta_4}(Q_Y)$ implies $\lambda|T_Y = \lambda_1$, and $\langle\lambda,\alpha_1\rangle = 2q$ for some p-power q.

By (6.9), $\langle\lambda,\alpha_2\rangle = 0$, $2q_0$, or $q_0$, for some p-power $q_0$. But then [8] and (1.27) imply $\dim V|A < \dim V|Y$. Thus, $L_Y'$ does not have type $A_2 \times A_1 \times A_4$. Consider now the case where $L_i \times L_j \times L_k$ has type $A_2 \times A_1 \times A_3$. Then (1.15) implies that if $L_k$ has type $A_3$, $V_{L_k}(-\beta_4)|L_A'$ is tensor decomposable. So if $Y = E_8$, $Q_Y/K_{\beta_8}$ has no $L_A'$ composition factor isomorphic to a twist of $Q_A{}^{\alpha_2}$; so $\langle\lambda,\beta_5+\beta_6+\beta_7+\beta_8\rangle = 0$. Moreover, the bound on $\dim V_{\beta_4}(Q_Y)$ implies that either $\langle\lambda,\beta_1\rangle = 1$ and $\langle\lambda,\beta_\ell\rangle = 0$ for $7 \geq \ell > 1$ or $\langle\lambda,\beta_7\rangle = 1$ and $\langle\lambda,\beta_\ell\rangle = 0$ for $\ell < 7$. In the first case, $Y = E_8$. But we may argue as in the previous case to see that $\dim V|A < \dim V|Y$. In the second case, previous remarks imply that $Y = E_7$, contradicting (8.1).

Finally, we must consider $L_i \times L_j \times L_k$ of type $A_2 \times A_1 \times A_2$. The bound on $\dim V_{\beta_4}(Q_Y)$ and (1.36) imply that $\langle\lambda,\beta_1\rangle = 1$ and $\langle\lambda,\beta_m\rangle = 0$ for $2 \leq m \leq 6$ or $\langle\lambda,\beta_6\rangle = 1$ and $\langle\lambda,\beta_m\rangle = 0$ for $1 \leq m \leq 5$. So if $Y = E_6$, $\dim V|Y = 27$ and the result follows from (8.1) and (1.10). If $Y = E_7$, then $Q_A \leq K_{\beta_7}$, as $Q_Y/K_{\beta_7}$ is a 3-dimensional irreducible $L_A'$-module. But then $\langle\lambda,\beta_k\rangle = 0$ for $k = 5, 6, 7$ and $\lambda|T_Y = \lambda_1$.

Suppose $Y = E_8$. Let $q_1$ (respectively, $q_2$, $q_3$) be the field twist on the embedding of $L_A'$ in $\langle U_{\pm\beta_1}, U_{\pm\beta_3}\rangle$ (respectively, $\langle U_{\pm\beta_2}\rangle$, $\langle U_{\pm\beta_5}, U_{\pm\beta_6}\rangle$). Then, (2.7) implies that $q_1$, $q_2$, $q_3$ are not all distinct. If $q_1 = q_2 \neq q_3$ or if $q_1 \neq q_2 = q_3$, the $L_A'$ composition factors of $Q_Y/K_{\beta_4}$ have dimensions 12 and 6. If $q_1 = q_3 \neq q_2$, the $L_A'$ compositions factors of $Q_Y/K_{\beta_4}$ have dimensions 10, 6, and 2. Thus, $q_1 = q_2 = q_3$. The $L_A'$ composition factors of $Q_Y/K_{\beta_4}$ have high weights $5q_1\mu_1$, $3q_1\mu_1$ and $q_1\mu_1$. Thus, $\beta_4|Z_A = q_1\alpha_2$. Examining the $T(L_A')$ weights in $Q_Y/K_{\beta_4}$ we see that $x_{-\alpha_2}(t) = x_{-\beta_2-\beta_4}(c_1t^{q_1})x_{-\beta_4-\beta_5}(c_2t^{q_1})x_{-\beta_3-\beta_4}(c_3t^{q_1})u_0$, for $c_i \in k$, $c_1$, $c_2$, $c_3$ not all zero, and $u_0 \in K_{\beta_4}$. Since $\beta_\ell|T_A = q_1\alpha_1$ for $\ell = 1, 2, 3, 5, 6$, $\beta_4|T_A = q_1(\alpha_2 - \alpha_1)$.

Let $L_i = \langle U_{\pm\beta_1}, U_{\pm\beta_3}\rangle$ and $L_j = \langle U_{\pm\beta_5}, U_{\pm\beta_6}\rangle$. Note that if $Q_A \nleq K_{\beta_7}$, $\dim(Q_Y/K_{\beta_7}) \geq 4$; so $\langle U_{\pm\beta_8}\rangle \leq L_Y'$ We first claim that $M_j$ is trivial. For suppose $M_j$ is nontrivial; in particular, $Q_A \nleq K_{\beta_7}$. Then (2.17) implies that the field twist on the embedding of $L_A'$ in $\langle U_{\pm\beta_8}\rangle$ is also $q_1$ and that $\beta_7|Z_A = q_1\alpha_2$. Thus, (2.5) and (2.6) imply $\langle\lambda,\beta_8\rangle = 0$. Now, the bound on $\dim V^2(Q_A)_{\lambda-q_1\alpha_2}$, of (1.22), implies $\lambda|T_Y = \lambda_6$. By (6.9), $\langle\lambda,\alpha_2\rangle = 0, 2q$, $q$, or $q + q_0$, for $q$ and $q_0$ distinct p-powers. However, (1.27) and (1.32) imply $\dim V|A < \dim V|Y$. Thus $M_j$ is trivial; so $M_i$ is nontrivial and $\langle\lambda,\beta_1\rangle = 1$, $\langle\lambda,\beta_\ell\rangle = 0$ for $2\le\ell\le6$. If $\lambda|T_Y = \lambda_1$, we may argue as in the $A_1XA_2XA_4$ case to produce a contradiction. Thus, $\langle\lambda,\beta_7+\beta_8\rangle \neq 0$. Argue as in the previous paragraph to get $\langle\lambda,\beta_8\rangle = 0$. But then the bound on $\dim V_{\beta_7}(Q_Y)$ is exceeded.

This completes the proof of (8.5).□

<u>(8.6).</u> Let $\gamma \in \Pi(Y)-\Pi(L_Y)$. Suppose there exists a unique pair $1\le i,j\le r$ with $(\Sigma L_i,\gamma) \neq 0 \neq (\Sigma L_j,\gamma)$ and $\dim(M_i\otimes M_j) > 1$. Then $L_iXL_j$ has type $A_1XA_2$ and only one of $M_i$ and $M_j$ is nontrivial. Moreover, if $\Pi(L_i) = \{\gamma_0\}$, $\Pi(L_j) = \{\gamma_1,\gamma_2\}$, with $(\gamma_1,\gamma) < 0$, then there does not exist $\delta \in \Pi(Y) - \Pi(L_Y)$ with $(\delta,\gamma_0) \neq 0$ (respectively, $(\delta,\gamma_2) \neq 0$) and $(\delta,\Sigma L_m) = 0$ for all $m \neq i$ (respectively, $m\neq j$).

<u>Proof</u>: By (8.3), each component, $L_k$, of $L_Y'$ has type $A_{m_k}$ for some $m_k\ge1$, so (1.5) implies $Z_A \le Z_Y$. Let $W_m$ denote the natural module for $L_m$, for $m = i, j$. By (7.1) of [12], if $M_m$ is nontrivial and $\mathrm{rank}(L_m) > 1$, $M_m \cong W_m$ or $W_m^*$, for $m = i$ or $j$. By (8.4), only one of $M_i$ and $M_j$ is nontrivial.

<u>Case I</u>: Suppose $V_{L_i}(-\gamma) \ncong W_i$ or $W_i^*$.

Then (1.15) and size restrictions imply that $L_iXL_j$ has type $A_3XA_\ell$, for $\ell = 1$ or 3, $A_4XA_\ell$, for $\ell = 1$ or 3, $A_5XA_1$ or $A_6XA_1$. If $L_iXL_j$ has type $A_mXA_1$ for $m = 4$, 5 or 6, the bound on $\dim V_\gamma(Q_Y)$, together with (1.36), implies that the $A_1$ component acts trivially on $V^1(Q_Y)$. But if $L_iXL_j$ has type $A_6XA_1$, the bound implies $\lambda|T_Y = \lambda_8$. Also, if $L_iXL_j$ has type $A_5XA_1$, then $Y = E_7$, else $Q_Y/K_{\beta_8}$ is a 6-dimensional irreducible $L_A'$-module containing a nontrivial image of $Q_A{}^{\alpha_2}$. The bound on

$\dim V_{\beta_3}(Q_Y)$ implies that $\lambda|T_Y = \lambda_7$, contradicting (8.1). Thus, $L_i \times L_j$ does not have type $A_6 \times A_1$ or $A_5 \times A_1$.

Consider now the case where $L_i \times L_j$ has type $A_4 \times A_1$ (so the $A_1$ component acts trivially on $V^1(Q_Y)$). The bound on $\dim V_\gamma(Q_Y)$ and (1.23) imply that if $Y = E_6$, $\lambda|T_Y = \lambda_2$; thus, $Y = E_7$ or $E_8$. If $Y = E_7$, $Q_Y/K_{\beta_7}$ is an irreducible $L_A'$-module of dimension 2 or 5; so $Q_A \leq K_{\beta_7}$. Thus, $\Pi(L_Y) = \{\beta_1,\beta_2,\beta_3,\beta_4,\beta_6\}$. By (2.10), $-\beta_7$ is not involved in $L_A'$. However, then (2.11) implies that there is a nontrivial image of $Q_A{}^{\alpha_2}$ in $Q_Y(\beta_5,\beta_7)$, an $L_A'$-module with no 4-dimensional, tensor indecomposable compositon factor. Hence, $Y = E_8$. Let $q_1$ (respectively, $q_2$) be the field twist on the embedding of $L_A'$ in $L_i$ (respectively, $L_j$). The $L_A'$ composition factors of $Q_Y/K_\gamma$ have high weights $(6q_1 + q_2)\mu_1$ anad $(2q_1 + q_2)\mu_1$; if p=5 and $5q_1 = q_2$, the high weights are $(2q_2 + q_1)\mu_1$, $q_1\mu_1$, and $(2q_1 + q_2)\mu_1$; if $q_1 = q_2$ the high weights are $7q_1\mu_1$, $5q_1\mu_1$, $3q_1\mu_1$ and $q_1\mu_1$. Thus, $q_1 = q_2$ and $\gamma|Z_A = q_1\alpha_2$.

Now standard arguments (using (2.5), (2.6), (2.17) and (1.22)) imply that $\Pi(L_i \times L_j) \neq \{\beta_k \mid k = 1,2,4,5,6\}$. We have, therefore, $\Pi(L_i \times L_j) = \{\beta_1,\beta_2, \beta_3,\beta_4,\beta_6\}$. As well, $\langle U_{\pm\beta_8}\rangle$ is a component of $L_Y'$, else we can argue as in $E_7$ to produce a contradiction. Also, (1.15) implies $Q_A \leq K_{\beta_7}$. If $-\beta_7$ is involved in $L_A'$, $\beta_7|Z_A = 0$. Otherwise, (2.11) implies $Q_Y(\beta_5,\beta_7)$ contains a nontrivial image of $Q_A{}^{\alpha_2}$. Arguing as with $Q_Y/K_{\beta_5}$ (in the previous paragraph), we see that the field twist on the embedding of $L_A'$ in $\langle U_{\pm\beta_8}\rangle$ is $q_1$ and $(\beta_5 + \beta_7)|Z_A = q_1\alpha_2$. So again $\beta_7|Z_A = 0$. Using the parabolic $P_Y\hat{}$ of (2.11), we see that $\lambda|T_Y = \lambda_1$, else the bound on $\dim V_{\beta_5}(Q_Y\hat{})$ is exceeded. By (6.9), $\langle\lambda,\alpha_2\rangle = 0$, 2q or q, for some p-power q. But [8] and (1.27) imply that $\dim V|A \neq \dim V|Y$. Thus, $L_i \times L_j$ does not have type $A_4 \times A_1$.

If $\Pi(L_i \times L_j) = \{\beta_k \mid k \neq 5\}$ in $E_8$, the bound on $\dim V_{\beta_5}(Q_Y)$ implies that $\lambda|T_Y = \lambda_1$ and $\langle\lambda,\alpha_1\rangle = 4q$ for some p-power q. Arguing as in the previous paragraph, we have $\dim V|A \neq \dim V|Y$. Thus, $L_i \times L_j$ does not have type $A_4 \times A_3$.

Consider the case where $\Pi(L_i) = \{\beta_2,\beta_3,\beta_4\}$ and $\Pi(L_j) = \{\beta_6,\beta_7,\beta_8\}$ in $E_8$. Note that $Q_A \not\leq K_{\beta_1}$. For otherwise, $\langle \lambda,\beta_k\rangle = 0$ for $1 \leq k \leq 4$ and the bound on $\dim V_{\beta_5}(Q_Y)$ implies that $\lambda|T_Y = \lambda_8$. Since $Q_A \not\leq K_{\beta_1}$, $W_i|L_A'$ is tensor indecomposable. Then (1.15) (applied to $Q_Y/K_{\beta_5}$) implies that $W_j|L_A'$ is also tensor indecomposable. Let $q_1$ (respectively, $q_2$) be the field twist on the embedding of $L_A'$ in $L_i$ (respectively, $L_j$). Then the $L_A'$ composition factors of $Q_Y/K_{\beta_5}$ have high weights $(4q_1 + 3q_2)\mu_1$ and $3q_2\mu_1$, if $q_1 \neq q_2$; if $q_1 = q_2$, the high weights are $7q_1\mu_1$, $5q_1\mu_1$, $3q_1\mu_1$ and $q_1\mu_1$. Thus, $\beta_5|Z_A = q_2\alpha_2$. Also, since $Q_Y/K_{\beta_1} \cong (Q_A{}^{\alpha_2})^{q_1}$, $\beta_1|T_A = q_1\alpha_2$. If $q_1 = q_2$, the bound on $\dim V^2(Q_A)_{\lambda - q_1\alpha_2}$, of (1.22), implies that $\lambda|T_Y = \lambda_8$. Thus $q_1 \neq q_2$. Examining $T(L_A')$ weight vectors in $Q_Y/K_{\beta_1}$ and in $Q_Y/K_{\beta_5}$, we have $x_{-\alpha_2}(t) = x_{-\beta_1}(at^{q_1})x_{-245}(b_1t^{q_2})\cdot x_{-345}(b_2t^{q_2})u$, where $a \in k^*$, $b_i \in k$, $u \in K_{\beta_1} \cap K_{\beta_5}$. In fact $b_2 \neq 0$, else there is a nontrivial contribution to the root group $U_{-\beta_4-\beta_5}$ in the expression for $[x_{\alpha_1}(t),x_{-\alpha_2}(t)]$. However, $b_2 \neq 0$ and $q_1 \neq q_2$ contradicts (2.8). Thus, $L_i \times L_j$ does not have type $A_3 \times A_3$.

We must now consider $L_i \times L_j$ of type $A_3 \times A_1$. Suppose $\Pi(L_i \times L_j) = \{\beta_2,\beta_3,\beta_4,\beta_6\}$. If $Q_A \leq K_{\beta_1}$, $-\beta_1$ is not involved in $L_A'$; else $\beta_1|Z_A = 0$, and using the parabolic $P_Y\hat{}$ of (2.11), we see that the bound on $\dim V_{\beta_5}(Q_Y\hat{})$ is exceeded. Thus, (2.11) implies $Q_A \not\leq K_{\beta_1}$, as $Q_Y(\beta_5,\beta_1)$ has no 4-dimensional $L_A'$ composition factor. So if $\Pi(L_i) = \{\beta_2,\beta_3,\beta_4\}$, $V_{L_i}(-\beta_1)$ is tensor indecomposable. We also note that the field twists on the embeddings of $L_A'$ in $L_i$ and $L_j$ are equal, else there is no 4-dimensional $L_A'$ composition factor of $Q_Y/K_{\beta_5}$. Call this twist q. Then the $L_A'$ composition factors of $Q_Y/K_{\beta_5}$ have high weights $5q\mu_1$, $3q\mu_1$ and $q\mu_1$. So $\beta_5|Z_A = q\alpha$. Then, examining the $T(L_A)$ weight vectors in $Q_Y/K_{\beta_1}$ and in $Q_Y/K_{\beta_5}$, we have

(1) $x_{-\alpha_2}(t) = x_{-\beta_1}(at^q)x_{-45}(a_1t^q)x_{-56}(a_2t^q)w$,

(2) $x_{-\alpha_1-\alpha_2}(t) = x_{-13}(bt^q)x_{-345}(b_1t^q)x_{-245}(b_2t^q)x_{-456}(b_3t^q)v$,

(3) $x_{-3\alpha_1-\alpha_2}(t) = x_{-1234}(ct^q)x_{-23456}(c_1t^q)x_{-\beta_2-\beta_3-2\beta_4-\beta_5}(c_2t^q)u$, where a, b, $c \in k^*$, $a_i$, $b_i$, $c_i \in k$, $a_1$ or $a_2$, $c_1$ or $c_2$ and some $b_i$ nonzero, u, v, $w \in K_{\beta_1} \cap K_{\beta_5}$,

Note that $a_2 \neq 0$ else there is a nontrivial contribution to the root group $U_{-\beta_5}$ in the expression for $[x_{-\alpha_2}(t), x_{\alpha_1}(u)]$. Then, this implies $b_3 \neq 0$, else no nonidentity element from $U_{-\beta_5-\beta_6}$ occurs in the factorization of $[x_{-\alpha_1-\alpha_2}(t), x_{\alpha_1}(u)]$. But if $b_3 \neq 0$, there is a nontrivial contribution to the root group $U_{-\beta_1-\beta_2-\beta_3-2\beta_4-\beta_5-\beta_6}$ in the factorization of $[x_{-\alpha_1-\alpha_2}(t), x_{-3\alpha_1-\alpha_2}(u)]$. Contradiction. Hence, $\Pi(L_i \times L_j) \neq \{\beta_2, \beta_3, \beta_4, \beta_6\}$.

It remains to consider the case where $\Pi(L_i \times L_j) = \{\beta_2, \beta_4, \beta_5, \beta_1\}$. The above argument implies that $Y = E_7$ or $E_8$ and $\langle U_{\pm\beta_7}\rangle \leq L_Y'$. Suppose $\langle U_{\pm\beta_7}\rangle$ is a component of $L_Y'$. Then $Q_Y/K_{\beta_6}$ has no 4-dimensional $L_A'$ composition factor, so $Q_A \leq K_{\beta_5}$. Also, if $Y = E_8$, $\dim(Q_Y/K_{\beta_8}) = 2$ implies $Q_A \leq K_{\beta_8}$. So $\langle \lambda, \beta_j\rangle = 0$ for $j = 2$, $j \geq 4$. In fact, the bound on $\dim V_{\beta_3}(Q_Y)$ implies $\lambda | T_Y = c\lambda_1$, where $c = 1$ or 2. If $c = 1$, $Y = E_8$ and by [8], $\dim V|Y = 3875$. However, referring to (6.9), we have $\langle \lambda, \alpha_2\rangle = 0, q_0$ or $2q_0$, for some p-power $q_0$. In each case, $\dim V|A < \dim V|Y$. Hence, $c = 2$. Then by (6.9), $\langle \lambda, \alpha_2\rangle = 0$ or $2q_0$. But (1.38) and (1.27) imply $\dim V|A < \dim V|Y$. Hence $\langle U_{\pm\beta_7}\rangle$ is not a component of $L_Y'$.

So $\Pi(L_Y) = \{\beta_1, \beta_2, \beta_4, \beta_5, \beta_7, \beta_8\}$. If $Q_A \leq K_{\beta_6}$, argue as in the preceding paragraph to produce a contradiction. Hence, $Q_A \nleq K_{\beta_6}$. Then (1.15) implies that $V_{L_i}(-\beta_6)|L_A'$ is tensor indecomposable, where $\Pi(L_i) = \{\beta_2, \beta_4, \beta_5\}$. Then previous remarks and (2.7) imply that the field twists on the embeddings of $L_A'$ in $L_i$, $L_j$ and $\langle U_{\pm\beta_7}, U_{\pm\beta_8}\rangle$ are equal, say q. So, $\langle \lambda, \beta_7+\beta_8\rangle = 0$. The $L_A'$ composition factors of $Q_Y/K_{\beta_6}$ have high weights $5q\mu_1$, $3q\mu_1$ and $q\mu_1$. Thus, $\beta_6|Z_A = q\alpha_2$. The bound on $\dim V^2(Q_A)_{\lambda-q\alpha_2}$ implies that $\lambda|T_Y = \lambda_2$ or $c\lambda_1$, for $c = 1$ or 2. If $\lambda|T_Y = c\lambda_1$, argue as above to produce a contradiction. If $\lambda|T_Y = \lambda_2$, refer to (6.9) to see that $\langle \lambda, \alpha_2\rangle = 0$, $q_1 + q_2$, $3q_1$, $2q_1$ or $q_1$, for $q_1$ and $q_2$ distinct p-powers. But by (1.27) and (1.38), dim $V|A <$ dim $V|Y$. Thus, this configuration cannot occur.

This completes the consideration of Case I.

Case II: Suppose $V_{L_k}(-\gamma) \cong W_k$ or $W_k{}^*$ for $k = i$ and $j$.

Then $Q_Y/K_\gamma$ has a 4-dimensional, tensor indecomposable $L_A'$ composition factor only if $L_i \times L_j$ has type $A_1 \times A_k$, for $k = 2$ or $4$ or $A_2 \times A_3$. Suppose that $L_i \times L_j$ has type $A_1 \times A_2$. Let $\gamma_0$, $\gamma_1$, and $\gamma_2$ be as in the statement of the result. If there exists $\delta \in \Pi(Y) - \Pi(L_Y)$ with $(\delta,\gamma_0) < 0$ and $(\delta,\Sigma L_m) = 0$ for all $m \neq i$, then $Q_Y/K_\delta$ is a 2-dimensional irreducible $L_A'$ module, so $Q_A \leq K_\delta$. By (2.10), $-\delta$ is not involved in $L_A'$, so (2.11) implies that there is a nontrivial image of $Q_A{}^{\alpha_2}$ in $Q_Y(\gamma,\delta)$. But $Q_Y(\gamma,\delta)$ is a 3-dimensional irreducible $L_A'$ module. Thus, no such $\delta$ exists. Arguing similarly, we show that there does not exist $\delta \in \Pi(Y) - \Pi(L_Y)$ with $(\delta,\gamma_2) \neq 0$ and $(\delta,\Sigma L_m) = 0$ for all $m \neq j$. In this case, we use $p > 2$ and (1.33) to see that if such a $\delta$ exists, we may assume $-\delta$ is not involved in $L_A'$. Thus, if $L_i \times L_j$ has type $A_1 \times A_2$, the result holds.

We now consider the case where $L_i$ (respectively, $L_j$) has type $A_1$ (respectively, $A_4$). Then (2.17) implies that there exists a p-power, q, which is the field twist on the embeddings of $L_A'$ in both $L_i$ and $L_j$ and such that $\gamma|Z_A = q\alpha_2$. Temporarily label as follows: $L_i = \langle U_{\pm\gamma_0} \rangle$, $L_j = \langle U_{\pm\gamma_k} \mid 1 \leq k \leq 4 \rangle$, with $(\gamma,\gamma_1) < 0$, $(\gamma_k,\gamma_{k+1}) < 0$ for $k = 1, 2, 3$. We first note that there does not exist $\delta \in \Pi(Y) - \Pi(L_Y)$ with $(\delta,\gamma_0) \neq 0$ and $(\delta,\Sigma L_k) = 0$ for all $k \neq i$. For if there exists such a $\delta$, $Q_Y/K_\delta$ is a 2-dimensional irreducible $L_A'$-module, so $Q_A \leq K_\delta$. By (2.10), $-\delta$ is not involved in $L_A'$, so (2.10) implies that there is a nontrivial image of $Q_A{}^{\alpha_2}$ in $Q_Y(\gamma,\delta)$, which is a 5-dimensional irreducible $L_A'$-module. Arguing similarly, we show that there does not exist $\delta \in \Pi(Y) - \Pi(L_Y)$ with $(\delta,\gamma_4) \neq 0$ and $(\delta,\Sigma L_k) = 0$ for all $k \neq i$. In this case, we must use $p > 3$ and (1.33) to see that we may assume $-\delta$ is not involved in $L_A'$. Also, if there exists $\delta \in \Pi(Y) - \Pi(L_Y)$ with $(\delta,\Sigma L_j) \neq 0$ and $(\delta,\Sigma L_\ell) = 0$ for all $\ell \neq j$, then

$Q_A \le K_\delta$. For neither $W_j|L_A'$ nor $(W_j \wedge W_j)|L_A'$ has a 4-dimensional tensor indecomposable composition factor. In particular, if such a $\delta$ exists, $M_j$ is trivial.

If $Y = E_7$, consider the action of $L_A'$ on the 56-dimensional irreducible kY-module, $V(\lambda_7)$. One checks that there is an $L_A'$ composition factor with high weight $6q\mu_1$ and one with high weight $4q\mu_1$. But there is no 56-dimensional kA-module which affords such an $L_A'$ composition series. Hence $Y = E_8$. Moreover, previous remarks imply that either $\Pi(L_i \times L_j) = \{\beta_2, \beta_k \mid 5 \le k \le 8\}$ or $\Pi(L_i \times L_j) = \{\beta_k \mid k = 2,4,5,6,8\}$. In the first configuration, we note that $L_Y' = L_i \times L_j$. For if $\langle U_{\pm\beta_1} \rangle$ is a component of $L_Y'$, $Q_Y/K_{\beta_3}$ is a 2-dimensional irreducible $L_A'$-module; so $Q_A \le K_{\beta_3}$. By (2.10), $-\beta_3$ is not involved in $L_A'$, so (2.11) implies that there is a nontrivial image of $Q_A{}^{\alpha_2}$ in $Q_Y(\beta_4, \beta_3)$, contradicting (1.15). Thus, $L_Y' = L_i \times L_j$ and (2.11) implies that $\beta_3|Z_A = 0$. Using the parabolic $P_Y\hat{}$ of (2.11), the bound on $\dim V_{\beta_4}(Q_Y\hat{})$, together with (1.36), implies that $\lambda|T_Y = \lambda_8$. Thus, the first configuration cannot occur.

Consider now the second configuration. If $Q_A \nleq K_{\beta_3}$, $\langle U_{\pm\beta_1} \rangle$ is a component of $L_Y'$. The work of Case I then implies that $\langle \lambda, \beta_j \rangle = 0$ for $j = 1,2,4,5,6$ and the bound on $\dim V_{\beta_3}(Q_Y)$ implies that $\langle \lambda, \beta_3 \rangle = 0$. If $Q_A \le K_{\beta_3}$, (2.3) implies $\langle \lambda, \beta_j \rangle = 0$ for $j \le 6$. As well, (1.36) and the bound on $\dim V_{\beta_7}(Q_Y)$ imply that $\lambda|T_Y = c\lambda_8$, for $1 < c \le 4$, and $\langle \lambda, \alpha_1 \rangle = c \cdot q$. By (6.9), $\langle \lambda, \alpha_2 \rangle = 0$ or $c \cdot q_0$, for some p-power $q_0$. However, (1.38) and (1.27) imply $\dim V|A < \dim V|Y$ in each case.

We must now consider $L_i \times L_j$ of type $A_2 \times A_3$. Temporarily label as follows: $L_i = \langle U_{\pm\gamma_1}, U_{\pm\gamma_2} \rangle$, $L_j = \langle U_{\pm\gamma_3}, U_{\pm\gamma_4}, U_{\pm\gamma_5} \rangle$, with $(\gamma_2, \gamma) \ne 0 \ne (\gamma, \gamma_3)$ and $(\gamma_k, \gamma_{k+1}) < 0$ for $k = 3, 4$. By (1.15), $V_{L_j}(-\gamma)|L_A'$ is tensor indecomposable. Let q be the field twist on the embeddings of $L_A'$ in $L_i$ and in $L_j$. (The twists are equal by (2.7).) Then the $L_A'$ composition factors of $Q_Y/K_\gamma$ have high weights $5q\mu_1$, $3q\mu_1$ and $q\mu_1$. Thus, $\gamma|Z_A = q\alpha_2$. Moreover, examining the $T(L_A')$ weight vectors in

$Q_Y/K_\gamma$, we see that $x_{-\alpha}(t) = x_{-\gamma_2-\gamma}(c_1t^q)x_{-\gamma-\gamma_3}(c_2t^q)u_1$, for $c_i \in k$, $c_1$ or $c_2$ nonzero, and $u_1 \in K_\gamma$. Since $\gamma_k|T_A = q\alpha_1$, for $1 \le k \le 5$, $\gamma|T_A = q(\alpha_2-\alpha_1)$. In particular, $\langle\lambda,\gamma\rangle = 0$.

We now point out various restrictions on the possible configurations which may arise with $L_i \times L_j$ as above. First, note that if there exists $\delta \in \Pi(Y)-\Pi(L_Y)$ with $(\delta,\Sigma L_i) \ne 0$ and $(\delta,\Sigma L_k) = 0$ for all $k \ne i$, then $Q_A \le K_\delta$. For $Q_Y/K_\delta$ is a 3-dimensional irreducible $L_A'$-module. In particular, if such a $\delta$ exists, $M_i$ is trivial and $\langle\lambda,\delta\rangle = 0$. Also, if there exists $\delta \in \Pi(Y)-\Pi(L_Y)$ with $(\delta,\gamma_4) \ne 0$ and $(\delta, \Sigma L_k) = 0$ for all $k \ne j$, then $Q_A \le K_\delta$. For $Q_Y/K_\gamma$ has $L_A'$ composition factors of dimensions 5 and 1. In particular, if such a $\delta$ exists, $M_j$ is trivial and $\langle\lambda,\delta\rangle = 0$. Now, the bound on $\dim V_\gamma(Q_Y)$ implies that $\langle\lambda,\gamma\rangle = 0$ and if $M_j$ is nontrivial, $\langle\lambda,\gamma_3+\gamma_4\rangle = 0$ and $\langle\lambda,\gamma_5\rangle = 1$. Also, note that if there exists $\delta \in \Pi(Y)-\Pi(L_Y)$ with $(\delta,\gamma_3) \ne 0$ or $(\delta,\gamma_5) \ne 0$ and $(\delta,\Sigma L_k) = 0$ for all $k \ne j$, then $Q_A \not\le K_\delta$. For otherwise, (2.10) implies that $-\delta$ is not involved in $L_A'$, and so by (2.11), there is a nontrivial image of $Q_A{}^{\alpha_2}$ in $Q_Y(\gamma,\delta)$, contradicting (1.15). Then since $Q_Y/K_\delta \cong (Q_A{}^{\alpha_2})^q$ as $L_A'$-modules and $Q_A \not\le K_\delta$, comparing high weight vectors we have $\delta|T_A = q\alpha_2$. Thus, if $\langle\lambda,\delta\rangle \ne 0$, the bound on $\dim V^2(Q_A)_{\lambda-q\alpha_2}$ implies $M_j$ is nontrivial and $(\delta,\gamma_5) \ne 0$.

Suppose $Y = E_7$. The above remarks and (8.1) imply $\lambda|T_Y = \lambda_3$ or $\lambda_6$ and $\langle\lambda,\alpha_1\rangle = 2q$. By (6.9), $\langle\lambda,\alpha_2\rangle = 0$, $q_0$ or $2q_0$ for some p-power $q_0$. But in each case, [8] implies $\dim V|A \ne \dim V|Y$. Thus, $Y = E_8$.

Suppose $\Pi(L_i) = \{\beta_1,\beta_3\}$ and $\Pi(L_j) = \{\beta_5,\beta_6,\beta_7\}$. We have $\beta_2|Z_A = 0$, by (2.11). Using the parabolic $P_Y\hat{}$ of (2.11), we see that if $M_i$ is nontrivial $\langle\lambda,\beta_1\rangle = 1$ and $\langle\lambda,\beta_3\rangle = 0$, else the bound on $\dim V_{\beta_4}(Q_Y\hat{})$ is exceeded. Previous remarks imply $\lambda|T_Y = \lambda_1$ and by (6.9), $\langle\lambda,\alpha_2\rangle = 0$, $q_0$ or $2q_0$ for some p-power $q_0$. But [8] and (1.27) imply $\dim V|A < \dim V|Y$ in each case. Thus, $M_j$ is nontrivial and $\lambda|T_Y = \lambda_7 + x\lambda_8$, for $x \ge 0$, and $\langle\lambda,\alpha_1\rangle = 3q$. Moreover, if $x \ne 0$, $0 \ne f_{\beta_8}v^+ \in V_{T_A}(\lambda - q\alpha_2)$, so in the p-adic expansion for $\langle\lambda,\alpha_2\rangle$, q has nonzero coefficient. By (6.9), $\langle\lambda,\alpha_2\rangle = 0$, $x\cdot q$, $q_0$ or $2q_0$, for $q_0$ some power of p. Then (1.27) and (1.32) imply $\dim V|A < \dim V|Y$, unless $x\neq 0$ and $\langle\lambda,\alpha_2\rangle = x\cdot q$. By the above remarks and (1.10), $\lambda|T_A = 3\mu_1 + x\mu_2$. Now, consider the subgroup $D \le Y$, $D = \langle U_{\pm\beta_k} \mid 2\le k\le 8\rangle$, of type $D_7$. The D composition factor of V afforded by $v^+$ has dimension strictly less than dim V|Y. Also, D has a natural subgroup of type $G_2$, say $A_0$ (found by letting $G_2$ act on its Lie algebra). Moreover, $v^+$ affords an $A_0$ composition factor of V with the same high weight as V|A, as $G_2$-module. Thus dim V|A < dim V|Y. Hence, $\Pi(L_i\times L_j) \neq \{\beta_1,\beta_3,\beta_5,\beta_6,\beta_7\}$.

Suppose $\Pi(L_i) = \{\beta_7,\beta_8\}$, so $\Pi(L_j) = \{\beta_2,\beta_4,\beta_5\}$ or $\{\beta_3,\beta_4,\beta_5\}$. The work of Case I, the general remarks of the $A_2 \times A_3$ work and the bound on $\dim V_\delta(Q_Y)$ for $\delta \in \Pi(Y) - \Pi(L_Y)$ imply $\lambda|T_Y = \lambda_7$. So $\langle\lambda,\alpha_1\rangle = 2q$ and (6.9) implies $\langle\lambda,\alpha_2\rangle = 0$, $q_0$ or $2q_0$ for some p-power $q_0$. But [8] and (1.32) imply $\dim V|A < \dim V|Y$. Similar arguments rule out $\Pi(L_i\times L_j) = \{\beta_6,\beta_7,\beta_1,\beta_3,\beta_4\}$.

The general remarks about $L_i\times L_j$ of type $A_2\times A_3$ imply that it remains to consider $\Pi(L_i) = \{\beta_2,\beta_4\}$ and $\Pi(L_j) = \{\beta_6,\beta_7,\beta_8\}$. Now $Q_A \not\le K_{\beta_3}$, else $\lambda|T_Y = \lambda_8$. Thus, $\langle U_{\pm\beta_1}\rangle$ is a component of $L_Y'$ and (2.17) implies that the field twist on the embedding of $L_A'$ in $\langle U_{\pm\beta_1}\rangle$ is also q. Thus, by (2.5) and (2.6), $\langle\lambda,\beta_1\rangle = 0$. Also, by (2.17), $\beta_3|Z_A = q\alpha_2$. Thus, the bound on $\dim V^2(Q_A)_{\lambda-q\alpha_2}$ implies $\lambda|T_Y = \lambda_2$, and $\langle\lambda,\alpha_1\rangle = 2q$. By (6.9), $\langle\lambda,\alpha_2\rangle = 0$, $3q_0$, $q_0 + q_1$, $2q_0$ or $q_0$, for $q_0$ and $q_1$ distinct p-powers. But in each case, (1.27) and (1.38) imply dim V|A < dim V|Y. Thus, $L_i\times L_j$ does not have type $A_2\times A_3$.

This completes the proof of (8.6).□

(8.7). Suppose there exists $1\le i\le r$ such that $L_i$ is separated from all other components of $L_Y'$ by more than one node of the Dynkin diagram. Then $M_i$ is trivial.

*Proof*: Suppose false; i.e., with i as given, suppose $\dim(M_i) > 1$. By

(8.3) and (1.5), $Z_A \leq Z_Y$ and $L_i$ has type $A_{k_i}$ for some $k_i$. Let $W_i$ denote the natural module for $L_i$. Arguing as in the proof of (6.7), we may reduce to $L_i$ of rank 3. Also, (8.2) and the working hypotheses imply $Y = E_7$ or $E_8$. Moreover, there does not exist $\delta \in \Pi(Y)-\Pi(L_Y)$ with $Q_Y/K_\delta \cong W_i \wedge W_i$. For $(W_i \wedge W_i)|L_A'$ has no 4-dimensional composition factor. Finally, note that $M_i$ is tensor indecomposable, as there exists $\delta \in \Pi(Y) - \Pi(L_Y)$ with $Q_Y/K_\delta \cong M_i$ or $M_i^*$ and $Q_A \nleq K_\delta$.

Consider now the possibility that $L_Y' = L_i \times L_j$, where $L_j$ has type $A_1$ and is separated by exactly 2 nodes of the Dynkin diagram from $L_i$. Temporarily label as follows: $L_i = \langle U_{\pm\gamma_1}, U_{\pm\gamma_2}, U_{\pm\gamma_3}\rangle$, $L_j = \langle U_{\pm\gamma_0}\rangle$, where $(\gamma_k, \gamma_{k+1}) < 0$ for $k = 1, 2$. Let $\delta_0$, $\delta_1 \in \Pi(Y)-\Pi(L_Y)$ with $(\delta_0, \delta_1) < 0$, $(\delta_0, \gamma_0) \neq 0 \neq (\delta_1, \gamma_1)$. Then $Q_A \leq K_{\delta_0}$, as $Q_Y/K_{\delta_0}$ is a 2-dimensional irreducible $L_A'$-module. Moreover, by (2.10), $-\delta_0$ is not involved in $L_A'$. Thus, (2.11) implies that there is a nontrivial image of $Q_A{}^{\alpha_2}$ in $Q_Y(\delta_1, \delta_0)$. However, this contradicts (1.15). Hence, $L_Y' \neq L_i \times L_j$ as described. In particular, $Y = E_8$.

Suppose $\Pi(L_i) = \{\beta_1, \beta_3, \beta_4\}$. Then (8.50, (8.6) and previous remarks imply $V^1(Q_Y) \cong M_i$. As well, since $\dim(Q_Y/K_{\beta_k}) < 4$ for $k = 6,7,8$, $\langle\lambda, \beta_6+\beta_7+\beta_8\rangle = 0$. So $\lambda|T_Y = \lambda_j + x\lambda_2 + y\lambda_5$, where $j = 1$ or $4$. In fact, either $x = 0$ or $y = 0$, else $f_{\beta_2}v^+$ and $f_{\beta_5}v^+$ are 2 linearly independent vectors in $V_{T_A}(\lambda - q\alpha_2)$, contradicting (1.31). Let $z = \langle\lambda, \beta_2+\beta_5\rangle$. If $z \neq 0$, $q$ has a nonzero coefficient in the p-adic expansion of $\langle\lambda, \alpha_2\rangle$. By (6.9), $\langle\lambda, \alpha_2\rangle = 0$, $zq$, $q_0$ or $2q_0$ for some p-power $q_0$. Now (1.32) and [8] imply $\dim V|A \neq \dim V|Y$ unless $z \neq 0$ and $\langle\lambda, \alpha_2\rangle = zq$; so by (1.10), $\lambda|T_A = 3\mu_1 + z\mu_2$. If $\lambda|T_Y = \lambda_4 + z\lambda_k$, $k = 2$ or $5$, then $V_{T_Y}(\lambda-\beta_4-\beta_2) \oplus V_{T_Y}(\lambda-\beta_4-\beta_5) \leq V_{T_A}(\lambda-\alpha_1-\alpha_2)$. So (1.35) implies $z = p-2$. But then $\dim V_{T_A}(\lambda-\alpha_1-\alpha_2) = 1$, also by (1.35), contradicting the above containment. So $\lambda|T_A = \lambda_1 + z\lambda_k$ for $k = 2$ or $5$. In this case, $z = 1$ else $V_{T_Y}(\lambda-2\beta_k-\beta_1) \oplus V_{T_Y}(\lambda-\beta_2-\beta_4-\beta_5) \oplus V_{T_Y}(\lambda-2\beta_k-\beta_4)$ is a 3-dimensional subspace of $V_{T_A}(\lambda-\alpha_1-2\alpha_2)$, contradicting (1.29). But now $\dim V|A < \dim V|Y$.

It remains to consider the case where $\Pi(L_i) = \{\beta_6,\beta_7,\beta_8\}$. We first claim that $\lambda|T_Y = x\lambda_5 + \lambda_k$, where k = 6 or 8. For if $\langle\lambda,\beta_j\rangle \neq 0$ for some $1 \leq j \leq 4$, either $\beta_j \in \Pi(L_Y)$ or $Q_Y/K_{\beta_j}$ has an $L_A'$ composition factor isomorphic to a twist of $Q_A{}^{\alpha_2}$. Then the work of this proof, (8.7) and (1.15) imply $\Pi(L_Y) = \{\beta_i \mid i \neq 4,5\}$ and $Q_A \nleq K_{\beta_4}$. So (2.17) implies that there is a power of p, say $q_0$, such that $q_0$ is the field twist on the embedding of $L_A'$ in $\langle U_{\pm\beta_1}, U_{\pm\beta_3}\rangle$ and in $\langle U_{\pm\beta_2}\rangle$, and such that $\beta_4|Z_A = q_0\alpha_2$. Then (2.8) implies that $q_0 = q$. Thus, by (2.5) and (2.6), $\langle\lambda,\beta_k\rangle = 0$ for k = 1, 2, 3, so $\langle\lambda,\beta_4\rangle \neq 0$. But then the bound on $\dim V_{\beta_4}(Q_Y)$ is exceeded. So $\lambda|T_Y = x\lambda_5 + \lambda_k$, for k = 6 or 8, as claimed. As well, if $x\neq 0$, $f_{\beta_5}v^+ \in V_{T_A}(\lambda - q\alpha_2)$, so q has a nonzero coefficient in the p-adic expansion of $\langle\lambda,\alpha_2\rangle$. By (6.9), $\langle\lambda,\alpha_2\rangle = 0$, $q_0$, $2q_0$, $q_0+q_1$ or $xq$, for distinct p-powers $q_0$ and $q_1$. Then by (1.27) and (1.32), $\dim V|A < \dim V|Y$ unless $x>1$ and $\langle\lambda,\alpha_2\rangle = xq$; so by (1.10) q = 1. Now (2.13) implies $\langle U_{\pm\beta_2}, U_{\pm\beta_3}\rangle \cap L_Y' \neq \{1\}$ and previous work of this proof implies $\dim(Q_Y/K_{\beta_4}) > 2$. If $\lambda|T_Y = x\lambda_5 + \lambda_6$, $\dim([V,Q_Y]/[V,Q_Y{}^3]) \geq 84$ if $x \neq p-2$, or $\geq 60$ if $x = p-2$. If $\lambda|T_Y = x\lambda_5+\lambda_8$, $\dim([V,Q_Y]/[V,Q_Y{}^3]) \geq 92$. (See Table 1 of [5].) But (1.20) and (1.35) imply $\dim([V,Q_A]/[V,Q_A{}^3]) \leq 80$ if $x\neq p-2$, and $\leq 55$ if $x = p-2$. Contradiction.

This completes the proof of (8.7).□

<u>(8.8).</u> Let $\gamma \in \Pi(Y)-\Pi(L_Y)$. Suppose there exists a unique pair $1\leq i,j\leq r$ with $(\Sigma L_i,\gamma) \neq 0 \neq (\Sigma L_j,\gamma)$. Assume $L_i$ is of type $A_2$ and $L_j$ is of type $A_1$. Then $M_i$ and $M_j$ are trivial.

<u>Proof</u>: Note that (1.5) implies $Z_A \leq Z_Y$. Also, if $Q_A \nleq K_\gamma$, (2.17) implies that there is a p-power $q_0$ which is the field twist on the embedding of $L_A'$ in $L_i$ and in $L_j$ and $\gamma|Z_A = q_0\alpha_2$.

<u>Claim 1:</u> If $L_Y'$ has components $L_k = \langle U_{\pm\beta_1}\rangle$ and $L_m = \langle U_{\pm\beta_2}, U_{\pm\beta_4}\rangle$, then $\dim V^1(Q_Y) = 1$.

Proof: Suppose $\dim V^1(Q_Y) > 1$. Let q be the field twist on the embedding of $L_A'$ in $L_k$ and $L_m$. If $\langle\lambda,\beta_k\rangle = 0$ for k = 1, 2, 4, then (8.6) and (8.7) imply $\langle U_{\pm\beta_6}\rangle$ is a component of $L_Y'$ and $\langle\lambda,\beta_6\rangle \neq 0$. Then $Y = E_6$, else $Q_Y/K_{\beta_7}$ has no $L_A'$ composition factor isomorphic to a twist $Q_A{}^{\alpha_2}$. Thus, by symmetry and (8.6), we may assume $\langle\lambda,\beta_k\rangle \neq 0$ for a unique $k \in \{1,2,4\}$.

Suppose $\langle U_{\pm\beta_6}\rangle \nleq L_Y'$. Then $Q_Y/K_{\beta_5}$ is a 3-dimensional irreducible $L_A'$ module, so $Q_A \leq K_{\beta_5}$ and $\langle\lambda,\beta_k\rangle = 0$ for k = 2,4,5. Hence, (1.23) implies $Y \neq E_6$. In fact, $\langle\lambda,\beta_k\rangle = 0$ for $k > 5$ also, as (8.6) and (8.7) imply $V^1(Q_Y) \cong M_k \otimes M_m$ and $\dim(Q_Y/K_{\beta_j}) < 4$ for $j > 5$. If $-\beta_5$ is involved in $L_A'$ then $\beta_5|Z_A = 0$. Otherwise, by (2.11) there is a nontrivial image of $Q_A{}^{\alpha_2}$ in $Q_Y(\beta_3,\beta_5)$. Since $Q_Y(\beta_3,\beta_5)$ has $L_A'$ composition factors with high weights $3q\mu_1$ and $q\mu_1$, $(\beta_3+\beta_5)|Z_A = q\alpha_2$ and again $\beta_5|Z_A = 0$. So we may use the parabolic $P_Y{}^\wedge$ of (2.11). The bound on $\dim V^2(Q_A)_{\lambda-q\alpha_2}$ implies $\langle\lambda,\beta_3\rangle = 0$ and $\langle\lambda,\beta_1\rangle \leq 2$. If $\langle\lambda,\beta_1\rangle = 1$, (6.9) implies that $\langle\lambda,\alpha_2\rangle = 0$, $q_0$, or $2q_0$ for some p-power $q_0$. In each case, $\dim V|A \neq \dim V|Y$, by (1.27) and [8]. So $\langle\lambda,\beta_1\rangle = 2$. Then $\langle\lambda,\alpha_2\rangle = 0$ or $2q_0$ and $\dim V|A < \dim V|Y$. Thus the assumption that $\langle U_{\pm\beta_6}\rangle \nleq L_Y'$ was incorrect.

Suppose $\langle U_{\pm\beta_6}\rangle$ is a component of $L_Y'$. Then $Q_A \nleq K_{\beta_5}$. For otherwise, (2.10) implies that $-\beta_5$ is not involved in $L_A'$, so by (2.11), there is a nontrivial image of $Q_A{}^{\alpha_2}$ in $Q_Y(\beta_3,\beta_5)$, contradicting (1.15). Now (2.17) implies that the field twist on the embedding of $L_A'$ in $\langle U_{\pm\beta_6}\rangle$ is also q. Moreover, $\beta_3|Z_A = q\alpha_2 = \beta_5|Z_A$. Examining the $T(L_A')$ weight vectors in $Q_Y/K_{\beta_3}$ and in $Q_Y/K_{\beta_5}$, we have $x_{-\alpha_2}(t) = x_{-\beta_3}(at^q)x_{-\beta_5}(bt^q)u_1$, $x_{-\alpha_1-\alpha_2}(t) = x_{-13}(a_1t^q)x_{-34}(a_2t^q)x_{-45}(b_1t^q)x_{-56}(b_2t^q)u_2$ and $x_{-3\alpha_1-\alpha_2}(t) = x_{-1234}(ct^q)x_{-2456}(dt^q)u_3$, where $a,b,c,d \in k^*$, $u_i \in K_{\beta_3} \cap K_{\beta_5}$, $a_i,b_i \in k$, $a_1$ or $a_2$ nonzero, $b_1$ or $b_2$ nonzero. In fact, $a_2 \neq 0$ as a nonidentity element from the root group $U_{-34}$ occurs in the factorization of $[x_{-\alpha_2}(t),x_{-\alpha_1}(u)]$ and $(-\beta_3-\beta_4)|T(L_A') = q\mu_1$. By examining $[x_{-\alpha_2}(t),x_{-3\alpha_1-\alpha_2}(t)]$, we see that $x_{-3\alpha_1-2\alpha_2}(t) = x_{-12345}(f_1(t))x_{-23456}(f_2(t))w$ where $0 \neq f_i(t) \in k[t]$ and $w \in \langle U_{-r} | r = \Sigma c_\gamma\gamma,\ \gamma \in \Pi(Y),\ c_\gamma \in \mathbb{Z}^+,\ c_{\beta_3}+c_{\beta_5} > 2 \text{ or } c_{\beta_7}+c_{\beta_8} > 0\rangle$.

But this gives a nontrivial contribution to the root group $U_{-\beta_1-\beta_2-2\beta_3-2\beta_4-\beta_5}$ in the expression for $[x_{-3\alpha_1-2\alpha_2}(t), x_{-\alpha_1-\alpha_2}(u)]$. Contradiction. Hence $\langle U_{\pm\beta_6}\rangle$ is not a component of $L_Y'$. (So $Y = E_7$ or $E_8$.)

If $\langle U_{\pm\beta_6}, U_{\pm\beta_7}\rangle$ is a component of $L_Y'$, then (1.15) implies $Q_A \leq K_{\beta_5}$ and $Q_A \leq K_{\beta_8}$. So $\langle\lambda,\beta_k\rangle = 0$ for $k = 2$ and $k \geq 4$. If $-\beta_5$ is involved in $L_A'$, $\beta_5|Z_A = 0$. If $-\beta_5$ is not involved in $L_A'$, (2.11) implies that there is a nontrivial image of $Q_A{}^{\alpha_2}$ in $Q_Y(\beta_3,\beta_5)$, so the field twist on the embedding of $L_A'$ in $\langle U_{\pm\beta_6}, U_{\pm\beta_7}\rangle$ is also q. Moreover, the $L_A'$ composition factors of $Q_Y(\beta_3,\beta_5)$ have high weights $5q\mu_1$, $3q\mu_1$ and $q\mu_1$, so $(\beta_3+\beta_5)|Z_A = q\alpha_2$. Hence, again $\beta_5|Z_A = 0$. But using the parabolic $P_Y{}^\wedge$ of (2.11), we see that the bound on $\dim V_{\beta_3}(Q_Y{}^\wedge)$ is exceeded.

Thus, $Y = E_8$ and $\Pi(L_Y) = \{\beta_1,\beta_2,\beta_4,\beta_6,\beta_7,\beta_8\}$. Now $-\beta_5$ is not involved in $L_A'$; else $\beta_5|Z_A = 0$ and we may again produce a contradiction. If the natural module for $\langle U_{\pm\beta_6}, U_{\pm\beta_7}, U_{\pm\beta_8}\rangle$ is a tensor decomposable $L_A'$ module, neither $Q_Y/K_{\beta_5}$ nor $Q_Y(\beta_3,\beta_5)$ has an $L_A'$ composition factor isomorphic to a twist of $Q_A{}^{\alpha_2}$, contradicting (2.11). So the natural module for $\langle U_{\pm\beta_6}, U_{\pm\beta_7}, U_{\pm\beta_8}\rangle$ is a tensor indecomposable $L_A'$ module; so (1.15) and (2.11) imply $Q_A \nleq K_{\beta_5}$. By (2.17), the field twist on the embedding of $L_A'$ in $\langle U_{\pm\beta_6}, U_{\pm\beta_7}, U_{\pm\beta_8}\rangle$ is also q and one checks that $\beta_5|Z_A = q\alpha_2$. Examining the $T(L_A')$ weight vectors in $Q_Y/K_{\beta_3}$ and $Q_Y/K_{\beta_5}$, we find that $x_{-\alpha_2}(t) = x_{-\beta_3}(at^q)x_{-45}(a_1t^q)x_{-56}(a_2t^q)u_1$ and $x_{-\alpha_1-\alpha_2}(t) = x_{-13}(b_1t^q)x_{-34}(b_2t^q x_{-245}(c_1t^q)x_{-456}(c_2t^q)x_{-567}(c_3t^q)u_2$, where $a \in k^*$, $a_i,b_i,c_i \in k$, $u_i \in K_{\beta_3} \cap K_{\beta_5}$. Also, $a_1$ or $a_2$, $b_1$ or $b_2$ and some $c_i$ is nonzero. In fact, $a_1 \neq 0 \neq a_2$, else there is a nontrivial contribution to the root group $U_{-\beta_5}$ in the expression for $[x_{-\alpha_2}(t), x_{\alpha_1}(t)]$. Also $b_1 \neq 0$, as a nonidentity element from the root group $U_{-13}$ occurs in the factorization of $[x_{-\alpha_2}(t), x_{-\alpha_1}(t)]$ and $(-\beta_1-\beta_3)|T(L_A') = q\mu_1$. However, we now see that there is a nontrivial contribution to the root group $U_{-1345}$ in the expression for $[x_{-\alpha_2}(t), x_{-\alpha_1-\alpha_2}(t)]$. Contradiction. This completes the proof of Claim 1.

Claim 2: If $L_\ell = \langle U_{\pm\beta_1}, U_{\pm\beta_3}\rangle$ and $L_m = \langle U_{\pm\beta_2}\rangle$ are components of $L_Y'$ with $\langle U_{\pm\beta_5}\rangle \not\leq L_Y'$, then $\dim V^1(Q_Y) = 1$.

Proof: Suppose false. Then (8.6) and (8.7) imply $\langle\lambda,\beta_k\rangle \neq 0$ for a unique $k \in \{1,2,3\}$ and $V^1(Q_Y) \cong M_\ell \otimes M_m$. Let q be the field twist on the embedding of $L_A'$ in $L_\ell$ and $L_m$, so $\beta_4|Z_A = q\alpha_2$. Suppose there exists $\delta \in \Pi(Y) - \Pi(L_Y)$, $\delta \neq \beta_4$, with $\langle\lambda,\delta\rangle \neq 0$. Since $Q_A \not\leq K_\delta$, $\dim(Q_Y/K_\delta) \geq 4$. By size restrictions and (1.15), $\delta = \beta_5$ and $\langle U_{\pm\beta_6}, U_{\pm\beta_7}, U_{\pm\beta_8}\rangle$ is a component of $L_Y'$ in Y of type $E_8$. But (2.8) implies $\beta_5|Z_A = q\alpha_2$, so the bound on $\dim V^2(Q_A)_{\lambda - q\alpha_2}$ is exceeded. Thus, no such $\delta$ exists and $\langle\lambda,\beta_k\rangle = 0$ for $k \geq 5$.

If $\langle U_{\pm\beta_6}\rangle \not\leq L_Y'$, (2.12) implies $\beta_5|Z_A = 0$. Using the parabolic $P_Y\hat{}$ of (2.11), and the bound on $\dim V_{\beta_4}(Q_Y\hat{})$, we have $\lambda|T_Y = \lambda_1$, $\lambda_3$ or $c\lambda_2$ for c = 1 or 2. Use (1.23), (6.9), (1.27), (1.32) and [8] to see that in every possible configuration, $\dim V|A \neq \dim V|Y$. Thus $\langle U_{\pm\beta_6}\rangle \leq L_Y'$. If $\langle U_{\pm\beta_6}\rangle$ is a component of $L_Y'$, then $Q_A \leq K_{\beta_5}$ and (2.10) and (2.11) produce a contradiction. So $Y = E_7$ or $E_8$ and $\langle U_{\pm\beta_6}, U_{\pm\beta_7}\rangle \leq L_Y'$. If $\langle U_{\pm\beta_6}, U_{\pm\beta_7}\rangle$ is a component of $L_Y'$, $Q_A \leq K_{\beta_5}$ and $Q_A \leq K_{\beta_8}$. So $\langle\lambda,\beta_k\rangle = 0$ for $k \geq 5$. If $-\beta_5$ is involved in $L_A'$, $\beta_5|Z_A = 0$. Otherwise, an application of (2.11) implies $\beta_5|Z_A = 0$. So in either case we may use the parabolic $P_Y\hat{}$ of (2.11) to see that the bound on $\dim V_{\beta_4}(Q_Y\hat{})$ implies $\lambda|T_Y = \lambda_1$. Again, use (6.9), (1.27) and [8] to see that $\dim V|A \neq \dim V|Y$.

Therefore, under the hypotheses of Claim 2, if $\dim V^1(Q_Y) > 1$, $Y = E_8$ and $L_Y' = L_\ell \times L_m \times \langle U_{\pm\beta_6}, U_{\pm\beta_7}, U_{\pm\beta_8}\rangle$. We claim that $Q_A \not\leq K_{\beta_5}$. For otherwise, since there is no $L_A'$ composition factor of $Q_Y(\beta_4,\beta_5)$ isomorphic to a twist of $Q_A{}^{\alpha_2}$, (2.11) would imply that $-\beta_5$ is involved in $L_A$. So $\beta_5|Z_A = 0$. But using the parabolic $P_Y\hat{}$ of (2.11) ,the bound on $\dim V_{\beta_4}(Q_Y\hat{})$ implies $\lambda|T_Y = \lambda_1$. Now we can argue as before to produce a contradiction. Since $Q_A \not\leq K_{\beta_5}$, (2.4) and (2.8) imply $Q_Y/K_{\beta_5}$ is a tensor indecomposable $L_A'$ module isomorphic to $(Q_A{}^{\alpha_2})^q$. Examining the $T(L_A')$ weight vectors in $Q_Y/K_{\beta_4}$ and $Q_Y/K_{\beta_5}$, we have $x_{-\alpha_2}(t) = x_{-\beta_4}(at^q)$.

$x_{-\beta_5}(bt^q)u_1$ and $x_{-\alpha_1-\alpha_2}(t) = x_{-\beta_3-\beta_4}(a_1t^q)x_{-\beta_2-\beta_4}(a_2t^q)\cdot$ $x_{-\beta_5-\beta_6}(b_1t^q)u_2$, where a, b, $b_1 \in k^*$, $u_i \in K_{\beta_4} \cap K_{\beta_5}$ and $a_i \in k$ with some $a_i$ nonzero. However, there is a nontrivial contribution to the root group $U_{-\beta_4-\beta_5-\beta_6}$ in the expression for $[x_{-\alpha_2}(t), x_{-\alpha_1-\alpha_2}(t)]$. Contradiction. This completes the proof of Claim 2.

Now suppose $L_i$ has type $A_2$ and $L_j$ has type $A_1$ with $\dim(M_i \otimes M_j) > 1$.

Case I: Suppose $M_i$ is nontrivial, so by (8.6) $M_j$ is trivial. Temporarily label as follows: $\Pi(L_i) = \{\gamma_1, \gamma_2\}$, $\Pi(L_j) = \{\gamma_0\}$, $(\gamma_1, \gamma) < 0$. We first note that there does not exist $\delta \in \Pi(Y) - \Pi(L_Y)$ with $(\delta, \Sigma L_i) \neq 0$ and $(\delta, \Sigma L_k) = 0$ for all $k \neq i$. For otherwise, $Q_Y/K_\delta$ is a 3-dimensional irreducible $L_A'$-module containing a nontrivial image of $Q_A{}^{\alpha_2}$. These remarks, together with Claim 1, imply $L_i \neq \langle U_{\pm\beta_2}, U_{\pm\beta_4}\rangle$. Also, (8.6) implies that there does not exist $\delta \in \Pi(Y) - \Pi(L_Y)$ with $(\delta, \Sigma L_j) \neq 0$ and $(\delta, \Sigma L_k) = 0$ for all $k \neq j$. Let q be the field twist on the embeddings of $L_A'$ in $L_i$ and in $L_j$, so $\gamma|Z_A = q\alpha_2$. Examining the $T(L_A')$ weight vectors in $Q_Y/K_\gamma$, we see that $x_{-\alpha_2}(t) = x_{-\gamma}(at^q)u$, where $a \in k^*$, $u \in K_\gamma$. Thus, $\gamma|T_A = q\alpha_2$.

Consider the possibility that $L_i = \langle U_{\pm\beta_1}, U_{\pm\beta_3}\rangle$. Then Claim 2 implies $L_j = \langle U_{\pm\beta_5}\rangle$ and previous remarks imply $Y = E_7$ or $E_8$ and $\langle U_{\pm\beta_7}\rangle \leq L_Y'$. Also, (2.12) implies that $\beta_2|T_A = 0$. Using the parabolic $P_Y{}^\wedge$ of (2.11), we see that $\langle \lambda, \beta_4\rangle = 0$, else the bound on $\dim V_{\beta_4}(Q_Y{}^\wedge)$ is exceeded. Also, $\langle \lambda, \beta_2\rangle = 0$, by (2.3). Suppose $L_Y' = L_i \times L_j \times \langle U_{\pm\beta_7}\rangle$. Then (1.15) implies $Q_A \leq K_{\beta_6}$. Thus, $\langle \lambda, \beta_k\rangle = 0$ for k = 5, 6, 7. If $Y = E_8$, $Q_A \leq K_{\beta_8}$, as $\dim(Q_Y/K_{\beta_8}) = 2$. So $\langle \lambda, \beta_8\rangle = 0$, also. Thus, $\lambda|T_Y = \lambda_1$ (with Y of type $E_8$) or $\lambda_3$. By (6.9), $\langle \lambda, \alpha_2\rangle = 0$, $q_0$ or $2q_0$, for some p-power $q_0$. However, (1.27), [8] and (1.32) imply $\dim V|A \neq \dim V|Y$.

Thus, if $L_i = \langle U_{\pm\beta_1}, U_{\pm\beta_3}\rangle$ and $L_j = \langle U_{\pm\beta_5}\rangle$, then $Y = E_8$ and $L_Y' = L_i \times L_j \times \langle U_{\pm\beta_7}, U_{\pm\beta_8}\rangle$. Using standard arguments, we reduce to $\lambda|T_Y = \lambda_1$ or $\lambda_3$. But then we proceed as before to produce a contradiction. Thus, $L_i \neq \langle U_{\pm\beta_1}, U_{\pm\beta_3}\rangle$.

Suppose now $L_i = \langle U_{\pm\beta_5}, U_{\pm\beta_6}\rangle$ in Y of type $E_8$. Then, (8.5), (8.6) and the general remarks at the beginning of Case I imply that $\langle U_{\pm\beta_2}\rangle$ and $\langle U_{\pm\beta_8}\rangle$ are components of $L_Y$; and $\langle U_{\pm\beta_3}\rangle \nleq L_Y'$. Also, the field twists on the embeddings of $L_A'$ in $L_i$, $\langle U_{\pm\beta_2}\rangle$ and $\langle U_{\pm\beta_8}\rangle$ are q and $\beta_4|Z_A = q\alpha_2 = \beta_7|Z_A$. Thus, the bound on $\dim V^2(Q_A)_{\lambda-q\alpha_2}$ implies $\langle\lambda,\beta_4+\beta_7\rangle = 0$. As usual, $\langle\lambda,\beta_2+\beta_8\rangle = 0$ and $\langle\lambda,\beta_1+\beta_3\rangle = 0$. Thus, $\lambda|T_Y = \lambda_5$ or $\lambda_6$ and $\langle\lambda,\alpha_1\rangle = 2q$. By (6.9), $\langle\lambda,\alpha_2\rangle = 0$, $2q_0$ or $q_0$, for some p-power $q_0$. However, (1.27) and (1.32) imply $\dim V|A < \dim V|Y$. Thus, $L_i \neq \langle U_{\pm\beta_5}, U_{\pm\beta_6}\rangle$, in Y of type $E_8$.

Reviewing the general remarks and the cases considered so far, we see that in Case I it remains to consider $L_i = \langle U_{\pm\beta_7}, U_{\pm\beta_8}\rangle$ in Y of type $E_8$; so $L_j = \langle U_{\pm\beta_5}\rangle$. Using (8.5), (8.6), (8.7), (2.17), (2.5) and (2.6), it is easy to see that $V^1(Q_Y) \cong M_i$. Suppose there exists $\delta \in \Pi(Y)-\Pi(L_Y)$, $\delta \neq \beta_6$, with $\langle\lambda,\delta\rangle \neq 0$. Then $Q_A \nleq K_\delta$ implies $\dim(Q_Y/K_\delta) \geq 4$. Thus, $\delta = \beta_4$ and $\langle U_{\pm\beta_2}, U_{\pm\beta_3}\rangle \cap L_Y' \neq 1$. The bound on $\dim V_{\beta_4}(Q_Y)$, of (1.25), implies that $L_Y' = L_i \times L_j \times \langle U_{\pm\beta_k}\rangle$ for k = 2 or 3. But then $Q_Y/K_{\beta_4}$ has no $L_A'$ composition factor isomorphic to a twist of $Q_A^{\alpha_2}$, contradicting (2.3). Thus, there exists no such $\delta$, and $\lambda|T_A = x\lambda_6 + \lambda_7$ or $x\lambda_6 + \lambda_8$, for $p>x\geq 0$. In fact, the bound on $\dim V_{\beta_6}(Q_Y)$ implies that $\lambda|T_Y = x\lambda_6+\lambda_7$, for $p>x\geq 0$. By (6.9), $\langle\lambda,\alpha_2\rangle = 0$, $xq_0$, $q_0$ or $2q_0$, for $q_0$ some p-power. Then, (1.27) and (1.32) imply $\dim V|A < \dim V|Y$ unless $\langle\lambda,\alpha_2\rangle = xq_0$ and $x\neq 0$. Since $\beta_6|T_A = q\alpha_2$, $0 \neq f_{\beta_6}v^+ \in V_{T_A}(\lambda-q\alpha_2)$, so in fact, $q_0 = q$. Then by (1.10), $\lambda|T_A = 2\mu_1+x\mu_2$.

Now, consider the subgroup $D \leq Y$ of type $D_4 \times D_4$ defined by $D = \langle U_{\pm\beta_2}, U_{\pm\beta_3}, U_{\pm(\beta_4+\beta_5+\beta_6)}, U_{\pm\beta_7}\rangle \circ \langle U_{\pm\beta_5}, U_{\pm(\beta_6+\beta_7+\beta_8)}, U_{\pm t_1}, U_{\pm t_2}\rangle$ (commuting product), where $t_1 = 2\beta_1+2\beta_2+3\beta_3+4\beta_4+3\beta_5+2\beta_6+\beta_7$ and $t_2 = \beta_2+\beta_3+2\beta_4+\beta_5$. Then $V|D$ is not irreducible by Theorem (4.1) of [12], so $\langle v^+\rangle$ affords a D-composition factor of $V|Y$ with dimension strictly less than $\dim V|Y$. Restrict this composition factor to the natural subgroup $G < D$ of type $G_2 \times G_2$. (The fixed point subgroup of $D_4$, under the graph automorphism of order three has type $G_2$.) Let $\Pi(G) = \Pi_1 \perp \Pi_2$ with $\Pi_1 = \{\gamma_1,\gamma_2\}$, $\Pi_2 = \{\gamma_3,\gamma_4\}$ and $(\gamma_1,\gamma_1) < (\gamma_2,\gamma_2)$, $(\gamma_3,\gamma_3) < (\gamma_4,\gamma_4)$. Let $\eta_i$ be the fundamental dominant weight corresponding to $\gamma_i$. Then, there is a G-composition factor of $V|Y$ afforded by $\langle v^+\rangle$ with high weight $(x\eta_2 + \eta_1) + ((x+1)\eta_4 + (2x+1)\eta_3)$. This composition factor has dimension at least $12\cdot\dim V(x\eta_2 + \eta_1)$. But since $V|A$ occurs as a compostion factor of the tensor product $V(\mu_1 + x\mu_2)\otimes V(\mu_1)$, $\dim V|A \leq 7\cdot\dim V(\mu_1 + x\mu_2) < 12\cdot\dim V(x\eta_2 + \eta_1) < \dim V|Y$. Contradiction.

This completes the consideration of Case I.

Case II: Suppose $M_j$ is nontrivial and $M_i$ is trivial.

By (8.6), there does not exist $\delta \in \Pi(Y)-\Pi(L_Y)$ with $(\delta,\Sigma L_j) \neq 0$ and $(\delta,\Sigma L_k) = 0$ for all $k\neq i$. Let q be as in Case I.

We now claim that $\Pi(L_j)$ corresponds to an end node of the Dynkin diagram. For, if not, the above remarks imply that there exists $k\neq i$ with $\Pi(L_k)$ separated from $\Pi(L_j)$ by exactly one node of the Dynkin diagram. By (8.5) and (8.6), $L_k$ is unique and has type $A_2$. Thus, $Y = E_8$ and $\Pi(L_Y) = \{\beta_1,\beta_3,\beta_5,\beta_7,\beta_8\}$. Also, by (8.6) and (2.3), $\langle\lambda,\beta_\ell\rangle = 0$ for $\ell = 1, 2, 3, 7, 8$. As in Case I, $\beta_4|T_A = q\alpha_2 = \beta_6|T_A$. Moreover, by (2.12), $\beta_2|Z_A = 0$. Now, using the parabolic $P_Y\hat{}$ of (2.11), we see that the bound on $\dim V^2(Q_A)_{\lambda-q\alpha_2}$ is exceeded. Thus, $\Pi(L_j)$ must correspond to an end node of the Dynkin diagram, as claimed.

Suppose $L_j = \langle U_{\pm\beta_1}\rangle$. Then Claim 1 implies $L_i = \langle U_{\pm\beta_4},U_{\pm\beta_5}\rangle$ and by (8.6), $\langle U_{\pm\beta_7}\rangle \leq L_Y'$. Now $Q_A \leq K_{\beta_2}$ and if $-\beta_2$ is involved in $L_A'$, $\beta_2|Z_A = 0$. Otherwise, (2.11) implies that there is a nontrivial image of $Q_A{}^{\alpha_2}$ in $Q_Y(\beta_3,\beta_2)$. But $Q_Y(\beta_3,\beta_2)$ has $L_A'$ composition factors with high weights $3q\mu_1$ and $q\mu_1$, so $(\beta_3 + \beta_2)|Z_A = q\alpha_2$ and again $\beta_2|Z_A = 0$. Using the parabolic $P_Y\hat{}$ of (2.11), and (1.36), we see that the bound on $\dim V_{\beta_3}(Q_Y\hat{})$ is exceeded unless $\langle\lambda,\beta_3\rangle = 0$ and $\langle\lambda,\beta_1\rangle \leq 2$.

Suppose $Q_A \nleq K_{\beta_6}$. Then (1.15) implies $\langle U_{\pm\beta_7}\rangle$ is a component of $L_Y'$, and by (2.17) the field twist on the embedding of $L_A'$ in $\langle U_{\pm\beta_7}\rangle$ is also q. Thus, by (2.5) and (2.6), $\langle\lambda,\beta_7\rangle = 0$. Also, the bound on $\dim V_{\beta_6}(Q_Y)$ implies $\langle\lambda,\beta_6\rangle = 0$. Finally, if $Y = E_8$, $Q_Y/K_{\beta_8}$ a 2-dimensional irreducible $L_A'$-module implies $Q_A \le K_{\beta_8}$, so $\langle\lambda,\beta_8\rangle = 0$. Thus, $\lambda|T_Y = c\lambda_1$, for $c\le 2$. Now, if $Q_A \le K_{\beta_6}$, (2.3) implies $\langle\lambda,\beta_6+\beta_7\rangle = 0$. If $Y = E_8$, either $\langle U_{\pm\beta_7}\rangle$ or $\langle U_{\pm\beta_7},U_{\pm\beta_8}\rangle$ is a component and we argue as above or apply (2.4) to get $\langle\lambda,\beta_8\rangle = 0$. Thus, if $Q_A \le K_{\beta_6}$, $\lambda|T_Y = c\lambda_1$ for $c\le 2$, as above. If $Y = E_7$, then $c \neq 1$. But now we argue as in Claim 1 to produce a contradiction. Thus, $L_j \neq \langle U_{\pm\beta_1}\rangle$.

Consider next the configuration where $L_j = \langle U_{\pm\beta_2}\rangle$. Then by Claim 2, the previous general remarks and symmetry, we may assume $L_i = \langle U_{\pm\beta_5}, U_{\pm\beta_6}\rangle$ and $Y = E_8$. A straightforward argument, using (2.10) and (2.11), implies that $\langle U_{\pm\beta_1},U_{\pm\beta_3}\rangle \cap L_Y' = 1$, $\langle\lambda,\beta_1+\beta_3\rangle = 0$ and $\beta_1|Z_A = 0 = \beta_3|Z_A$, by (2.12). Using the parabolic $P_Y\hat{}$ of (2.12) and (1.36), we see that the bound on $\dim V_{\beta_4}(Q_Y\hat{})$ is exceeded. Thus, $L_j \neq \langle U_{\pm\beta_2}\rangle$.

The opening remarks of the proof and the cases considered thus far allow us to reduce, finally, to the case where $Y = E_8$ and $L_i = \langle U_{\pm\beta_5},U_{\pm\beta_6}\rangle$, $L_j = \langle U_{\pm\beta_8}\rangle$. We first claim that $V^1(Q_Y) \cong M_j$. For otherwise, (8.5), (8.6) and (8.7) imply that there exists a unique $1\le k\le r$, $k \neq i, j$, with $L_k$ of type $A_1$, $(\Sigma L_k, \beta_4) \neq 0$, and $M_k$ nontrivial. However, (2.17) implies that the field twist on the embedding of $L_A'$ in $L_k$ is also q, contradicting (2.5) and (2.6). Thus, $V^1(Q_Y) \cong M_j$, as claimed. Suppose there exists $\delta \in \Pi(Y)-\Pi(L_Y)$, $\delta \neq \beta_7$ with $\langle\lambda,\delta\rangle \neq 0$. Then $Q_A \not\le K_\delta$ implies $\dim(Q_Y/K_\delta) \ge 4$. Thus $\delta = \beta_4$ and $\langle U_{\pm\beta_2},U_{\pm\beta_3}\rangle \cap L_Y' \neq 1$. However, the bound on $\dim V_{\beta_4}(Q_Y)$, of (1.25), is exceeded. Thus, if $\delta \in \Pi(Y)-\Pi(L_Y)$ with $\langle\lambda,\delta\rangle \neq 0$, $\delta = \beta_7$. So $\lambda|T_Y = x\lambda_7 + c\lambda_8$, for $p>x\ge 0$ and $p>c>0$, and $\langle\lambda,\alpha_1\rangle = c\cdot q$. Recall $\lambda|T_Y \neq \lambda_8$. By (6.9), $\langle\lambda,\alpha_2\rangle = 0$, $cq_0$ or $xq_0$ for some p-power $q_0$. In fact, if $x\neq 0$, $0 \neq f_{\beta_7}v^+ \in V_{T_A}(\lambda-q\alpha_2)$ implies that in the p-adic expansion for $\langle\lambda,\alpha_2\rangle$, q has nonzero coefficient, so by (1.10) $q = q_0 = 1$.

Let $D_0 \le Y$ be the subgroup of type $D_4$ defined by $D_0 = \langle U_{\pm\beta_2}, U_{\pm\beta_3}, U_{\pm(\beta_4+\beta_5+\beta_6+\beta_7)}, U_{\pm\beta_8} \rangle$. Then, if $G \le D_0$ is the fixed point subgroup of the graph automorphism of $D_0$ of order 3, G has type $G_2$. Now, $D_0$ is the Levi factor of a proper parabolic of Y, so the $D_0$ composition factor of $V|Y$ afforded by $v^+$ is not all of $V|Y$. Moreover, the G-composition factor afforded by $v^+$ has high weight $c\mu_1 + x\mu_2$, as $G_2$-module . Thus, if $\lambda|T_A = c\mu_1 + x\mu_2$, $\dim V|A < \dim V|Y$. Hence, $\langle\lambda,\alpha_2\rangle = 0$ or $cq_0$.

We now claim that if $x \neq 0$, $x=c$. Let $P \ge B_A^-$ be the parabolic subgroup of A with Levi factor $L = \langle U_{\pm\alpha_2}\rangle T_A$. Since $V|A$ is basic, (1.9) implies that there is a parabolic subgroup $P_0$ of Y with $P \le P_0$, $Q = R_u(P) \le Q_0 = R_u(P_0)$ and $L \le L_0$, a Levi factor of $P_0$. Moreover, since $T_A \le T_Y$, we may take $T_Y \le L_0$. Thus, there is a subsystem $\Sigma_0 \subseteq \Sigma(Y)$ with $L_0 = T_Y\langle U_{\pm\gamma} \mid \gamma \in \Sigma_0^+ \rangle$. Write $L_0' = L_{01} \times \cdots \times L_{0d}$, a product of simple algebraic groups, with $L_{01}v^+ \neq v^+$ and $L_{0\ell}v^+ = v^+$ for $\ell \neq 1$. (This is possible since $V|A$ is basic.) Then $V|A$ basic and Theorem (7.1) of [12] imply that $L_{01}$ has type $A_k$ for some k and that if $k>1$, $V^1(Q_0)$ is isomorphic to $W_{01}$, the natural module for $L_{01}$ (or to $W_{01}^*$). Note that if $k>1$, there does not exist $K_\gamma \le Q_0$ such that $Q_0/K_\gamma \cong W_{01}$ or $W_{01}^*$, and $Q_A \nleq K_\gamma$. For otherwise, $Q_0/K_\gamma$ is an irreducible L' module on which $Z(L')^\circ$ induces scalars. So $Q_AK_\gamma/K_\gamma$ is an L' submodule of $Q_0/K_\gamma$. But $\dim(Q_AK_\gamma/K_\gamma) < \dim(Q_0/K_\gamma)$. Hence, $L_{01}$ has type $A_1$. So there exists $\tau \in \Sigma_0^+$ with $L_{01} = \langle U_{\pm\tau}\rangle$, $\langle\lambda,\tau\rangle = c$ and $\langle\lambda,\eta\rangle = 0$ for all $\eta \in \Sigma_0$ with $\eta \neq \pm\tau$. But, by the earlier work in this configuration, $x_{-\beta_7}(at^q)$ occurs in the factorization of $x_{-\alpha_2}(t)$, and $\langle\lambda,\beta_7\rangle \neq 0$. Since $L \le L_0$, $\beta_7 = \tau$ and $x = \langle\lambda,\beta_7\rangle = \langle\lambda,\tau\rangle = c$, as claimed. But as in the preceeding paragraph, $\dim V|A < \dim V|Y$. Thus, in fact, $x=0$. Moreover, the preceding argument with $D_0$ implies that $\langle\lambda,\alpha_2\rangle \neq 0$.

Suppose $q_0 = q$, where $\langle\lambda,\alpha_2\rangle = cq_0$. Let $X \le Y$ be the subgroup of type $D_7$ defined by $X = \langle U_{\pm t}, U_{\beta_k} \mid 3 \le k \le 8,\ t = \beta_1+2\beta_2+2\beta_3+3\beta_4+2\beta_5+\beta_6\rangle$.

Then X has a natural subgroup, $G \le X$, of type $G_2$. Moreover, the G–composition factor of $V|Y$ afforded by $v^+$ has the same dimension, as $G_2$ module, as does $V(q(c\mu_1+c\mu_2))$. However, X is contained in a proper parabolic of Y and so the X–composition factor of V afforded by $v^+$ is not all of V. Thus, $\dim V(q(c\mu_1+c\mu_2)) < \dim V|Y$.

Thus, we have $\lambda|T_Y = c\lambda_8$, $c > 1$ and $\lambda|T_A = cq\mu_1+cq_0\mu_2$, with $q_0 \neq q$. Note that $3c = \dim V_{\beta_7}(Q_Y) \le \dim V^2(Q_A)_{\lambda-q\alpha_2} \le 3c$. Hence, $\beta_4|Z_A \neq 0$; in particular, $-\beta_4$ is not involved in $L_A{}'$. Previous remarks imply that $\langle U_{\pm\beta_2}, U_{\pm\beta_3}\rangle \cap L_Y{}' \neq \{1\}$. If $\Pi(L_Y) = \{\beta_2,\beta_3,\beta_5,\beta_6,\beta_8\}$ or $\{\beta_1,\beta_3,\beta_5,\beta_6,\beta_8\}$, (1.15) implies $Q_A \le K_{\beta_4}$ and (2.11) implies $\beta_4|Z_A = 0$. If $\Pi(L_Y) = \{\beta_1,\beta_2,\beta_3,\beta_5,\beta_6,\beta_8\}$, (2.11) and (1.15) imply $Q_A \nleq K_{\beta_4}$. Examining the $L_A{}'$ composition factors and $T(L_A{}')$ weight vectors in $Q_Y/K_{\beta_4}$, we see that the field twist on the embedding of $L_A{}'$ in $\langle U_{\pm\beta_2}\rangle$ and in $\langle U_{\pm\beta_1}, U_{\pm\beta_3}\rangle$ is also q and $\beta_4|T_A = q(\alpha_2 - \alpha_1)$. But then $(V_{T_Y}(\lambda-\beta_4-\beta_5-\beta_6-\beta_7-\beta_8) \oplus V_{T_Y}(\lambda-2\beta_7-2\beta_8) + [V,Q_A{}^3])/[V,Q_A{}^3]$ is a 2–dimensional subspace of $(V_{T_A}(\lambda-q(2\alpha_1+2\alpha_2)) + [V,Q_A{}^3]/[V,Q_A{}^3]$. However, using (1.28) and the description of commutator subspaces given in (1.21), we see that the latter weight space has dimension 1. Hence, $L_Y{}' \ge \langle U_{\pm\beta_2}\rangle$ but $\langle U_{\pm\beta_3}\rangle \nleq L_Y{}'$, or $L_Y{}' = L_i \times L_j \times \langle U_{\pm\beta_3}\rangle$. In either case, (2.11) implies $Q_A \nleq K_{\beta_4}$ and arguing as usual, we find that $\beta_4|T_A = q\alpha_2$. Say $\langle U_{\pm\beta_2}\rangle \le L_Y{}'$ and $\langle U_{\pm\beta_3}\rangle \nleq L_Y{}'$. (The other case is handled similarly.) Using (1.21) and (1.29), we have $\dim(V_{T_A}(\lambda-q(4\alpha_1+2\alpha_2)) + [V,Q_A{}^3])/[V,Q_A{}^3] \le 5$. However, $[V_{T_Y}(\lambda-2\beta_7-4\beta_8) \oplus V_{T_Y}(\lambda-\beta_6-2\beta_7-3\beta_8) \oplus V_{T_Y}(\lambda-\beta_5-\beta_6-2\beta_7-2\beta_8) \oplus V_{T_Y}(\lambda-2\beta_6-2\beta_7-2\beta_8) \oplus V_{T_Y}(\lambda-\beta_4-\beta_5-\beta_6-\beta_7-2\beta_8) \oplus V_{T_Y}(\lambda-\beta_2-\beta_4-\beta_5-\beta_6-\beta_7-\beta_8) + [V,Q_A{}^3]]/[V,Q_A{}^3]$ is a 6–dimensional subspace lying in $(V_{T_A}(\lambda-q(4\alpha_1+2\alpha_2)) + [V,Q_A{}^3])/[V,Q_A{}^3]$, unless c = 2 or 3. But if c = 2 or 3, (1.38) and (1.27) imply $\dim V|A < \dim V|Y$. Contradiction. This completes the consideration of Case II and the proof of (8.8).□

(8.9). $\langle \lambda,\alpha_1 \rangle \neq 0$.

Proof: Suppose false; i.e., suppose $\langle \lambda,\alpha_1 \rangle = 0$. Then $\langle \lambda,\alpha_2 \rangle \neq 0$. Let $P \geq B_A^-$ be the parabolic subgroup of A with Levi factor $L = \langle U_{\pm\alpha_2} \rangle T_A$. Let $P_0$ be a parabolic subgroup of Y with $P \leq P_0$, $Q = R_u(P) \leq R_u(P_0) = Q_0$ and such that $P_0$ is minimal with these properties. Let $L_0$ be the Levi factor of $P_0$. Let $T_0$ be a maximal torus of $L_0$ with $T_A \leq T_0$. Fix a base $\Pi_0(Y)$ of the root system $\Sigma_0(Y)$, such that $U_A \cap L \leq Q_0(U_0 \cap L_0)$, where $U_0$ is the product of the $T_0$ root subgroups corresponding to roots in $\Sigma_0^+(Y)$ and $Q_0$ is the product of the $T_0$ root subgroups corresponding to the roots in $\Sigma_0^-(Y) - \Sigma_0(L_0)$. Let $\Pi_0(Y) = \{\gamma_1 \ldots,\gamma_n\}$. Let $\nu_i$ be the fundamental dominant weight corresponding to $\gamma_i$ and $\langle w^+ \rangle$ be the unique 1-space of V invariant under $B_0 = \langle U_r \mid r \in \Sigma_0^+(Y) \rangle$. Let $L_0' = D_1 \times \cdots \times D_s$, with $D_i$ a simple algebraic group. Note that each $D_i$ is of classical type by (7.1) of [12] and Remark (6.2), so (1.5) implies $Z = Z(L)^\circ \leq Z_0 = Z(L_0)^\circ$.

Now, referring to (6.9) and using (1.27), (8.1), (1.38), (1.32), [8] and $p>3$, we see that $\dim V|A \neq \dim V|Y$ unless one of the following holds:

(i) $Y = E_8$, $\lambda|T_Y = x\nu_7 + c\nu_8$, for $p>x\geq 0$, $p>c>0$ and $\langle \lambda,\alpha_2 \rangle = c \cdot q$, for some p-power q.

(ii) The hypotheses of (6.7) hold.

The configurations of (i) and (ii) may be described as follows: $\langle \lambda,\alpha_2 \rangle = c$, for $0<c<p$ (by (1.10)), $D_1$ has type $A_1$, $\dim M_1 = c+1$ and $M_i$ is trivial for $i \neq 1$. Now, one checks that $\dim V^2(Q) = c$ and $\dim V^3(Q) \leq 2c$. Thus, $D_1$ must be separated by more than one node of the Dynkin diagram from all other components of $L_0'$, else $\dim V^2(Q_0) > c$. However, there do not exist $\delta_1, \delta_2 \in \Pi_0(Y)-\Pi(L_0)$ with $(\delta_1,\Sigma D_1) \neq 0$, $(\delta_1,\delta_2) < 0$ and $(\delta_2,\Sigma L_0) = 0$. For otherwise (2.14) implies $\langle \lambda,\alpha_1 \rangle \neq 0$. Thus, there exists $1<i\leq s$ with $D_i$ separated from $D_1$ by exactly 2 nodes of the Dynkin diagram. Let $\gamma,\delta \in \Pi_0(Y)-\Pi(L_0)$ with $(\gamma,\Sigma D_1) \neq 0 \neq (\delta,\Sigma D_i)$ and $(\gamma,\delta) < 0$. If $-\delta$ is involved in $L_0'$, $\delta|Z = 0$. Otherwise, if $Q_A \leq K_\delta$, (2.11) applies to give $(\gamma+\delta)|Z = \alpha_1$, so again $\delta|Z = 0$. Using the parabolic $P_0^\wedge$ of (2.11), we see that the bound on dim $V^2(Q)$ is exceeded. Thus, $Q_A \nsubseteq K_\delta$ and $-\delta$ is not involved in $L_0'$. Now $\langle \lambda, \gamma \rangle = 0$, else dim $V^2(Q_0) > c$; but if $\Pi(D_1) = \{\tau_1\}$, $f_{\tau_1+\gamma}w^+$ affords an $L_0'$ composition factor in $V^2(Q_0)$ of dimension c. Hence, $[V,Q_0{}^2] = [V,Q^2]$. Now, $f_{\tau_1+\gamma}{}^2w^+$ and $f_{\tau_1+\gamma+\delta}w^+$ afford distinct L' composition factors in $V^3(Q_0)$ of dimensions c−1 and nc, where $n = \dim V_{L_0'}(-\delta) \geq 2$. But dim $V^3(Q) \leq 2c$ implies c=1 and $\dim V|A = 14 <$ $\dim V|Y$. Contradiction.□

The results (8.2) – (8.9) form a complete proof of Theorem (8.0)(b) in case $p > 3$.□

# CHAPTER 9: SPECIAL CASES

In this chapter, we consider the remaining special cases which will complete the proof that the only possible triples $(A,Y,V)$ are those described in the Main Theorem. In particular, we consider the cases where $Y$ has type $E_n$ and rank$A = 2$ with $p=2$, or $(A,p)$ has type $(G_2,3)$ or where $Y$ is exceptional and $A$ is a non-simple, semisimple algebraic group. We find that the only possible configuration is that of (8.0)(a).

## Section I: Rank(A) = 2 and p = 2.

Before establishing the notation to be used throughout this section of Chapter 9, we prove a preliminary lemma.

(9.1). Let $p=2$, $X = SL_2$, $Y = SO_8$. If $X < Y$, acting irreducibly on a rational $kY$-module, $V$, then $X$ acts irreducibly on two of the three fundamental, restricted 8-dimensional irreducible $kY$-modules.

Proof: If $X<Y$ is as in Theorem (7.1) (d) of [12], then it is clear that $X$ acts irreducibly on the two fundamental spin modules for $Y$, since $X < SO_7 < Y$ and each of these representations restrict to the same irreducible representation of $SO_7$.

Consider now the possibility that $V|Y$ is the "natural" module for $Y$. Let $D$ be a group of type $D_4$ and let $X < D$ be as in (7.1)(d) of [12]. In particular, if $\Pi(D) = \{\gamma_1,\gamma_2,\gamma_3,\gamma_4\}$ is labelled as throughout and if $\mu_i$ is the fundamental dominant weight corresponding to $\gamma_i$, $X$ acts irreducibly on the $kD$-module, $W$, with high weight $\mu_3$. Also, if $\Pi(X) = \{\alpha\}$, $h_\alpha(c) =$ $h_{\gamma_3}(c^{q_1})h_{\gamma_4}(c^{q_1})h_{\gamma_2+\gamma_3}(c^{q_2})\, h_{\gamma_2+\gamma_4}(c^{q_2})h_{\gamma_1+\gamma_2+\gamma_3}(c^{q_3})h_{\gamma_1+\gamma_2+\gamma_4}(c^{q_3})$, for $c \in k^*$, $q_1$, $q_2$, and $q_3$ distinct p-powers and $U_\alpha \leq \langle U_{\gamma_i} | 1 \leq i \leq 4 \rangle$. Now, by (1.11), D fixes a form on W. Since p=2, we may take the form on W to be symplectic. So if $\varphi: D \longrightarrow SL(W)$ is the kD-representation corresponding to W, $\varphi(D) \leq Sp(W)$. Then $\varphi(D) < Sp(W)$ is a maximal rank configuration, so by Theorem 4.1 and Table 4.1 of [12], $\varphi(D) = \langle U_r \mid r \in \Sigma(Sp(W)), r \text{ short} \rangle$. Now, choose a base $\Pi(\varphi(D)) = \{\eta_1, \eta_2, \eta_3, \eta_4\}$ such that the Dynkin diagram is labelled as throughout and such that $W|\varphi(D)$ has high weight $\nu_1$ (where $\nu_i$ is the fundamental dominant weight corresponding to $\eta_i$). Since p=2, $X \cong \varphi(X)$ and $D \cong \varphi(D)$ as abstract groups, and $\varphi(X) < \varphi(D) < SL(W)$ is the desired embedding; i.e. $W|\varphi(D)$ is the "natural" module for $\varphi(D)$. Moreover,we may take $\varphi(h_\alpha(c)) = h_{\eta_1}(c^{q_1}) h_{\eta_3}(c^{q_1}) h_{\eta_1+\eta_2}(c^{q_2}) h_{\eta_2+\eta_3}(c^{q_2}) \cdot h_{\eta_1+\eta_2+\eta_4}(c^{q_3}) h_{\eta_2+\eta_3+\eta_4}(c^{q_3})$. Also $\varphi(U_\alpha) \leq \langle U_{\eta_i} | 1 \leq i \leq 4 \rangle$. Hence, $\varphi(X)$ acts irreducibly on the $\varphi(D)$ modules with high weights $\nu_1$ and $\nu_3$.

By Theorem (7.1) of [12] the above considerations are sufficient and the proof of (9.1) is complete.□

For the remainder of this section, we suppose (A,Y,V) is an example in the Main Theorem, with A a rank two simple algebraic group, Y of type $E_n$ and p = 2. Adopt Notation and Hypothesis (2.0); in addition, choose $P_A = Q_A L_A$ with $L_A = \langle U_{\pm\beta} \rangle T_A$ such that $\langle \lambda, \beta \rangle \neq 0$. (Then $\Pi(A) - \Pi(L_A) = \{\alpha\}$ as usual.) Note that (7.1) of [12], the minimality of $P_Y$ and induction imply that if $L_j$, a component of $L_Y'$, has type $A_{k_j}$ for some $k_j$, then $k_j = 1$, 3 or 7. Finally, one checks, using [8] and (1.35) that $\dim V|A = 4^k$, $3^k 8^\ell$, or $6^k 14^\ell 64^m$, for $k, \ell, m \in \mathbb{Z}^+$.

<u>(9.2)</u>. Suppose $\dim(M_i) > 1$ for some i.

(i) If $L_i$ has type $A_3$ and $\dim(M_j) > 1$ for some $j \neq i$, then $\mathrm{rank}(L_j) = 1$.

(ii) $L_i$ has type $A_1$ or $A_3$.

<u>Proof:</u> By induction and (7.1) of [12], $L_i$ has type $D_k$ for

$k = 4,5,6$ or $7$ or $L_i$ has type $A_1$, $A_3$, or $A_7$. Consider first the case where $L_i$ has type $A_3$ and suppose there exists $j \neq i$ with $M_j$ nontrivial and $\mathrm{rank}(L_j) > 1$. Rank restrictions imply $L_j$ has type $A_3$ and by (1.5), $Z_A \leq Z_Y$. Also, (2.5) and (2.7) imply that $\Pi(L_Y) = \{\beta_2,\beta_3,\beta_4,\beta_6,\beta_7,\beta_8\}$. The bound on $\dim V_{\beta_5}(Q_Y)$ implies $\langle \lambda,\beta_j \rangle = 0$ for $j = 5,6,7$. However, $0 \neq w_1 \in V_{T_Y}(\lambda-\beta_5-\beta_6-\beta_7-\beta_8)$ and $0 \neq w_2 \in V_{T_Y}(\lambda-\beta_2-\beta_4-\beta_5) \oplus V_{T_Y}(\lambda-\beta_3-\beta_4-\beta_5)$ afford $L_Y'$ composition factors of $V_{\beta_5}(Q_Y)$ of dimensions 20 and 56, respectively. Hence, the bound is still exceeded and the first statement of the result holds.

Suppose (9.2)(ii) is false. Previous remarks and (1.5) imply that $Z_A \leq Z_Y$. If $L_i$ has type $A_7$, the bound on $\dim V_{\beta_2}(Q_Y)$ implies that $\lambda|T_Y = \lambda_8$; thus, $L_i$ has type $D_k$ for some $k$. If $k > 4$, $M_i$ is one of the two irreducible, restricted spin modules for $L_i$. Thus, $\langle U_{\pm\beta_1} \rangle \leq L_i$ and $L_i$ has type $D_5$; else $Q_Y/K_{\beta_1}$ is a $2^{k-1}$ – dimensional irreducible $L_A'$ module containing a nontrivial image of $Q_A{}^{\alpha}$. Now, the same argument forces $Y$ to be of type $E_7$ or $E_8$ with $\langle U_{\pm\beta_7} \rangle$ a component of $L_Y'$. By (2.5), (2.6) and (2.7), $\langle \lambda,\beta_7 \rangle = 0$. However, the bound on $\dim V_{\beta_6}(Q_Y)$ is exceeded in every possible configuration. So we have reduced to $L_i$ having type $D_4$.

By (9.1), $L_A'$ acts irreducibly on 2 of the three fundamental restricted 8-dimensional irreducible $L_i$ modules. But $Q_A \nleq K_{\beta_1}$, so $Q_Y/K_{\beta_1}$ is a reducible $L_A'$ module. Using (7.1) of [12] to obtain a precise description of the embedding of $L_A'$ in $L_i$, we see that $L_A'$ acts on $Q_Y/K_{\beta_1}$ with composition factors of dimensions 2 and 1. Thus, if $A = G_2$, then $\beta$ must be long. Also, since $V_{L_i}(-\beta_6)$ is an irreducible $L_A'$ module, $Y = E_7$ or $E_8$ and $\langle U_{\pm\beta_7} \rangle$ is a component of $L_Y'$. Moreover, by (2.5), (2.6) and (2.7), $\langle \lambda,\beta_7 \rangle = 0$. The above remarks about $Q_Y/K_{\beta_1}$ imply that $\langle \lambda,\beta_3 \rangle = 0$; and by (7.1) of [12], $\langle \lambda,\beta_4 \rangle = 0$. Moreover, the bound on $\dim V_{\beta_j}(Q_Y)$, for $j = 1$ or $6$, implies that $\langle \lambda,\beta_k \rangle = 0$ for $k = 1,5,6$. (We have used Table 1 of [5] and (1.35).) Then, [8] implies that $Y = E_8$, else $\dim V|A \neq \dim V|Y$. Suppose $\lambda|T_Y = \lambda_2$. The remarks of the previous paragraph imply

$\langle\lambda,\alpha\rangle = 0$, q, $q+q_0$ or $q+q_0+q_1$ for q, $q_0$ and $q_1$ distinct p-powers, unless $A = G_2$, in which case $\langle\lambda,\alpha\rangle \neq q+q_0+q_1$. Hence, $\dim V|A \leq 14^3 \cdot 6^2 < \dim V|Y$. A variation of the method described in [8] was run on the VAX at the University of Oregon to determine the multiplicity of the subdominant weight $\lambda-\beta_1-3\beta_2-3\beta_3-5\beta_4-4\beta_5-3\beta_6-2\beta_7-\beta_8$. This computation, together with [8], implies $\dim V|Y > 14^3 \cdot 6^2$. If $\lambda|T_Y = \lambda_2 + \lambda_8$, previous remarks and part (i) imply $\langle\lambda,\alpha\rangle = 0$, q, $q+q_0$ or $q+q_0+q_1$, for $q,q_0$ and $q_1$ distinct p-powers. Then $\dim V|A \leq 14^3 \cdot 6^3 < \dim V|Y$, by (1.38). Contradiction.

This completes the proof of (9.2).□

<u>(9.3).</u> Assume $\beta$ is long if $A = G_2$. If $\dim(M_i) > 1$ then $\text{rank}(L_i) = 1$.

<u>Proof:</u> Suppose false. Then by (9.2), $L_i$ has type $A_3$, so by (1.5), $Z_A \leq Z_Y$. Since $L_A'$ acts irreducibly on W, the natural module for $L_i$, (2.3) implies that there does not exist $\gamma \in \Pi(Y) - \Pi(L_Y)$ with $(\gamma,\Sigma L_i) \neq 0$ and $Q_Y/K_\gamma \cong W$ or $W^*$. Consider the case where $\Pi(L_i) = \{\beta_2,\beta_4,\beta_5\}$, so $Y = E_7$ or $E_8$ and $\langle U_{\pm\beta_7}\rangle$ is a component of $L_Y'$. Now, $\langle\lambda,\beta_7\rangle = 0$, by (2.5), (2.6) and (2.7), and $\langle\lambda,\beta_6\rangle = 0$, else the bound on $\dim V_{\beta_6}(Q_Y)$ is exceeded. Also, the bound on $\dim V_{\beta_3}(Q_Y)$ implies $\langle\lambda,\beta_3\rangle = 0$. If $\langle U_{\pm\beta_1}\rangle$ is a component of $L_Y'$, $\langle\lambda,\beta_1\rangle = 0$, else $0 \neq w_1 \in V_{T_Y}(\lambda-\beta_1-\beta_3)$ and $0 \neq w_2 \in V_{T_Y}(\lambda-\beta_2-\beta_3-\beta_4) \oplus V_{T_Y}(\lambda-\beta_3-\beta_4-\beta_5)$ afford distinct $L_Y'$ composition factors of $V_{\beta_3}(Q_Y)$, exceeding the given bound. Now, [8] implies that $\lambda|T_Y \neq \lambda_2$ or $\lambda_5$ if $Y = E_7$. Hence, $Y = E_8$. If $\lambda|T_Y = \lambda_2$ or $\lambda_5$, (9.2) implies $\langle\lambda,\alpha\rangle = 0$, q or $q + q_0$, for q and $q_0$ distinct p-powers. So $\dim V|A \leq 14^2 \cdot 6^2 < \dim V|A$ by (1.38). Thus, $\lambda|T_Y = \lambda_2 + \lambda_8$ or $\lambda_5 + \lambda_8$. Then (9.2) implies $\langle\lambda,\alpha\rangle = 0$, $q_1$, $q_1+q_2$, or $q_1+q_2+q_3$, for $q_1$, $q_2$, $q_3$ distinct p-powers. Hence, $\dim V|A \leq 14^2 \cdot 6^3 < \dim V|Y$, by (1.38). Thus, $\Pi(L_i) \neq \{\beta_2,\beta_4,\beta_5\}$.

Consider the case where $\Pi(L_i) = \{\beta_6,\beta_7,\beta_8\}$. Then $\langle U_{\pm\beta_4}\rangle \leq L_Y'$. If $L_Y'$ has a second component of type $A_3$, (9.2) implies $V^1(Q_Y) \cong M_i$. But in this case the bound on $\dim V_\gamma(Q_Y)$, for $\gamma \in \Pi(Y) - \Pi(L_Y)$, implies

$\lambda|T_Y = \lambda_8$. Thus, $\langle U_{\pm\beta_4}\rangle$ is a component of $L_Y'$. By (2.5), (2.6) and (2.7), $\langle\lambda,\beta_4\rangle = 0$ and the bound on $\dim V_{\beta_5}(Q_Y)$ implies $\langle\lambda,\beta_5\rangle = 0$. Suppose $L_A'$ acts on the natural module for $L_i$ with high weight $(q_1+q_2)\mu_\beta$. Then, by (2.6) and (2.7) we may assume that the field twist on the embedding of $L_A'$ in $\langle U_{\pm\beta_4}\rangle$ is $q_1$. If $\langle U_{\pm\beta_1}\rangle$ is a component of $L_Y'$ with $\langle\lambda,\beta_1\rangle \neq 0$, then $Q_A \nsubseteq K_{\beta_3}$ implies that the field twist on the embedding of $L_A'$ in $\langle U_{\pm\beta_1}\rangle$ is also $q_1$. But this contradicts (2.5). If $\langle U_{\pm\beta_1}\rangle \nsubseteq L_Y'$, then $\langle\lambda,\beta_1\rangle = 0$ by (2.3). So $\lambda|T_Y = x\lambda_2 + y\lambda_3+\lambda_6$ or $x\lambda_2+y\lambda_3+\lambda_8$, where $x, y \in \{0,1\}$. By (9.2), $\dim V|A \leq 14^2\cdot 6^2$ if $x=0=y$ and $\dim V|A \leq 14^2\cdot 6^4$ if $x+y\neq 0$. Then (1.38) and (1.32) imply $\dim V|A < \dim V|Y$. Hence, $\Pi(L_i) \neq \{\beta_6,\beta_7,\beta_8\}$.

We have reduced to the configuration where $Y = E_7$ and $\Pi(L_i) = \{\beta_5,\beta_6,\beta_7\}$. If $\Pi(L_Y) = \{\beta_2,\beta_3,\beta_5,\beta_6,\beta_7\}$, (2.5), (2.7) and the bound on $\dim V_{\beta_4}(Q_Y)$ imply that $\langle\lambda,\beta_j\rangle = 0$ for $2 \leq j \leq 6$. So $\lambda|T_Y = \lambda_7$ or $\lambda_1 + \lambda_7$. But then [8] implies that $\dim V|A \neq \dim V|Y$. The above remarks imply that $\Pi(L_Y) = \Pi(L_i) \cup \{\beta_2\}$, $\Pi(L_i) \cup \{\beta_3\}$, or $\Pi(L_i) \cup \{\beta_1,\beta_2\}$, $V_1(Q_Y) \cong M_i$ in the first two cases, and in each case, $\langle\lambda,\beta_2+\beta_4\rangle = 0$. Now, $\lambda|T_Y \neq \lambda_5$ or $\lambda_7$, by [8], so $\langle\lambda,\beta_1+\beta_3\rangle > 0$; in particular, $\Pi(L_Y) \neq \{\beta_2,\beta_5,\beta_6,\beta_7\}$. If $A = A_2$ or $B_2$, $\dim V|A \leq 2^{12} < \dim V|Y$, by (1.32) and [8]. So $A = G_2$. If $L_Y' = L_i \times \langle U_{\pm\beta_3}\rangle$, then previous remarks imply that $\lambda|T_Y = \lambda_1+\lambda_5$ and $\dim V|A \leq 6^3\cdot 14^2 < \dim V|Y$, by (1.38). In the remaining case, we consider the action of $L_A'$ on the 56-dimensional irreducible $kY$-module $V(\lambda_7)$. One checks that there are at least five 4-dimensional $L_A'$ composition factors of $V(\lambda_7)$. However, there is no 56-dimensional $kA$-module affording such an $L_A'$ composition series. This completes the proof of (9.3).□

<u>(9.4).</u> A does not have type $A_2$ or $B_2$.

<u>Proof:</u> Suppose false. If $\langle\lambda,\alpha\rangle = 0$ or $q$ and $\langle\lambda,\beta\rangle = 0$ or $q_0$ for some p-powers $q$ and $q_0$, $\dim V|A = 3,8,9,4$ or $16 < \dim V|Y$. So we may choose a parabolic $P_A$ with Levi factor $L_A = \langle U_{\pm\beta}\rangle T_A$ such that $\langle\lambda,\beta\rangle$ has more than one p-power in its p-adic expansion. Moreover, choose $\beta$ such

that the number of distinct p-powers in the p-adic expansion of $\langle\lambda,\beta\rangle$ is greater than or equal to the number of p-powers in the p-adic expansion of $\langle\lambda,\alpha\rangle$. Let $P_Y$ be as before. Then (9.3) implies that $L_Y'$ has at least two components of type $A_1$, so (1.5) implies $Z_A \leq Z_Y$. Also, $\langle\lambda,\beta_i\rangle = 1 = \langle\lambda,\beta_j\rangle$ for some $1 \leq i,j \leq \text{rank}(Y)$, with $(\beta_i,\beta_j) = 0$. Recall, $\dim V|A = 4^k$ or $3^k 8^\ell$. Suppose $\langle\lambda,\beta\rangle = q_1+q_2$, for $q_1$ and $q_2$ distinct p-powers. Then, $\dim V|A \leq 2^8$. But counting only the conjugates of $V_{T_Y}(\lambda)$ in $V|Y$, we have $\dim V|Y > \dim V|A$. Hence, $\langle\lambda,\beta\rangle \neq q_1+q_2$. Suppose $\langle\lambda,\beta\rangle = q_1+q_2+q_3$, for $q_1$, $q_2$, and $q_3$ distinct p-powers. Then $\langle\lambda,\beta_i\rangle = 1 = \langle\lambda,\beta_j\rangle = \langle\lambda,\beta_k\rangle$ for some $i\neq j\neq k$ with $(\beta_i,\beta_j) = 0 = (\beta_j,\beta_k) = 0 = (\beta_i,\beta_k)$. Also, $\dim V|A \leq 2^{12}$. Once again, counting only the conjugates of $V_{T_Y}(\lambda)$ in $V|Y$, we have $\dim V|Y > \dim V|A$, unless $Y = E_6$ and $\lambda|T_Y = \lambda_1+\lambda_2+\lambda_6$. However, by (1.38), $\dim V|A < \dim V|Y$ here also. By rank restrictions, it remains to consider the case where $\langle\lambda,\beta\rangle = q_1+q_2+q_3+q_4$. Arguing exactly as above, we find that $\dim V|A < \dim V|Y$. This completes the proof of (9.4).□

<u>(9.5).</u> Let A have type $G_2$, with $\beta$ long. Suppose $M_i$ and $M_j$ are nontrivial for some $i \neq j$.

(1) If $(\Sigma L_i,\gamma) \neq 0 \neq (\gamma,\Sigma L_j)$ for some $\gamma \in \Pi(Y) - \Pi(L_Y)$, then $\Pi(L_i \times L_j) \subset \{\beta_2,\beta_3,\beta_5\} \subset \Pi(L_Y)$. Moreover, $\langle\lambda,\beta_2+\beta_3+\beta_5\rangle = 2$.

(2) If $(\Pi(L_i),\gamma_i) \neq 0 \neq (\gamma_i,\gamma_j) \neq 0 \neq (\gamma_j,\Pi(L_j))$ for some $\gamma_i \neq \gamma_j \in \Pi(Y) - \Pi(L_Y)$, then $\beta_4 \in \{\gamma_i,\gamma_j\}$ and $|\{\beta_2,\beta_3,\beta_5\} \cap \Pi(L_Y)| = 2$.

(3) $\langle\lambda,\beta\rangle = 0$, $q$, $q+q_0$ or $q+q_0+\hat{q}$, for $q$, $q_0$ and $\hat{q}$ distinct p-powers. If $Y = E_6$, $\langle\lambda,\beta\rangle \neq q+q_0+\hat{q}$.

<u>Proof:</u> (1) follows from (2.5) and (2.7). For (2), let $\Pi(L_i) = \{\beta_i\}$ and $\Pi(L_j) = \{\beta_j\}$ and take $\ell \in \{i,j\}$. Note that if $\gamma \in \Pi(Y) - \Pi(L_Y)$ with $(\gamma,\beta_\ell) \neq 0$ and $(\gamma,\Sigma L_k) = 0$ for all $k\neq\ell$, then $Q_Y/K_\gamma \cong (Q_A{}^\alpha)^q$, where q is the field twist on the embedding of $L_A'$ in $L_\ell$. Hence, $x_{-\alpha}(t) = x_{-\gamma}(at^q)u$, for some $a \in k^*$, $u \in K_\gamma$ and $\gamma|T_A = q\alpha$. The result of (2) then follows from (2.5), (2.6) and (2.8). Then (9.3), (9.5)(1) and (9.5)(2) imply (3).□

(9.6). There are no examples in the Main Theorem with A a rank two simple algebraic group, Y of type $E_n$ and p=2.

Proof: Suppose false. Then (9.4) implies that $A = G_2$. Let $\Pi(A) = \{\alpha_1,\alpha_2\}$, with $\alpha_1$ short. If $P_A$ is as before with $\dim V^1(Q_A) > 1$, then all components of $L_Y'$ have classical type. For otherwise, $Y = E_8$, $\Pi(L_Y) = \{\beta_j \mid j \neq 7\}$ and by induction, $\lambda|T_Y = \lambda_7+\lambda_8$. But (9.2), (9.3) and (1.32) imply $\dim V|A \leq 6^2 \cdot 14 < \dim V|Y$. So all components of $L_Y'$ have classical type and by (1.5), $Z_A \leq Z_Y$.

We first make a few notes about the case where $L_A' = \langle U_{\pm\alpha_1}\rangle$ and $\dim(M_i) > 1$ with $L_i$ of type $A_3$. (1) If $\dim(M_j) > 1$ for some $j \neq i$, (9.2) implies $L_j$ has type $A_1$. (2) There does not exist $\gamma \in \Pi(Y) - \Pi(L_Y)$ with $(\gamma,\Sigma L_i) \neq 0$, $(\gamma,\Sigma L_m) = 0$ for all $m \neq i$ and $\dim(Q_Y/K_\gamma) = 6$. For otherwise, the $L_A'$ composition factors of $Q_Y/K_\gamma$ would have dimensions 1 and 2, contradicting (2.3). (3) By (9.2), (2.5) and (2.7), there does not exist $\gamma \in \Pi(L_Y) - \Pi(L_i)$, $\delta \in \Pi(Y) - \Pi(L_Y)$ with $\langle\lambda,\gamma\rangle \neq 0$, $(\gamma,\delta) \neq 0 \neq (\delta,\Sigma L_i)$ and $\dim(Q_Y/K_\gamma) = 8$. (4) If $\gamma \in \Pi(Y) - \Pi(L_Y)$ such that $V_\gamma(Q_Y) \neq 0$, then (2.3) implies $\dim(Q_Y/K_\gamma) \geq 4$.

Note that $\langle\lambda,\alpha_1\rangle \neq q_1+q_2+q_3+q_4$, for $q_1$, $q_2$, $q_3$, and $q_4$ distinct p-powers. For otherwise, previous remarks imply $Y = E_8$ and $\langle\lambda,\beta_6\rangle = 1 = \langle\lambda,\beta_8\rangle = 1 = \langle\lambda,\beta_2+\beta_3\rangle$. But then (9.5) implies $\dim V|A \leq 6^4 \cdot 14^3 < \dim V|Y$, by (1.32). Now, (9.2) implies that if there are r distinct p-powers in the p-adic expansion of $\langle\lambda,\alpha_1\rangle$, there exist $\{\beta_{j_1},\dots,\beta_{j_{r-1}}\} \subset \Pi(Y)$ with $\langle\lambda,\beta_{j_k}\rangle = 1$ and $(\beta_{j_k},\beta_{j_\ell}) = 0$ for $k \neq \ell$. Previous work implies that $\langle\lambda,\alpha_1\rangle = 0$, $q_1$, $q_1+q_2$, or $q_1+q_2+q_3$, for distinct p-powers $q_1$, $q_2$, and $q_3$. We next claim that if $\langle\lambda,\alpha_1\rangle = q_1+q_2+q_3$, $Y = E_7$ or $E_8$. For otherwise, taking $P_A$ as before, with $L_A' = \langle U_{\pm\alpha_1}\rangle$, we may assume $\Pi(L_Y) = \{\beta_2,\beta_3,\beta_4,\beta_6\}$. Also, $L_A'$ acts on the natural module for $\langle U_{\pm\beta_2},U_{\pm\beta_3},U_{\pm\beta_4}\rangle$ with high weight $(q_1+q_2)\mu_\beta$ and the field twist on the embedding of $L_A'$ in $\langle U_{\pm\beta_6}\rangle$ is $q_3 \neq q_i$ for i = 1,2. Consider the action of $L_A'$ on the 27-dimensional irreducible $L_A'$ module $V(\lambda_6)$. There is an $L_A'$ composition factor with high weight $(\lambda-\beta_1-\beta_2-\beta_3-2\beta_4-2\beta_5-\beta_6)|T(L_A') = (q_1+q_2+q_3)\mu_1$. But there is no 27-dimensional $kA$-module affording such an $L_A'$ composition factor.

Previous remarks, [8] and (1.32) imply that $\lambda|T_Y \neq \lambda_\ell$ for any $\ell$, else $\dim V|A \neq \dim V|Y$. Note that (9.3) and (9.5) imply that for each distinct p-power $q_j$ in the p-adic expansion of $\langle\lambda,\alpha_2\rangle$, there exists $\beta_j \in \Pi(Y)$ with $\langle\lambda,\beta_j\rangle = 1$ and $(\beta_j,\beta_k) = 0$ for any $\beta_k$ corresponding to a different p-power $q_k$. As well, previous remarks, [8], (1.26), (1.32) and (9.5) imply that $\langle\lambda,\alpha_1\rangle \neq 0 \neq \langle\lambda,\alpha_2\rangle$, else $\dim V|A < \dim V|Y$.

Using the above remarks and the usual dimension arguments, it is straightforward to show that $\langle\lambda,\alpha_2\rangle \neq q$ and if $\langle\lambda,\alpha_2\rangle = q+q_0$, then $Y \neq E_8$. (Here $q$ and $q_0$ are distinct p-powers.)

Suppose $\langle\lambda,\alpha_2\rangle = q+q_0$ and $Y = E_6$. Then $\dim V|A \leq 6^2\cdot 14^2$. Also there exist $\beta,\eta \in \Pi(Y)$ with $(\beta,\eta) = 0$ and $\langle\lambda,\beta\rangle = 1 = \langle\lambda,\eta\rangle$. Then [8] and (1.32) imply $\langle\lambda,\alpha_1\rangle = q_1+q_2$, else $\dim V|A < \dim V|Y$. Let $P_A$ be as before, with $L_A' = \langle U_{\pm\alpha_1}\rangle$. By previous remarks, symmetry and (1.23), we may assume that either $\Pi(L_Y) = \{\beta_2,\beta_3,\beta_4,\beta_6\}$ with $\lambda|T_Y = \lambda_3+\lambda_5$ or $\Pi(L_Y) = \{\beta_1,\beta_4,\beta_6\}$ with $\langle\lambda,\beta_1\rangle = 1 = \langle\lambda,\beta_6\rangle$ and $\langle\lambda,\beta_4+\beta_2\rangle = 0$. In the first case, the bound on $\dim V_{\beta_5}(Q_Y)$ is exceeded. In the second case, [8] implies that $\lambda|T_Y \neq \lambda_1+\lambda_6$, and (1.38) implies that $\dim V|A < \dim V|Y$ in the remaining cases.

So if $\langle\lambda,\alpha_2\rangle = q+q_0$, $Y = E_7$. Also, $\langle\lambda,\alpha_1\rangle = q_1+q_2+q_3$, else $\dim V|A \leq 6^2\cdot 14^2$ and using (1.32), [8] and (1.38), we see that $\dim V|A < \dim V|Y$. Let $P_A$ be as before with $L_A' = \langle U_{\pm\alpha_1}\rangle$. Then either (a) $\Pi(L_Y) = \{\beta_2,\beta_3,\beta_5,\beta_6,\beta_7\}$, with $\langle\lambda,\beta_1+\beta_3+\beta_6\rangle = 0$ and $\langle\lambda,\beta_2\rangle = 1 = \langle\lambda,\beta_5+\beta_7\rangle$ or (b) $L_Y'$ has components $\langle U_{\pm\beta_1}\rangle$ and $\langle U_{\pm\beta_2}, U_{\pm\beta_4}, U_{\pm\beta_5}\rangle$, with $\langle\lambda,\beta_1\rangle = 1 = \langle\lambda,\beta_2+\beta_5\rangle$ and $\langle\lambda,\beta_4+\beta_7\rangle = 0$ or (c) $L_Y'$ has components $\langle U_{\pm\beta_2}\rangle$, $\langle U_{\pm\beta_5}\rangle$, and $\langle U_{\pm\beta_7}\rangle$, with $\langle\lambda,\beta_j\rangle = 1$, for $j = 2,5,7$ and $\langle\lambda,\beta_1+\beta_3\rangle = 0$. In the last case, we need only count the conjugates in $V|Y$ of $V_{T_Y}(\lambda)$ to see that $\dim V|A < \dim V|Y$. In case (a), the bound on

$\dim V_{\beta_4}(Q_Y)$ implies that $\langle\lambda,\beta_4+\beta_7\rangle = 0$. So $\lambda|T_Y = \lambda_2+\lambda_5$ and (1.38) implies that $\dim V|A < \dim V|Y$. Finally, to see that the configurations described in (b) do not occur, we consider the action of $L_A'$ on the 56-dimensional irreducible kY module, $V(\lambda_7)$. There is an $L_A'$ composition factor with high weight $\lambda-\beta_3-\beta_4-\beta_5-\beta_6-\beta_7|T(L_A') = (q_1+q_2+q_3)\mu_\beta$. But there is no 56-dimensional kA module which affords such an $L_A'$ composition factor. Thus, in fact, $\langle\lambda,\alpha_2\rangle \neq q+q_0$.

By (9.5), it remains to consider the case where $\langle\lambda,\alpha_2\rangle = q+q_0+q^\wedge$ and $Y = E_7$ or $E_8$. So there exist $\beta,\eta,\varepsilon \in \Pi(Y)$ such that $\langle\lambda,\beta\rangle = 1 = \langle\lambda,\eta\rangle = \langle\lambda,\varepsilon\rangle$ and $0 = (\beta,\eta) = (\beta,\varepsilon) = (\eta,\varepsilon)$. Also, $\dim V|A \leq 6^4 \cdot 14^3$. If $Y = E_8$, (1.32) and (1.38) imply $\dim V|A < \dim V|Y$. So $Y = E_7$. Let $P_A$ be as in the previous notation with $L_A' = \langle U_{\pm\alpha_2}\rangle$. Then (9.3) and (9.5) imply that either $\Pi(L_Y) = \{\beta_1,\beta_2,\beta_5,\beta_7\}$, with $\langle\lambda,\beta_1\rangle = 1$ or $\Pi(L_Y) = \{\beta_2,\beta_3,\beta_5,\beta_7\}$, with $\langle\lambda,\beta_k\rangle = 1$, for $k = 2,3,7$ and $\langle\lambda,\beta_5\rangle = 0$. But in the first case (2.7) implies that the field twists on the embeddings of $L_A'$ in $\langle U_{\pm\beta_2}\rangle$, $\langle U_{\pm\beta_5}\rangle$ and $\langle U_{\pm\beta_7}\rangle$ are all equal, contradicting (2.5). In the second case, the field twists on the embeddings of $L_A'$ in $\langle U_{\pm\beta_5}\rangle$ and in $\langle U_{\pm\beta_7}\rangle$ are equal, while the field twists on the embeddings of $L_A'$ in $\langle U_{\pm\beta_2}\rangle$, $\langle U_{\pm\beta_3}\rangle$ and $\langle U_{\pm\beta_7}\rangle$ are distinct, contradicting (2.7) (with $\gamma = \beta_4$).

This completes the proof of (9.6).□

## Section II: $A = G_2$ and $p = 3$.

Let $(A,p) = (G_2,3)$ and $\Pi(A) = \{\alpha,\beta\}$. Choose $P_A$ such that $L_A = \langle U_{\pm\beta}\rangle T_A$ and $\langle\lambda,\beta\rangle \neq 0$. Note that $I_\alpha$ is a 2-dimensional irreducible $L_A'$ module. (See (2.2) for the definition of $I_\alpha$.) Adopt the remaining notation of (2.0). Note that $\dim V|A = 7^k \cdot 27^\ell$, for $k,\ell \in \mathbb{Z}^+$.

<u>(9.7).</u> If $M_i$ is nontrivial, $L_i$ has type $A_1$ or $A_2$.

<u>Proof:</u> Suppose false. Then (7.1) of [12] and rank restrictions imply

that all components of $L_Y'$ have classical type. Hence, by (1.5), $Z_A \leq Z_Y$. We first note that $L_i$ has type $A_{k_i}$ for some $k_i$. For otherwise, (7.1) of [12] implies that $L_i$ has type $D_6$ and $M_i$ is isomorphic to the natural module for $L_i$. As well, the bound on $\dim V_{\beta_1}(Q_Y)$ and $\dim V_{\beta_8}(Q_Y)$ implies $Y = E_7$ and $\lambda|T_Y = \lambda_7$. But then $\dim V|A > \dim V|Y$. Thus, $L_i$ has type $A_{k_i}$ as claimed. Also note that $L_i$ does not have type $A_7$. For otherwise the bound on $\dim V_{\beta_2}(Q_Y)$ implies $\lambda|T_Y = \lambda_8$.

Suppose $L_i$ has type $A_5$. If $Y = E_6$, the bound on $\dim V_{\beta_2}(Q_Y)$ implies $\lambda|T_Y = \lambda_1$ or $\lambda_6$ and $\dim V|Y = 27$. But $\dim V|A > 27$. Hence, $Y = E_7$ or $E_8$. Note that $L_A'$ acts irreducibly on W, the natural module for $L_i$, so there does not exist $\gamma \in \Pi(Y) - \Pi(L_Y)$ with $(\gamma, \Sigma L_i) \neq 0$ and $Q_Y/K_\gamma \cong W$ or $W^*$. Hence, if $Y = E_7$, $\Pi(L_i) = \{\beta_k \mid k \neq 1,3\}$. The bound on $\dim V_{\beta_3}(Q_Y)$ implies $\lambda|T_Y = \lambda_2$ or $\lambda_7$. However, then [8] implies $\dim V|A \neq \dim V|Y$. Thus, $Y = E_8$ and $\Pi(L_Y) = \{\beta_k \mid k \neq 2,7\}$. But (1.15) implies $Q_A \leq K_{\beta_7}$, contradicting (2.3).

The above remarks imply $L_i$ has type $A_3$. If W is the natural module for $L_i$, $W|L_A'$ is a 4-dimensional tensor decomposable irreducible and $M_i \cong W$ or $W^*$. If $M_i|L_A'$ has high weight $(q_1+q_2)\mu_\beta$, the $L_A'$ composition factors of $W \wedge W$ have high weights $2q_1\mu_\beta$ and $2q_2\mu_\beta$. Hence, there does not exist $\delta \in \Pi(Y) - \Pi(L_Y)$ with $(\delta, \Sigma L_i) \neq 0$ and $Q_Y/K_\delta \cong W, W^*$, or $W \wedge W$. So if $\Pi(L_i) = \{\beta_2, \beta_4, \beta_5\}$, then $Y = E_7$ or $E_8$ and $\langle U_{\pm\beta_1}, U_{\pm\beta_7}\rangle \leq L_Y$. The above remarks about $W \wedge W$, (2.5) and (2.7) imply $\langle \lambda, \beta_1+\beta_7\rangle = 0$. Also, if $\langle U_{\pm\beta_8}\rangle \leq L_Y$, $\langle \lambda, \beta_8\rangle = 0$. The bound on $\dim V_{\beta_3}(Q_Y)$ and $\dim V_{\beta_6}(Q_Y)$ implies $\lambda|T_Y = \lambda_2 + x\lambda_8$ or $\lambda_5 + x\lambda_8$, where $x \in \{0,1,2\}$. Now, [8] implies $\dim V|A \neq \dim V|Y$ if $Y = E_7$. If $x = 0$, the work of this result and (1.38) imply $\dim V|A \leq 7^4 < \dim V|Y$. Hence $x \neq 0$. But then previous work and (1.32) imply $\dim V|A \leq 7^6 < \dim V|Y$. So, $\Pi(L_i) \neq \{\beta_2, \beta_4, \beta_5\}$.

Consider now the case where $Y = E_8$ and $\Pi(L_i) = \{\beta_6, \beta_7, \beta_8\}$. Then (1.15) and previous remarks imply $\langle U_{\pm\beta_4}\rangle$ is a component of $L_Y'$. By (2.5) and (2.7), $\langle \lambda, \beta_4\rangle = 0$. The bound on $\dim V_{\beta_5}(Q_Y)$ implies $\langle \lambda, \beta_5\rangle = 0$.

If $\langle U_{\pm\beta_1}\rangle \nleq L_Y$, then $\langle\lambda,\beta_1+\beta_3\rangle = 0$, by (2.3) and (2.13). If $\langle U_{\pm\beta_1}\rangle \leq L_Y$, (1.15) implies $Q_A \leq K_{\beta_3}$ and again $\langle\lambda,\beta_1+\beta_3\rangle = 0$. Hence, $\lambda|T_Y = x\lambda_2+\lambda_j$, where $x\in\{0,1,2\}$ and $j = 6$ or $8$. If $\lambda|T_Y = \lambda_2+\lambda_6$, $\lambda_2+\lambda_8$, $2\lambda_2+\lambda_6$ or $2\lambda_2+\lambda_8$, $\dim V|A \leq 7^6 < \dim V|Y$, by (1.38). If $\lambda|T_Y = \lambda_6$, $\dim V|A \leq 7^4 < \dim V|Y$ by (1.38). Hence $\Pi(L_i) \neq \{\beta_6,\beta_7,\beta_8\}$.

Finally, consider the case where $Y = E_7$ and $\Pi(L_i) = \{\beta_5,\beta_6,\beta_7\}$. Applying (1.15), we see that $\Pi(L_Y) = \Pi(L_i) \cup \mathcal{S}$, where $\mathcal{S}$ is (a) $\{\beta_3\}$, (b) $\{\beta_2\}$, (c) $\{\beta_2,\beta_1\}$, or (d) $\{\beta_2,\beta_1,\beta_3\}$. In (d), (2.5), (2.7) and the bound on $\dim V_{\beta_4}(Q_Y)$ imply $\lambda|T_Y = \lambda_7$ or $\lambda_1+\lambda_7$. But then [8] implies $\dim V|A \neq \dim V|Y$. In each of the remaining cases, we consider the action of $L_A'$ on the 56-dimensional irreducible $kY$ module, $V(\lambda_7)$. With $q_1$ and $q_2$ as above, we may assume that the field twist on the embedding of $L_A'$ in $\langle U_{\pm\beta_k}\rangle$ is $q_1$, where $k = 3$ in (a) and $k = 2$ in (b) and (c). In case (c), we find that the field twist on the embedding of $L_A'$ in $\langle U_{\pm\beta_1}\rangle$ is either $q_1$ or $q_2$, else there is an 8-dimensional $L_A'$ composition factor of $V(\lambda_7)$ with high weight $\lambda-\beta_2-\beta_3-2\beta_4-\beta_5-\beta_6-\beta_7|T(L_A')$. But there is no 56-dimensional $kA$ module affording such an $L_A'$ compositon factor. Now, in cases (a) – (c), there is a 6-dimensional $L_A'$ composition factor of $V(\lambda_7)$. This implies that there is a 49-dimensional $A$ composition factor of $V(\lambda_7)$. Also, there are less than seven 1-dimensional $L_A'$ composition factors, so there must also be a 7-dimensional $A$ composition factor of $V(\lambda_7)$. But this contradicts the following information about $V(\lambda_7)$: If (a) holds, there are four trivial $L_A'$ composition factors of $V(\lambda_7)$; if (b) holds, there are eight 4-dimensional $L_A'$ composition factors; if (c) holds, there are either exactly 2 trivial $L_A'$ composition factors or exactly six 4-dimensional $L_A'$ composition factors.

This completes the proof of (9.7).□

<u>Remarks (9.8).</u> Assume $Z_A \leq Z_Y$. (1) If $i\neq j$ with $M_i$ and $M_j$ nontrivial and $(\Sigma L_i,\gamma) \neq 0 \neq (\Sigma L_j,\gamma)$ for some $\gamma\in\Pi(Y) - \Pi(L_Y)$, then (9.7)

and (2.5) – (2.7) imply $\gamma = \beta_4$ and $\{\beta_2,\beta_3,\beta_5\} \subset \Pi(L_Y)$.

(2) Suppose $\langle U_{\pm\beta_\ell}\rangle$ and $\langle U_{\pm\beta_m}\rangle$ are components of $L_Y'$ and $\gamma, \delta \in \Pi(Y) - \Pi(L_Y)$ with $(\beta_\ell,\gamma) \neq 0 \neq (\gamma,\delta) \neq 0 \neq (\delta,\beta_m)$ and $\dim(Q_Y/K_\gamma) = 2 = \dim(Q_Y/K_\delta)$. If $Q_A \nleq K_\gamma$ and $Q_A \nleq K_\delta$, then the field twists on the embeddings of $L_A'$ in $\langle U_{\pm\beta_\ell}\rangle$ and $\langle U_{\pm\beta_m}\rangle$ are equal. For one may check that there is a nontrivial contribution to some root group of $Q_Y(\gamma,\delta)$ in the factorization of nonidentity elements of $Q_A'$. Since $Z_A \leq Z_Y$ and $Q_A' \leq K_\gamma \cap K_\delta$, the image of $Q_A'$ in $Q_Y(\gamma,\delta)$ is a 1-dimensional $L_A'$ submodule of $Q_Y(\gamma,\delta)$. Such a submodule can exist only if the twists are equal.

(3) If $L_\ell$ and $L_m$ have type $A_1$ with $(\Sigma L_\ell,\gamma) \neq 0 \neq (\Sigma L_m,\gamma)$ for some $\gamma \in \Pi(Y) - \Pi(L_Y)$ and $\dim(Q_Y/K_\gamma) = 4$, then (1.15) implies that $Q_A \leq K_\gamma$.

<u>(9.9).</u> Suppose $M_i$ is nontrivial and $L_i$ has type $A_2$. Then $Y$ has type $E_6$, $\lambda|T_Y = \lambda_1$ or $\lambda_6$ and $\lambda|T_A = 2\mu_1$.

<u>Proof:</u> Since all components of $L_Y'$ are necessarily of classical type, $Z_A \leq Z_Y$. Also, there does not exist $\gamma \in \Pi(Y) - \Pi(L_Y)$ with $(\gamma,\Sigma L_i) \neq 0$ and $(\gamma,\Sigma L_k) = 0$ for $k \neq i$. For otherwise, $Q_Y/K_\gamma$ is a 3-dimensional irreducible $L_A'$ module, contradicting (2.4).

Consider first the case where $\Pi(L)_i = \{\beta_2,\beta_4\}$. The above remarks imply $\langle U_{\pm\beta_1}\rangle$ is a component of $L_Y'$ and $\langle U_{\pm\beta_6}\rangle \leq L_Y$. In fact, p=3 and (1.15) imply $\langle U_{\pm\beta_6}\rangle$ is a component of $L_Y'$. By (2.6) and (2.7), the field twists on the embeddings of $L_A'$ in $L_i$, $\langle U_{\pm\beta_1}\rangle$ and $\langle U_{\pm\beta_6}\rangle$ are all equal. Hence, (2.5) implies $\langle \lambda,\beta_1+\beta_6\rangle = 0$. If $\langle U_{\pm\beta_8}\rangle$ is a component of $L_Y'$, (9.8) (3) implies that $\langle \lambda,\beta_7+\beta_8\rangle = 0$. Otherwise, (2.3) and (2.13) imply the same conclusion. Note that $\langle \lambda,\beta_3+\beta_5\rangle \neq 0$. For otherwise, (9.7), the work of this result so far, [8] and (1.38) imply $\dim V|A \neq \dim V|Y$. Since $\langle \lambda,\beta_3\rangle \neq 0$ or $\langle \lambda,\beta_5\rangle \neq 0$, the bound on $\dim V_{\beta_k}(Q_Y)$, for k = 3 or 5, implies $\langle \lambda,\beta_4\rangle = 1$ and $\langle \lambda,\beta_2\rangle = 0$. Moreover, (1.35) implies $\langle \lambda,\beta_k\rangle = 1$ when $\langle \lambda,\beta_k\rangle \neq 0$, for k = 3 or 5. If $\lambda|T_Y = \lambda_3+\lambda_4$ or $\lambda_4+\lambda_5$,

$\dim V|A \le 27^2$. Counting only the conjugates of $V_{T_Y}(\lambda)$ in $V|Y$, we have $\dim V|A < \dim V|Y$. A similar argument rules out $\lambda|T_Y = \lambda_3+\lambda_4+x\lambda_7$, $\lambda_4+\lambda_5+x\lambda_7$ and $\lambda_3+\lambda_4+\lambda_5+x\lambda_7$ when $x \ne 0$ and $\lambda|T_Y = \lambda_3+\lambda_4+\lambda_5$ when $Y = E_7$ or $E_8$. Finally, if $Y = E_6$ and $\lambda|T_Y = \lambda_3+\lambda_4+\lambda_5$, $\dim V|A \le 27^3 < \dim V|Y$, by (1.38). Hence, $\Pi(L_i) \ne \{\beta_2,\beta_4\}$. Similar arguments show that $\Pi(L_i) \ne \{\beta_6,\beta_7\}$.

<u>Claim:</u> If $\langle U_{\pm\beta_2}\rangle$, $\langle U_{\pm\beta_1},U_{\pm\beta_3}\rangle$ and $\langle U_{\pm\beta_5},U_{\pm\beta_6}\rangle$ are components of $L_Y'$, then the statement of (9.9) holds.

Reason: First note that $Y \ne E_7$, else previous remarks and the bound on $\dim V_{\beta_4}(Q_Y)$ imply $\lambda|T_Y = \lambda_1$. Now $\Pi(L_i) = \{\beta_1,\beta_3\}$ or $\{\beta_5,\beta_6\}$. Suppose $M_j$ is nontrivial for some $j \ne i$ with $(\Sigma L_j,\beta_4) \ne 0$. Then (2.5), (2.7) and the bound on $\dim V_{\beta_4}(Q_Y)$ imply $\langle\lambda,\beta_1\rangle = 1 = \langle\lambda,\beta_6\rangle$ and $\langle\lambda,\beta_k\rangle = 0$ for $2 \le k \le 5$. If $Y = E_6$, $\dim V|A \ne \dim V|Y$, by [8]. Previous remarks then imply $Y = E_8$ and $\langle U_{\pm\beta_8}\rangle$ is a component of $L_Y'$. By (2.7), $\langle\lambda,\beta_8\rangle = 0$. The bound on $\dim V_{\beta_7}(Q_Y)$ and (1.35) imply that if $\langle\lambda,\beta_7\rangle \ne 0$, then $\langle\lambda,\beta_7\rangle = 1$. In any case, $\dim V|A \le 27^4 < \dim V|Y$, by (1.32). Thus, $\langle\lambda,\beta_1+\beta_2+\beta_3+\beta_5+\beta_6\rangle = 1$. The bound on $\dim V_{\beta_4}(Q_Y)$ implies that $\langle\lambda,\beta_4\rangle = 0$ and $V_{L_i}(-\beta_4) \cong M_i^*$.

Suppose $Y = E_8$. If $\langle\lambda,\beta_6\rangle = 1$, $\langle U_{\pm\beta_8}\rangle$ is a component of $L_Y'$, and by (2.7), $\langle\lambda,\beta_8\rangle = 0$. So $\dim V|A \le 27^2 < \dim V|Y$ by (1.32). Thus, $\langle\lambda,\beta_1\rangle = 1$. If $\langle\lambda,\beta_7+\beta_8\rangle \ne 0$, previous remarks, the bound on $\dim V_{\beta_7}(Q_Y)$ and (1.35) imply $\langle U_{\pm\beta_8}\rangle$ is a component of $L_Y'$ and $\langle\lambda,\beta_8\rangle \ne 0$. If $\langle\lambda,\beta_8\rangle = 2$, $\dim V|A \le 27^4 < \dim V|Y$ by (1.38). If $\langle\lambda,\beta_8\rangle = 1$, the bound on $\dim V_{\beta_7}(Q_Y)$ implies $\langle\lambda,\beta_7\rangle = 0$, and then $\dim V|A \le 7\cdot 27^3 < \dim V|Y$, by (1.38).

Thus, $Y = E_6$, $\dim V|Y = 27$; so $\langle\lambda,\alpha\rangle = 0$. Hence by (1.10), $\lambda|T_A = 2\mu_1$ or $2q\mu_2$. If $\lambda|T_A = 2q\mu_2$, (4.1) of [12] implies that there is a closed subgroup $B < A$ of type $A_2$ such that $V|B$ is irreducible. Moreover, (1.10) and knowledge of the embedding of B in A implies that $q = 1$. Thus, V is a restricted kA module and hence irreducible for L(A), by (1.1) But now, the proof of (11.1) in [15] shows that the span of the $e_r$ and $f_r$, for short roots $r \in \Sigma(A)$, is a noncommutative ideal which lies in the kernel of the action of

L(A) on V. But the action of L(Y) on V has no such ideal in its kernel. Hence, $\lambda|T_A = 2\mu_1$, and the Claim holds.

Consider now the case where $\Pi(L_i) = \{\beta_1,\beta_3\}$. By the preceding Claim, previous remarks and (1.15), we may assume that either $L_Y'$ has component $\langle U_{\pm\beta_j}\rangle$ and $\langle U_{\pm\beta_k}\rangle \nleq L_Y'$ for {j,k} = {2,5} or $\Pi(L_Y) = \{\beta_1,\beta_3,\beta_2,\beta_5,\beta_6,\beta_7\}$. In either case, $\langle\lambda,\beta_2\rangle = 0$; as well, in the first case $\langle\lambda,\beta_j\rangle = 0$ and in the second case $\langle\lambda,\beta_5+\beta_6+\beta_7\rangle = 0$. (We have used (9.7) and the bound on $\dim V_{\beta_4}(Q_Y)$.) Now $Y \neq E_6$; else (1.23) and previous remarks imply that $\lambda|T_Y = \lambda_1+\lambda_6$, $\lambda_3+\lambda_5$, or $\lambda_3+\lambda_4+\lambda_5$. But by [8], $\dim V|A \neq \dim V|Y$ in the first case and in the latter cases, $\dim V|A \leq 27\cdot 7^2 < \dim V|Y$, by (1.38).

Suppose $\Pi(L_i) = \{\beta_1,\beta_3\}$ and $Y = E_7$. We first note that $\langle\lambda,\beta_5+\beta_6+\beta_7\rangle \neq 0$. For otherwise, the previous work of this result implies $\lambda|T_Y = \lambda_1$, $\lambda_3$ or $\lambda_3+\lambda_4$. But [8] rules out the first two possibilities and in the last case, $\dim V|A \leq 27^2 < \dim V|Y$ by (1.38). Now examining all possible configurations, using previous work, (2.3), (9.8) and (2.13), we see that either (a) $\Pi(L_Y) = \{\beta_1,\beta_3,\beta_2,\beta_6\}$, or (b) $\Pi(L_Y) = \{\beta_1,\beta_3,\beta_2,\beta_7\}$ and $\langle\lambda,\beta_7\rangle \neq 0$. Consider the action of $L_A'$ on the 56-dimensional irreducible kY module, $V(\lambda_7)$. In case (a), if the field twist on the embedding of $L_A'$ in $\langle U_{\pm\beta_6}\rangle$ is not equal to the twist on the embedding in $\langle U_{\pm\beta_2}\rangle$ (and $L_i$), there are exactly two 6-dimensional and two 4-dimensional $L_A'$ composition factors of $V(\lambda_7)$. In case (b), the $L_A'$ composition series of $V(\lambda_7)$ has the same properties. But there is no 56-dimensional kA module affording such an $L_A'$ composition series. Hence, (2.5), (2.6), and (2.7) imply that $L_Y'$ is as in (a) with $\langle\lambda,\beta_6\rangle = 0$. So $\langle\lambda,\beta_5+\beta_7\rangle \neq 0$ and $\dim V|A \leq 7\cdot 27^2$. (We have used (9.8).) But [8], (1.32) and (1.38) then imply $\dim V|A \neq \dim V|Y$.

Now suppose $\Pi(L_i) = \{\beta_1,\beta_3\}$ and $Y = E_8$. Note that if $L_Y'$ has component $\langle U_{\pm\beta_5},U_{\pm\beta_6},U_{\pm\beta_7}\rangle$ (respectively, $\langle U_{\pm\beta_6},U_{\pm\beta_7},U_{\pm\beta_8}\rangle$, $\langle U_{\pm\beta_6},U_{\pm\beta_7}\rangle$), then $\langle\lambda,\beta_5+\beta_6+\beta_7+\beta_8\rangle = 0$, as $Q_Y/K_{\beta_8}$ (respectively,

$Q_Y/K_{\beta_5}$, $Q_Y/K_{\beta_8}$) is a 3- or 4-dimensional irreducible $L_A'$ module. If $\langle U_{\pm\beta_7}, U_{\pm\beta_8}\rangle$ is a component of $L_Y'$, $\langle\lambda,\beta_5+\beta_6+\beta_7+\beta_8\rangle = 0$, also. For if $\langle U_{\pm\beta_5}\rangle \nleq L_Y'$, we may argue as above. Otherwise, use (2.5) – (2.7) and the bound on $\dim V_{\beta_6}(Q_Y)$. But if $\langle\lambda,\beta_5+\beta_6+\beta_7+\beta_8\rangle = 0$, $\dim V|A \leq 27^2 < \dim V|Y$, by [8] and (1.32).

Now, if $\lambda|T_Y = \lambda_1+\lambda_8$ or $\lambda_3+x\lambda_4+\lambda_8$, for $x \in \{0,1\}$, $\dim V|A \leq 27^2 \cdot 7^2 < \dim V|Y$, by (1.38) and (1.32). If $\lambda|T_Y = \lambda_1+2\lambda_8$ or $\lambda_3+x\lambda_4+2\lambda_8$, x as above, then $\dim V|A \leq 27^4 < \dim V|Y$, by (1.38). Also, if $\lambda|T_Y = \lambda_1+z\lambda_5$ or $\lambda_3+x\lambda_4+z\lambda_5$ with $0 < z \leq 2$ and x as above, $\dim V|A \leq 27^2 \cdot 7 < \dim V|Y$, by (1.32). Applying the restrictions imposed on $\lambda|T_Y$ by (2.3), (2.13), (9.8) and the above remarks, we find that one the following holds:

(a) $L_Y' = L_i \times \langle U_{\pm\beta_5}\rangle \times \langle U_{\pm\beta_8}\rangle$ and $\langle\lambda,\beta_6+\beta_7\rangle \neq 0$,

(b) $L_Y' = L_i \times \langle U_{\pm\beta_2}\rangle \times \langle U_{\pm\beta_6}\rangle$ and $\langle\lambda,\beta_6\rangle \neq 0 = \langle\lambda,\beta_7+\beta_8\rangle$, or

(c) $L_Y' = L_i \times \langle U_{\pm\beta_2}\rangle \times \langle U_{\pm\beta_7}\rangle$ and $\langle\lambda,\beta_7\rangle \neq 0 = \langle\lambda,\beta_5+\beta_6\rangle$.

In case (a), (9.8), (2.5) and (2.7) imply that at most one of $\langle\lambda,\beta_6\rangle$ and $\langle\lambda,\beta_8\rangle$ is nonzero. If $\langle\lambda,\beta_8\rangle \neq 0$, (so $\langle\lambda,\beta_7\rangle \neq 0$) $\dim V|A \leq 27^4$. But counting only the conjugates of $V_{T_Y}(\lambda)$ and $V_{T_Y}(\lambda-\beta_7-\beta_8)$ in $V|Y$, we have $\dim V|Y > 27^4$. So $\langle\lambda,\beta_8\rangle = 0$ and $\dim V|A \leq 27^3 < \dim V|Y$ by (1.32). In case (b), $\dim V|A \leq 27^4 < \dim V|Y$ by (1.32). In case (c), if $\langle\lambda,\beta_7\rangle = 1$, $\dim V|A \leq 27^3 \cdot 7$; if $\langle\lambda,\beta_7\rangle = 2$, $\dim V|A \leq 27^4$. But we need only count the conjugates of $V_{T_Y}(\lambda)$ when $\langle\lambda,\beta_7\rangle = 1$, and in addition, the conjugates of $V_{T_Y}(\lambda-\beta_7)$ when $\langle\lambda,\beta_7\rangle = 2$, to see that $\dim V|Y > \dim V|A$. Hence, we have shown that if $L_i = \langle U_{\pm\beta_1}, U_{\pm\beta_3}\rangle$, then the result holds.

Consider the case where $\Pi(L_i) = \{\beta_5,\beta_6\}$. By symmetry and previous work of this result, we may assume $Y = E_8$ and $\langle U_{\pm\beta_j}\rangle$ is a component of $L_Y'$ with $\langle\lambda,\beta_j\rangle = 0$, for j = 2 or 3, and $\langle U_{\pm\beta_2}, U_{\pm\beta_3}\rangle \nleq L_Y'$. Also, $\langle U_{\pm\beta_8}\rangle \leq L_Y'$ and $\langle\lambda,\beta_8\rangle = 0$. If $\langle U_{\pm\beta_3}\rangle$ is a component of $L_Y'$, $\langle\lambda,\beta_2\rangle = 0$. Moreover, earlier remarks which apply to the bound on $\dim V_{\beta_k}(Q_Y)$ for k = 4 or 7, imply $\lambda|T_Y = y\lambda_1+x\lambda_4+\lambda_5$ or $y\lambda_1+\lambda_6+x\lambda_7$, where $x \in \{0,1\}$ and $0 \leq y \leq 2$.

If $y = 0$, $\dim V|A \leq 27^2$ and if $y \neq 0$, $\dim V|A \leq 27^3$. But counting only the conjugates of $V_{T_Y}(\lambda)$ in $V|Y$, we have $\dim V|Y > \dim V|A$. Thus, $\langle U_{\pm\beta_2}\rangle$ is a component of $L_Y'$. Also, $\lambda|T_Y = y\lambda_1+z\lambda_3+x\lambda_4+\lambda_5$ or $y\lambda_1+z\lambda_3+\lambda_6+x\lambda_7$, where $x$ and $y$ are as above and $0 \leq z \leq 2$. If $y = 0 = z$, $\dim V|A \leq 27^2$, if $y = 0 \neq z$, $\dim V|A \leq 27^3$ and otherwise, $\dim V|A \leq 27^4$. But in each case (1.32) implies that $\dim V|A < \dim V|Y$. Hence, $L_i \neq \langle U_{\pm\beta_5}, U_{\pm\beta_6}\rangle$.

Finally, consider the case where $\Pi(L_i) = \{\beta_7,\beta_8\}$. Previous remarks, the bound on $\dim V_{\beta_6}(Q_Y)$ and (1.15) imply $\langle U_{\pm\beta_5}\rangle$ is a component of $L_Y'$ with $\langle\lambda,\beta_5\rangle = 0$. We first claim that there exists $j\neq i$ with $M_j$ nontrivial. For suppose not. Then, (a) if $\langle\lambda,\beta_k\rangle = 0$ for $1\leq k\leq 4$, $\dim V|A \leq 27^2$, (b) if $\langle\lambda,\beta_1+\beta_3+\beta_4\rangle > 0$ and $\langle\lambda,\beta_2\rangle = 0$ or if $\langle\lambda,\beta_2\rangle \neq 0$ and $\langle\lambda,\beta_1+\beta_3+\beta_4\rangle = 0$, $\dim V|A \leq 27^3$, and (c) $\dim V|A \leq 27^4$ otherwise. (We have used (9.8) and the previous work of this result.) Recall that $\lambda|T_Y \neq \lambda_8$. In (a) and (b), we need only count the conjugates of $V_{T_Y}(\lambda)$ in $V|Y$ to see that $\dim V|Y > \dim V|A$. For (c), we may assume $\langle\lambda,\beta_1+\beta_3+\beta_4\rangle > 0$ and $\langle\lambda,\beta_2\rangle > 0$. But again, as in (a) and (b) we show that $\dim V|A < \dim V|Y$. Thus, there exists a $j\neq i$ with $M_j$ nontrivial.

The previous work of this result and (9.8) imply that either (a) $\Pi(L_Y) = \{\beta_1,\beta_2,\beta_5,\beta_7,\beta_8\}$, (b) $\Pi(L_Y) = \{\beta_1,\beta_5,\beta_7,\beta_8\}$ or (c) $\Pi(L_Y) = \{\beta_2,\beta_3,\beta_5,\beta_7,\beta_8\}$. If (a) or (b) holds, (2.3), (2.13) and (9.8) imply that $\langle\lambda,\beta_2+\beta_4\rangle = 0$. So $\lambda|T_Y = c\lambda_1+x\lambda_3+\lambda_8$ or $c\lambda_1+x\lambda_3+y\lambda_6+\lambda_7$ for $c\in\{1,2\}$, $0\leq x\leq 2$ and $y\in\{0,1\}$. Moreover, $\dim V|A \leq 7\cdot 27^3$ if $c=1$ and $\dim V|A \leq 27^4$ if $c=2$. In each case, (1.38) and (1.32) imply $\dim V|A < \dim V|Y$. Suppose $L_Y'$ is as in (c). By (2.7), the set of field twists on the embeddings of $L_A'$ in $\langle U_{\pm\beta_k}\rangle$ for $k=2,3,5$ consists of at most 2 distinct primes. Recall that the field twist on the embedding of $L_A'$ in $\langle U_{\pm\beta_5}\rangle$ equals that on the embedding in $L_i$. And since $\langle\lambda,\beta_2+\beta_3\rangle \neq 0$, (2.5) implies that there are exactly two distinct field twists on the embeddings of $L_A'$ in the triple of $A_1$'s, $\langle U_{\pm\beta_k}\rangle$, $k=2,3,5$, and exactly one of $\langle\lambda,\beta_2\rangle$ and $\langle\lambda,\beta_3\rangle$ is nonzero. Moreover, the bound on $\dim V_{\beta_4}(Q_Y)$ implies $\langle\lambda,\beta_4\rangle = 0$.

So in case (c), $\lambda|T_Y = x\lambda_1+c\lambda_k+\lambda_8$ or $x\lambda_1+c\lambda_k+y\lambda_6+\lambda_7$ for $k = 2$ or $3$, $0\le x\le 2$, $c\in\{1,2\}$ and $y\in\{0,1\}$. Referring to cases (a) and (b), (9.8) and the previous work of this result, we see that $\dim V|A \le 7^3\cdot 27^2$ if $c=1$ and $\dim V|A \le 27^4\cdot 7$ if $c=2$. Again (1.32) and (1.38) imply $\dim V|A < \dim V|Y$. Hence, $L_i \ne \langle U_{\pm\beta_7}, U_{\pm\beta_8}\rangle$.

This completes the proof of (9.9).□

<u>(9.10).</u> Let $\gamma\in\Pi(Y)-\Pi(L_Y)$ and $i\ne j$ such that $(\Sigma L_i,\gamma)\ne 0\ne(\gamma,\Sigma L_j)$. Then $M_i$ or $M_j$ is trivial.

<u>Proof:</u> Suppose false; i.e., suppose $M_i$ and $M_j$ are both nontrivial. Then (9.7) – (9.9) imply that $L_i$ and $L_j$ each have type $A_1$ and there exists $k\ne i,j$ with $(\Sigma L_k,\gamma)\ne 0$. Moreover, (1.15), $p=3$ and the minimality of $P_Y$ imply that $L_k$ has type $A_1$ also. Let $\beta_i$, $\beta_i$, $\beta_k$ be such that $L_\ell = \langle U_{\pm\beta_\ell}\rangle$ for $\ell = i,j,k$. So $\langle\lambda,\beta_i\rangle\ne 0\ne\langle\lambda,\beta_j\rangle$. Let $q_\ell$ be the field twist on the embedding of $L_A{}'$ in $L_\ell$ for $\ell = i,j,k$. Then we may assume $q_k = q_j \ne q_i$. The $L_A{}'$ composition factors of $Q_Y/K_\gamma$ have high weights $(2q_j+q_i)\mu_\beta$ and $q_i\mu_\beta$. Hence, $\dim(Q_AK_\gamma/K_\gamma) = 2$. Now since $V_{T_Y}(\lambda-\beta_i-\gamma)\ne 0$, a nonidentity element from the set $U_{-\gamma}\cdot U_{-\gamma-\beta_i}$ must occur in the factorization of some element of $Q_AK_\gamma/K_\gamma$. However, $\gamma$ (respectively, $-\gamma-\beta_i$) affords $T(L_A{}')$ weight $(q_i+2q_j)\mu_\beta$ (respectively, $(2q_j-q)\mu_\beta$). While the weights in $Q_AK_\gamma/K_\gamma$ are $q_i\mu_\beta$ and $-q_i\mu_\beta$, contradicting (2.4). This completes the proof of (9.10).□

<u>(9.11).</u> If $(A,Y,V)$ is an example in the Main Theorem, with $(A,p) = (G_2,3)$ and $Y$ of type $E_n$, then $\lambda|T_Y = \lambda_1$ or $\lambda_6$ and $\lambda|T_A = 2\mu_1$.

<u>Proof:</u> Suppose $\langle\lambda,\gamma\rangle = 0$ for some $\gamma\in\Pi(A)$. Then (4.1) of [12] implies that there exists $B < A$ of type $A_2$ such that $V|B$ is irreducible. Then (6.0) implies that $Y = E_6$, $\lambda|T_Y = \lambda_1$ or $\lambda_6$ and $\lambda|T_A = 2q\mu_1$ or $2q\mu_2$ for some p-power q. Hence, the hypotheses of (9.9) hold (for some choice of $\beta\in\Pi(A)$), and we have the result of (9.11). Thus, we may assume

$\langle\lambda,\gamma\rangle \neq 0$ for all $\gamma \in \Pi(A)$ and that the hypotheses of (9.9) do not hold. Thus, for any choice of $\beta \in \Pi(A)$ with $P_A$ and $P_Y$ as before, if $M_i$ is nontrivial then $L_i$ has type $A_1$. So we will choose $\beta \in \Pi(A)$ such that the number of p-powers in the p-adic expansion of $\langle\lambda,\beta\rangle$ is as large as possible. And we note that there must be at least two such p-powers; i.e., $V^1(Q_A)$ is tensor decomposable. Otherwise, $\langle\lambda,\beta\rangle = q$ and $\dim V|A = 49$ or 189 or $\langle\lambda,\beta\rangle = 2q$, $\langle\lambda,\beta_\ell\rangle = 2$ for some $1 \leq \ell \leq \mathrm{rank} Y$ and $\dim V|A = 189$ or 729. But in each case, [8] and (1.32) imply $\dim V|A \neq \dim V|Y$. Thus, there exists $1 \leq k \neq \ell \leq \mathrm{rank} Y$ with $(\beta_k,\beta_\ell) = 0$ and $\langle\lambda,\beta_k\rangle \neq 0 \neq \langle\lambda,\beta_\ell\rangle$.

Suppose $Y = E_6$. The opening remarks of the proof, (9.8), (9.10) and symmetry imply that either $\Pi(L_Y) = \{\beta_1,\beta_2,\beta_5,\beta_6\}$ with $\langle\lambda,\beta_5+\beta_6\rangle = 0$ and $\langle\lambda,\beta_1\rangle \neq 0 \neq \langle\lambda,\beta_2\rangle$, or $L_Y'$ has components $\langle U_{\pm\beta_1}\rangle$ and $\langle U_{\pm\beta_6}\rangle$ with $\langle\lambda,\beta_k\rangle \neq 0$ for $k = 1$ and 6. The first configuration is ruled out by (1.23). So suppose $L_Y'$ has components $\langle U_{\pm\beta_k}\rangle$ with $\langle\lambda,\beta_k\rangle \neq 0$ for $k = 1$ and 6. If $\langle U_{\pm\beta_2}\rangle$ is also a component of $L_Y'$, (9.8) implies that $Q_A \leq K_{\beta_4}$; else the field twists on the embeddings of $L_A'$ in $\langle U_{\pm\beta_1}\rangle$, $\langle U_{\pm\beta_2}\rangle$ and $\langle U_{\pm\beta_6}\rangle$ are all equal, contradicting (2.5) and (2.6). But $-\beta_4$ is not involved in $L_A'$, by (2.10). So there is a nontrivial image of $Q_A{}^\alpha$ in $Q_Y(\beta_3,\beta_4)$ and in $Q_Y(\beta_5,\beta_4)$ again implying that the field twists are equal. Hence, (2.7) implies $\Pi(L_Y) = \{\beta_1,\beta_6\}$. Let $q_k$ be the field twist on the embedding of $L_A'$ in $\langle U_{\pm\beta_k}\rangle$. Now, there are two 4-dimensional $L_A'$ composition factors of the 27-dimensional $kY$- module $V(\lambda_1)$. However, there is no 27-dimensional $kA$ module affording such an $L_A'$ composition series. Hence $Y \neq E_6$.

Suppose $Y = E_8$. Choose i and j such that $M_i$ and $M_j$ are nontrivial. So $L_i$ and $L_j$ each have type $A_1$ and one of the following holds:

(a) $\Pi L_i$ and $\Pi L_j$ are separated by more than 2 nodes of the Dynkin diagram.

(b) $\{\Pi L_i,\Pi L_j\} = \{\beta_1,\beta_2\}$ and $\langle U_{\pm\beta_5},U_{\pm\beta_6}\rangle \leq L_Y'$.

(c) $\{\Pi L_i,\Pi L_j\} = \{\beta_2,\beta_6\}$ and $\langle U_{\pm\beta_1},U_{\pm\beta_3}\rangle \leq L_Y'$.

This follows from (9.8), (9.10) and (1.15). Rank restrictions, (9.8) and (9.10) imply that if there exists $k \neq i, j$ with $M_k$ nontrivial, then (b) holds and $L_Y' = L_i \times L_j \times \langle U_{\pm\beta_5}, U_{\pm\beta_6}\rangle \times \langle U_{\pm\beta_8}\rangle$, with $\langle \lambda, \beta_8\rangle \neq 0$. But (2.7) implies that the field twists on the embeddings of $L_A'$ in $\langle U_{\pm\beta_2}\rangle$, $\langle U_{\pm\beta_5}, U_{\pm\beta_6}\rangle$ and $\langle U_{\pm\beta_8}\rangle$ must all be equal, contradicting (2.5). Hence, $V^1(Q_Y) \cong M_i \otimes M_j$, and by the choice of $\beta$, $\dim V|A \leq 27^4$.

Now we may assume $\lambda|T_Y = c\lambda_r + d\lambda_t$ for $0 < c,d < 3$. For if $W_0$, the stabilizer of $\lambda$ in W has rank at most 5, the number of distinct conjugates of $V_{T_Y}(\lambda)$ in V|Y exceeds $27^4$, unless $W_0$ has type $D_5$. But if we count also the conjugates of a maximal subdominant weight we find that $\dim V|Y > 27^4$. So $\lambda|T_Y = c\lambda_r + d\lambda_t$ and by (9.10), $(\beta_r, \beta_t) = 0$. Now, $|W:W_0| > 27^4$ unless $W_0$ has type $D_5 \times A_1$, $D_6$, $A_5 \times A_1$ or $A_6$. If $2 \in \{c,d\}$, we count the conjugates of $V_{T_Y}(\lambda - \beta_r)$ or $V_{T_Y}(\lambda - \beta_t)$, whichever is dominant, in addition to the conjugates of $V_{T_Y}(\lambda)$, in order to see that $\dim V|Y > \dim V|A$ in most cases. For the remaining cases, refer to (1.38) for the same conclusion.

Finally, consider the case where $Y = E_7$. Applying (9.8), (9.10) and previous work of this result, we see that $L_Y'$ has exactly 2 nontrivially acting components, $\langle U_{\pm\beta_k}\rangle$, $\langle U_{\pm\beta_\ell}\rangle$ where $\{k,\ell\} = \{7,j\}$, $j = 1, 2$ or $3$, $\{1,2\}$, $\{1,6\}$ or $\{2,6\}$. If $\{k,\ell\} = \{3,7\}$, (9.8) implies (i) $\Pi(L_Y) = \{\beta_3, \beta_7\}$. If $\{k,\ell\} = \{2,7\}$, (9.8) implies that either (ii) $\Pi(L_Y) = \{\beta_2, \beta_7\}$ or (iii) $\Pi(L_Y) = \{\beta_1, \beta_2, \beta_7\}$ or (iv) $\Pi(L_Y) = \{\beta_1, \beta_3, \beta_2, \beta_7\}$. If $\{k,\ell\} = \{1,2\}$, (9.8) implies (v) $\Pi(L_Y) = \{\beta_1, \beta_2, \beta_5, \beta_6\}$ or (vi) $\Pi(L_Y) = \{\beta_1, \beta_2, \beta_5, \beta_6, \beta_7\}$. If $\{k,\ell\} = \{2,6\}$, (1.15) and (9.8) imply (vii) $\Pi(L_Y) = \{\beta_1, \beta_3, \beta_2, \beta_6\}$. If $\{k,\ell\} = \{1,6\}$, previous remarks about Y of type $E_6$ imply (viii) $\Pi(L_Y) = \{\beta_1, \beta_6\}$. Finally, suppose $\{k,\ell\} = \{1,7\}$. Then (1.15) implies $\langle U_{\pm\beta_k}\rangle$ is not a component of $L_Y'$, for $k = 4$ or $5$. Also, $\langle U_{\pm\beta_4}, U_{\pm\beta_5}\rangle$ is not a component of $L_Y'$, else (2.7) implies that the field twists on the embeddings of $L_A'$ in $\langle U_{\pm\beta_1}\rangle$, $\langle U_{\pm\beta_4}, U_{\pm\beta_5}\rangle$ and $\langle U_{\pm\beta_7}\rangle$ are all equal, contradicting (2.5). So (ix) $\Pi(L_Y) = \{\beta_1, \beta_2, \beta_4, \beta_5, \beta_7\}$ or (x) $\Pi(L_Y) = \{\beta_1, \beta_2, \beta_4, \beta_7\}$, or (xi) $\Pi(L_Y) = \{\beta_1, \beta_7\}$, or $L_Y'$ is as in (iii).

Recall that the field twists on the embedding of $L_A'$ in $\langle U_{\pm\beta_k}\rangle$ and in $\langle U_{\pm\beta_\ell}\rangle$ are not equal. In case (iii), (9.8), (2.10) and (2.11) imply that the field twists on the embeddings of $L_A'$ in $\langle U_{\pm\beta_1}\rangle$ and in $\langle U_{\pm\beta_2}\rangle$ must be equal. In cases (iv), (v), (vii) and (x), if $L_m$ is the component of type $A_2$, (2.7) implies that the field twist on the embedding of $L_A'$ in $L_m$ is equal to the twist on the embedding of $L_A'$ in $\langle U_{\pm\beta_k}\rangle$, where $(\beta_k,\gamma) \neq 0 \neq (\gamma,\Sigma L_m)$, for some $\gamma \in \Pi(Y) - \Pi(L_Y)$. In case (vi) (respectively, case (ix)), if $L_m$ is the component of type $A_3$ with natural module W, say $W|L_A'$ has high weight $(q_1+q_2)\mu_\beta$. By (2.7), we may assume that the field twist on the embedding of $L_A'$ in $\langle U_{\pm\beta_2}\rangle$ (respectively, $\langle U_{\pm\beta_7}\rangle$) is $q_1$. Moreover, in case (ix), the field twist on the embedding of $L_A'$ in $\langle U_{\pm\beta_1}\rangle$ is $q_2$, else $Q_Y/K_{\beta_3}$ has no 2-dimensional $L_A'$ composition factor.

We now consider the action of $L_A'$ on the 56-dimensional irreducible kY module, $V(\lambda_7)$. We note that there is no 56-dimensional kA module affording an 8-dimensional $L_A'$ composition factor, nor exactly three 6-dimensional $L_A'$ composition factors. As well, any 56-dimensional kA module affording exactly two 4-dimensional $L_A'$ composition factors must also afford no 6-dimensional and six 3-dimensional $L_A'$ composition factors. One checks that these restrictions rule out all configurations except that of case (iii). In this case, $V(\lambda_7)$ has no 6-dimensional, exactly four 4-dimensional, and exactly two 3-dimensional $L_A'$ composition factors. However again, there is no 56-dimensional kA module affording such an $L_A'$ composition series. This completes the proof of (9.11).□

## Section III: A < Y, A non-simple

In this section, we consider the case where (A,Y,V) is an example in the main theorem with A a non-simple, semisimple algebraic group and Y a simple algebraic group of exceptional type. Theorem 4.1 of [12] implies that rankA < rankY. Let $A = H_1 \circ H_2 \circ \cdots \circ H_m$ be a commuting product of simple algebraic groups $H_i$, with a fixed maximal torus $T_A$. Let $P_A = L_AQ_A$ be a parabolic subgroup of A with Levi factor $L_A = H_1 \cdot T_A$ and $R_u(P_A) = Q_A$. Adopt the remaining notation of (2.0). We first make a few general

Remarks (9.12): (1) If (A,Y,V) is as above, we may assume, after a suitable reordering that $\mathrm{rank}(H_1) > 1$. For if $\mathrm{rank}(H_i) = 1$ for all i, there exists $B \leq A$, B a simple algebraic group of type $A_1$ such that $V|B$ is irreducible. But this contradicts (7.1) of [12]. In particular, $\mathrm{rank}(Y) > \mathrm{rank}(A) > 2$. As well, since A is an actual subgroup of Y, $\dim V^1(Q_A) > 1$.

(2) Since $Z(L_A') \leq Z(A) \leq Z(Y)$, if $Z(Y) = 1$, then $Z(L_A') = 1$.

(3) Note that all $L_A'$ composition factors of V are isomorphic to $V^1(Q_A)$. So if $\mu$ is a weight in $V|Y$ which affords the high weight of an $L_Y'$ (and hence of an $L_A'$) composition factor, then $\mu|T(L_A') = \lambda|T(L_A')$. In particular, for $\gamma \in \Pi(Y) - \Pi(L_Y)$ with $(\gamma, \Sigma L_Y) \neq 0$ and $\gamma|T(L_A') \neq 0$, $\langle\lambda,\gamma\rangle = 0$. Otherwise, $\mu = \lambda-\gamma$ fails to satisfy the above condition.

(4) Given $\alpha \in \Pi(A) - \Pi(L_A)$, $Q_A/K_\alpha$ is a 1-dimensional, irreducible $L_A'$ module. Assume $Z_A \leq Z_Y$ (as will be the case under the hypotheses of (1.5)). Let $\gamma \in \Pi(Y) - \Pi(L_Y)$ with $V_\gamma(Q_Y) \neq 0$. Then the proof of (3.3)(ii) in [12] implies that there exists $\alpha_0 \in \Pi(A) - \Pi(L_A)$ such that $U_{-\alpha_0} \nleq K_\gamma$. Hence, $Q_Y/K_\gamma$ has a 1-dimensional $L_A'$ composition factor. Also, if $U_{-\alpha_0} \nleq K_\gamma$, $U_{-\alpha_0} \leq \langle U_{-r} \leq Q_Y \mid r|T(L_A') = 0\rangle K_\gamma$. See (2.4).

(9.13): Y has type $E_n$ for some n.

Proof: Suppose false. Then (9.12) (1) implies that $Y = F_4$, $\mathrm{rank}(A) = 3$ and $\mathrm{rank}(L_A') = 2$. As well, (1.5) implies $Z_A \leq Z_Y$. If $L_A'$ has type $B_2$, the Main Theorem of [12] implies that $L_Y' = \langle U_{\pm\beta_2}, U_{\pm\beta_3}\rangle$. Since $Q_Y/K_{\beta_4}$ is an irreducible $L_A'$ module, (9.12) (4) implies that $V_{\beta_4}(Q_Y) = 0$. Hence, p = 2. By (1.7) we may assume that $V|Y$ is either a basic or p-basic module. So [8] implies that $\dim V|A = 26, 246$ or $4096$. As well, $\dim V^1(Q_A) | \dim V|A$. So $\dim V|A = 4096$, $\lambda|T_Y = \lambda_1+\lambda_2$ or $\lambda_3+\lambda_4$ and

$\dim V^1(Q_A) = 4$. But $A = B_2 \times A_1$, so by induction, $\dim V|A \leq 32 < \dim V|Y$. Thus, $L_A' \neq B_2$.

Suppose $L_A'$ has type $A_2$. The Main Theorem of [12] implies $\Pi(L_Y) = \{\beta_1,\beta_2,\beta_3\}$ and $p = 3$, or $L_Y'$ has type $A_2$. If $L_Y'$ has type $A_2$, (9.12)(4) implies that $p = 2$ and $L_Y' = \langle U_{\pm\beta_1}, U_{\pm\beta_2}\rangle$. As well, we may assume that $V|Y$ is a p-basic module, so $\lambda|T_A = \lambda_1$, $\lambda_2$ or $\lambda_1+\lambda_2$, $\dim V|Y = 26$, 246 or 4096, respectively, and $\dim V^1(Q_A) = 3,3$ or 8, respectively. Since $\dim V^1(Q_A) | \dim V|A$, $\lambda|T_Y = \lambda_2$ or $\lambda_1+\lambda_2$. But by induction, $\dim V|A \leq 64 < \dim V|Y$. So if $L_A'$ has type $A_2$, $\Pi(L_Y) = \{\beta_1,\beta_2,\beta_3\}$ and the Main Theorem of [12] implies $\lambda|T_Y = \lambda_1+x\lambda_4$ or $2\lambda_1+x\lambda_4$. By (9.12) (3), $x = 0$ in either case. (It is necessary to compute the embedding of $T(L_A')$ in $T(L_Y')$ to see this.) Since $\lambda|T_Y \neq \lambda_1$, we have $\lambda|T_Y = 2\lambda_1$. However, $\mu = \lambda-\beta_1-\beta_2-\beta_3-\beta_4$ contradicts (9.12) (3). Thus, $L_A'$ does not have type $A_2$.

It remains to consider the case where $L_A' = G_2$. By the Main Theorem of [12], $L_Y'$ has type $B_3$ or $p = 2$ and $L_Y'$ has type $C_3$. If $p = 2$, $\dim V|A = 6^k \cdot 14^\ell \cdot 64^m \cdot 2^n$, for $k, \ell, m, n \in \mathbb{Z}^+$. Since we may assume $V|Y$ is tensor indecomposable, $V|Y$ is either basic or p-basic. (See (1.7).) Thus, [8] implies $\lambda|T_Y = \lambda_1+\lambda_2$ or $\lambda_3+\lambda_4$ and $\dim V|Y = 2^{12}$. By induction, $\lambda|T_Y = \lambda_3+\lambda_4$ and $\dim V^1(Q_A) = 64$. If $P_0 \geq B_A^-$ is the parabolic of $A$ with Levi factor $H_2 \cdot T_A$, $H_2$ of type $A_1$, then $\dim V^1(R_u(P_0)) = 64$. But this contradicts (1.19) and (7.1) of [12]. Thus, if $L_A' = G_2$, $L_Y'$ has type $B_3$ and $p \neq 2$. One checks that if $\Pi(L_A) = \{\alpha_1,\alpha_2\}$, with $\alpha_1$ short, then $h_{\alpha_1}(c) = h_{\beta_1}(c^q) \cdot h_{\beta_3}(c^q)$ and $h_{\alpha_2}(c) = h_{\beta_2}(c^q)$, where $q$ is the field twist on the embedding of $L_A'$ in $L_Y'$. So $\beta_4 | T(L_A') \neq 0$ and by (9.12) (3), $\langle \lambda,\beta_4\rangle = 0$. For $\lambda|T_Y = k\lambda_1$, let $\mu = \lambda=\beta_1-\beta_2-2\beta_3-2\beta_4$, for $\lambda|T_Y = x\lambda_2+y\lambda_3$, let $\mu = \lambda-\beta_3-\beta_4$ and for $\lambda|T_Y = y\lambda_1+x\lambda_2$, let $\mu = \lambda-\beta_2-\beta_3-\beta_4$. In each case, $\mu$ contradicts (9.12) (3). But by the Main Theorem of [12], these are the only possible configurations.

This completes the proof of (9.13).□

<u>(9.14).</u> If Y has type $E_n$, then $L_Y'$ is not a quasisimple algebraic group.

<u>Proof:</u> Suppose false. We first note that $L_A' \not\cong L_Y'$, otherwise there exists some $\gamma \in \Pi(Y) - \Pi(L_Y)$ which contradicts (9.12) (4). Suppose $L_Y'$ has type $A_k$ for some k. Then by (1.5), $Z_A \leq Z_Y$. Since $L_A'$ acts irreducibly on W, the natural module for $L_Y'$, (9.12) (4) implies that there does not exist $\gamma \in \Pi(Y) - \Pi(L_Y)$ with $Q_Y/K_\gamma \cong W$ or $W^*$. Hence, $k \geq 4$. In fact, $k > 4$, else $p > 2$, $L_A' = B_2$ and there exists $\gamma \in \Pi(Y) - \Pi(L_Y)$ with $V_\gamma(Q_Y) \neq 0$ and $Q_Y/K_\gamma$ a nontrivial irreducible $L_A'$ module ($\cong W \wedge W$ or $W^* \wedge W^*$), contradicting (9.12) (4).

Consider the case where $L_Y' = A_5$. By induction, $L_A'$ has type $A_2$, $A_3$, $C_3$, or $p = 2$ and $L_A'$ has type $G_2$ or $B_3$. Previous remarks imply that either $Y = E_6$ or $Y = E_7$ and $\Pi(L_Y) = \{\beta_k \mid k \neq 1,3\}$. If $Y = E_6$, $V_{\beta_2}(Q_Y) \neq 0$ and $Q_Y/K_{\beta_2} \cong W \wedge W \wedge W$. But there is no 1-dimensional $L_A'$ composition factor of $W \wedge W \wedge W$, contradicting (9.12)(4). Hence, the second configuration holds. Then $V_{\beta_3}(Q_Y) \neq 0$ and $Q_Y/K_{\beta_3} \cong W \wedge W$ or $W^* \wedge W^*$, which has a 1-dimensional $L_A'$ composition factor only if $L_A'$ has type $C_3$ or $p = 2$ and $L_A'$ has type $G_2$, $A_3$ or $B_3$. Also, one checks that $\beta_3|T(L_A') \neq 0$, so (9.12) (3) implies $\langle \lambda, \beta_3 \rangle = 0$. In fact, $\langle \lambda, \beta_1 \rangle = 0$, as well. (Consider the $L_A'$ composition factor afforded by $\lambda - \beta_1 - \beta_3$.) In the cases where $p = 2$, $\lambda|T_Y = \lambda_7$, else $\mu = \lambda - \beta_2 - \beta_3 - \beta_4$ contradicts (9.12) (3). But $6 = \dim V^1(Q_A) \nmid \dim V = 56$. Contradiction. Hence, $L_A'$ has type $C_3$. Examining the $T(L_A')$ weight vector decomposition of $Q_Y/K_{\beta_3}$, we find that for $\alpha \in \Pi(A) - \Pi(L_A)$ such that $U_{-\alpha} \not\leq K_{\beta_3}$, $U_{-\alpha} \leq (U_{-34567} \cdot U_{-23456} \cdot U_{(0,1,1,2,1,0,0)})K_{\beta_3}$. This fact, together with the equality $[V,Q_A] = [V,Q_Y]$, restricts the possible weights in $[V,Q_Y]$, and hence the labellings of $V^1(Q_Y)$. In fact, referring to the Main Theorem of [12], we find that $\langle \lambda, \beta_m \rangle = 0$ for $m = 2,4,5,6$ and $\langle \lambda, \beta_7 \rangle = c$. So $\lambda|T_Y = c\lambda_7$. But then $\mu = \lambda - \beta_1 - \beta_2 - 2\beta_3 - 2\beta_4 - \beta_5 - \beta_6 - \beta_7$ contradicts (9.12) (3). So $L_Y'$ does not have type $A_5$.

Consider now the case where $L_Y{}' = A_6$; so $L_A{}'$ has type $G_2$ or $B_3$ or $p = 3$ and $L_A{}'$ has type $A_2$. If $\Pi(L_Y) = \{\beta_j \mid j = 1, 3 \le j \le 7\}$, so $Y = E_7$, then $\beta_2|T(L_A) \neq 0$. So $\langle\lambda,\beta_2\rangle = 0$. Also, $Q_Y/K_{\beta_2}$ has a 1-dimensional $L_A{}'$ composition factor only if $L_A{}'$ has type $G_2$ or $A_2$. In each of these cases $V^1(Q_A) \cong W$ or $W^*$. Now, $\lambda|T_Y \neq \lambda_1$, so $\lambda|T_Y = \lambda_7$. However, $\mu = \lambda-\beta_2-\beta_4-\beta_5-\beta_6-\beta_7$ contradicts (9.12) (3). Thus, if $L_Y{}' = A_6$, then $Y = E_8$ and $\Pi(L_Y) = \{\beta_j \mid j \neq 1,3\}$. One checks that $W \wedge W$ has a 1-dimensional $L_A{}'$ composition factor only if $L_A{}' = A_3$. So $V^1(Q_A) \cong W$ or $W^*$. Now, $\langle\lambda,\beta_1+\beta_3\rangle = 0$ as $\beta_3|T(L_A{}') \neq 0$. So $\lambda|T_Y = \lambda_2$. But $\mu = \lambda-\beta_2-\beta_3-\beta_4$ contradicts (9.12) (3).

Thus, for $L_Y{}'$ of type $A_k$, we have reduced to $L_Y{}'$ of type $A_7$ in $E_8$. So $L_A{}'$ has type $A_2$, $B_3$, $C_4$ or $D_4$ or $p = 2$ and $L_A{}'$ has type $C_3$ or $B_4$. If $L_A{}'$ has type $C_4$, $B_4$ or $D_4$, $Q_Y/K_{\beta_2}$ has no 1-dimensional $L_A{}'$ composition factor. In the remaining cases, $V^1(Q_Y) \cong W$ or $W^*$. Now, $\beta_2|T(L_A{}') \neq 0$, so by (9.12) (3), $\langle\lambda,\beta_2\rangle = 0$ and $\lambda|T_Y = \lambda_1$. But $8 = \dim V^1(Q_A) \mid \dim V|A$, contradicting [8]. Hence, $L_Y{}'$ does not have type $A_k$.

Suppose $L_Y{}'$ has type $D_k$, for some $k \ge 4$. Again (1.5) implies $Z_A \le Z_Y$. Now $L_A{}'$ must act reducibly on the 2 fundamental spin modules for $L_Y{}'$, as there exists $\gamma \in \Pi(Y) - \Pi(L_Y)$ with $V_\gamma(Q_Y) \neq 0$ and $Q_Y/K_\gamma$ isomorphic to one of these. (See (9.12)(4).) The Main Theorem of [12] then implies that $L_A{}'$ must act irreducibly on $W$, the natural module for $L_Y{}'$. Hence, there does not exist $\gamma \in \Pi(Y) - \Pi(L_Y)$ with $Q_Y/K_\gamma \cong W$. Thus, the triple $(L_A{}',L_Y{}',p)$ is one of $(A_2,D_4,p)$, $(B_2,D_5,5)$, $(B_2,D_7,3)$, $(C_3,D_7,3)$, or $(C_3,D_7,7)$. In the first case, $L_A{}'$ acts irreducibly on all three of the fundamental 8-dimensional irreducible $L_Y{}'$ modules, so there is no 1-dimensional $L_A{}'$ composition factor of $Q_Y/K_{\beta_1}$. If $L_Y{}' = D_7$, (9.12) (3) implies $\langle\lambda,\beta_1\rangle = 0$, but then $\lambda|T_Y = \lambda_8$. Thus, $(L_A{}',L_Y{}',p)$ has type $(B_2,D_5,5)$. Also, there does not exist $\gamma \in \Pi(Y) - \Pi(L_Y)$ with $Q_Y/K_\gamma \cong W$, so we may assume $\Pi(L_Y) = \{\beta_j \mid 1 \le j \le 5\}$. However, there is no 1-dimensional composition factor of $Q_Y/K_{\beta_6}$.

Hence, if $L_Y'$ is quasisimple, $L_Y'$ has type $E_m$ for some m. By induction and the previous work of this paper, the pair $(L_A', L_Y')$ is one of the following: $(A_2,E_6)$, $(G_2,E_6)$, $(C_4,E_6)$ or $(F_4,E_6)$. In the first three cases, $L_A'$ acts irreducibly on $Q_Y/K_{\beta_7}$; so $Z_A$ induces scalars on $Q_Y/K_{\beta_7}$. But this forces $Q_AK_{\beta_7}/K_{\beta_7}$ to be an $L_A'$ submodule of $Q_Y/K_{\beta_7}$, contradicting (9.12) (4). In the last case, (9.12) (3) implies $\langle\lambda,\beta_7\rangle = 0$. It is now a check to see that in every configuration afforded by induction, there exists a weight $\mu$ (of $Q_Y$ level 1) which contradicts (9.12) (3).□

<u>(9.15).</u> There are no examples (A,Y,V) in the main theorem with A non-simple and Y of type $E_n$.

<u>Proof:</u> Suppose false. Let $P_A$, $P_Y$ be as before. Then (9.14) implies that $L_Y'$ has more than one component. In particular, $\text{rank}(L_A') \leq 3$. Also rank $L_A' > 1$ and (1.5) imply $Z_A \leq Z_Y$. If $\text{rank}(L_A') = 3$, rank restrictions imply $L_A' = A_3$. Since $A_3$ has no 5-dimensional irreducible representation, $L_Y'$ has type $A_3 \times A_3$ in $E_8$. Now, $Q_Y/K_{\beta_5}$ has a trivial $L_A'$ composition factor only if $\Pi(L_Y) = \{\beta_j \mid j \neq 2,5\}$. Moreover, if $L_A' = \langle U_{\pm\alpha_i} \mid 1 \leq i \leq 3\rangle$, labelled as throughout, we may assume $h_{\alpha_1}(c) = h_{\beta_1}(c^q)h_{\beta_6}(c^q)$, $h_{\alpha_2}(c) = h_{\beta_3}(c^q)h_{\beta_7}(c^q)$ and $h_{\alpha_3}(c) = h_{\beta_4}(c^q)h_{\beta_8}(c^q)$, for some p-power q. Also, since $Q_Y/K_{\beta_2}$ is a 4-dimensional irreducible $L_A'$ module, (9.12) (4) implies $\langle\lambda,\beta_i\rangle = 0$ for $1 \leq i \leq 4$. The $T(L_A')$ 0-weight space of $Q_Y/K_{\beta_5}$ is spanned by the root groups $U_{-1345}$, $U_{-5678}$, $U_{-4567}$ and $U_{-3456}$. Hence, if $\alpha \in \Pi(A) - \Pi(L_A)$ with $U_{-\alpha} \nleq K_{\beta_5}$,
$U_{-\alpha} \leq \langle U_{-1345}\cdot U_{-5678}\cdot U_{-4567}\cdot U_{-3456}\rangle K_{\beta_5}$. This restricts the possible $T_Y$ weights in $[V,Q_Y] = [V,Q_A]$. In particular, $\langle\lambda,\beta_5+\beta_6+\beta_7\rangle = 0$, so $\lambda|T_Y = x\lambda_8$, for some $x > 1$. But this contradicts (9.12) (2). Thus, $\text{rank}(L_A') < 3$. Note that rank restrictions imply $L_A' \neq G_2$.

Suppose $L_A' = B_2$. Then $L_Y'$ has a component of type $A_3$ and (9.12) (2) implies p = 2. Since $L_A'$ has no 5-dimensional irreducible representation, $L_Y'$ has type $A_3 \times A_3$. As in the previous case, we reduce to

$\Pi(L_Y) = \{\beta_j \mid j \neq 2,5\}$, and $\langle \lambda, \beta_k \rangle = 0$ for $1 \leq k \leq 4$. Also, $\beta_5 | T(L_A') \neq 0$, so $\langle \lambda, \beta_5 \rangle = 0$. The Main Theorem of [12] implies $\lambda | T_Y = \lambda_6$. But then $\mu = \lambda - \beta_5 - \beta_6$ contradicts (9.12) (3). Thus, $L_A' \neq B_2$.

It remains to consider the case where $L_A' = A_2$. The minimality of $P_Y$, the Main Theorem of [12] and rank restrictions imply that $L_Y' = L_1 X L_2$, where $L_i$ has type $A_2$ or $D_4$. Suppose $L_i = A_2$ for $i = 1,2$. If there exists $\gamma \in \Pi(Y) - \Pi(L_Y)$ with $(\gamma, \Sigma L_j) \neq 0$ and $(\gamma, \Sigma L_m) = 0$ for $m \neq j$, then $Q_Y / K_\gamma$ is a 3-dimensional irreducible $L_A'$ module. So (9.12) (4) implies $V_\gamma(Q_Y) = 0$. Thus, one of the following holds:

(a) $\Pi(L_Y) = \{\beta_1, \beta_3, \beta_5, \beta_6\}$ and $\langle \lambda, \beta_5 + \beta_6 + \beta_7 \rangle = 0$ if $Y = E_7$ or $E_8$.

(b) $\Pi(L_Y) = \{\beta_2, \beta_4, \beta_6, \beta_7\}$ in $Y$ of type $E_7$ and $\langle \lambda, \beta_k \rangle = 0$ for $k = 2,3,4$.

(c) $\Pi(L_Y) = \{\beta_3, \beta_4 \beta_6, \beta_7\}$ in $Y$ of type $E_7$ and $\langle \lambda, \beta_k \rangle = 0$ for $1 \leq k \leq 4$.

(d) $\Pi(L_Y) = \{\beta_4, \beta_5, \beta_7, \beta_8\}$ and $\langle \lambda, \beta_k \rangle = 0$ for $2 \leq k \leq 5$.

Now for $\gamma \in \Pi(Y) - \Pi(L_Y)$ such that $(\gamma, \Sigma L_1) \neq 0 \neq (\gamma, \Sigma L_2)$, $\gamma | T(L_A) \neq 0$, so (9.12) (3) implies $\langle \lambda, \gamma \rangle = 0$. In fact, (9.12) (3) implies $\langle \lambda, \beta_2 + \beta_8 \rangle = 0$, in case (a) and $\langle \lambda, \beta_1 \rangle = 0$ in cases (b) and (d).

Temporarily label as follows: $\Pi(L_Y) = \{\gamma_1, \gamma_2, \gamma_3, \gamma_4\}$, $\gamma \in \Pi(Y) - \Pi(L_Y)$ and $\Pi(L_A) = \{\alpha_1, \alpha_2\}$, where $\Pi(L_1) = \{\gamma_1, \gamma_2\}$, $\Pi(L_2) = \{\gamma_3, \gamma_4\}$ and $(\gamma_2, \gamma) \neq 0 \neq (\gamma, \gamma_3)$. Since $Q_Y / K_\gamma$ must have a 1-dimensional $L_A'$ composition factor, the field twists on the embeddings of $L_A'$ in $L_1$ and $L_2$ must be equal, so $V^1(Q_Y)$ is tensor indecomposable. As well, we may assume that $h_{\alpha_1}(c) = h_{\gamma_1}(c^q) h_{\gamma_3}(c^q)$ and $h_{\alpha_2}(c) = h_{\gamma_2}(c^q) h_{\gamma_4}(c^q)$, for some p-power q. Then, the $T(L_A')$ 0 – weight space in $Q_Y / K_\gamma$ is spanned by $U_{-\gamma_1 - \gamma_2 - \gamma}$ , $U_{-\gamma_2 - \gamma - \gamma_3}$ and $U_{-\gamma - \gamma_3 - \gamma_4}$. This restricts the possible $T_Y$ weights in $[V, Q_Y] = [V, Q_A]$. For instance, suppose $\langle \lambda, \gamma_3 + \gamma_4 \rangle = 0$. Then, the factorization of elements in $Q_A$ implies that $\langle \lambda, \gamma_2 \rangle = 0$, so $\langle \lambda, \gamma_1 \rangle \neq 0$. Similarly, if $\langle \lambda, \gamma_1 + \gamma_2 \rangle = 0$, then $\langle \lambda, \gamma_3 \rangle = 0$. In case (a) (respectively, (b), (c), (d)),

$\mu = \lambda-\beta_1-\beta_2-\beta_3-2\beta_4-\beta_5$ (respectively, $\mu = \lambda-\beta_3-\beta_4-\beta_5-\beta_6-\beta_7$, $\mu = \lambda-\beta_2-\beta_4-\beta_5-\beta_6-\beta_7$, $\mu = \lambda-\beta_2-\beta_4-\beta_5-\beta_6-\beta_7-\beta_8$) contradicts (9.12) (3).

Thus, it remains to consider the case where $L_A' = A_2$ and $L_Y' = L_1 \times L_2$ with $L_1$ of type $D_4$ and $L_2$ of type $A_2$; so $Y = E_8$. One checks that $L_A'$ acts on each of the three fundamental 8-dimensional representations of $L_1$ with composition factors of dimensions 1 and 7, when $p = 3$, and otherwise $L_A'$ acts irreducibly. It is then easy to see that there is no 1-dimensional $L_A'$ composition factor of $Q_Y/K_{\beta_6}$, contradicting (9.12) (4). This completes the proof of (9.15).□

TABLE 1

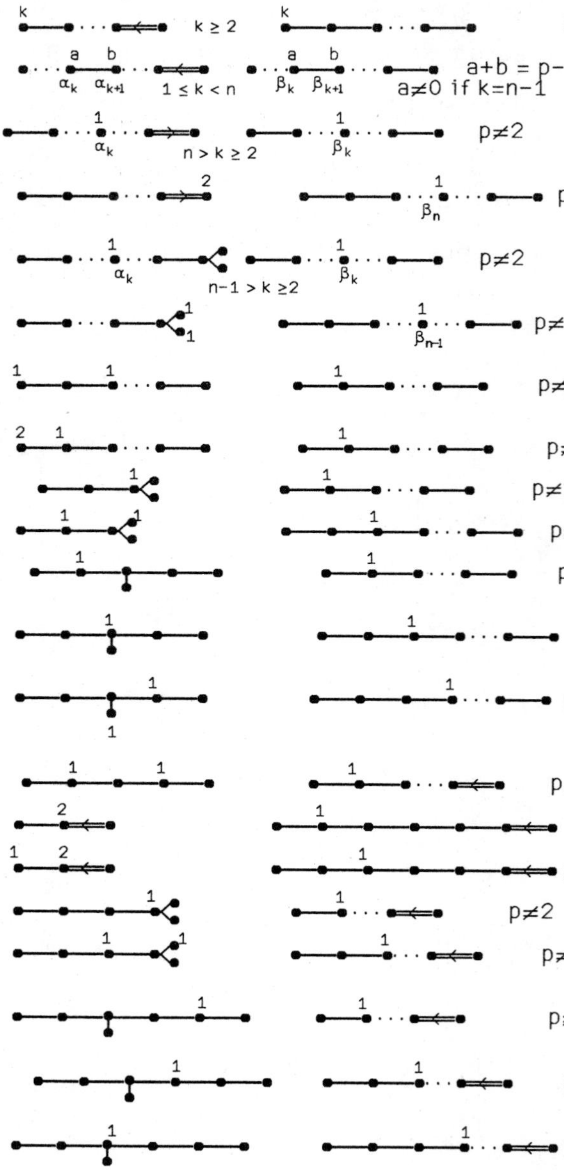

| no. | A < Y | W\|A | V\|A | V\|Y | p |
|---|---|---|---|---|---|
| $I_1$ | $C_n < A_{2n-1}$, $n \geq 2$ | $\mu_1$ | $k \geq 2$ | | $k > 1$ |
| $I_1'$ | $C_n < A_{2n-1}$, $n \geq 2$ | $\mu_1$ | $1 \leq k < n$ | | $a+b = p-1 > 1$, $a \neq 0$ if $k=n-1$ |
| $I_2$ | $B_n < A_{2n}$, $n \geq 3$ | $\mu_1$ | $n > k \geq 2$ | | $p \neq 2$ |
| $I_3$ | $B_n < A_{2n}$, $n \geq 2$ | $\mu_1$ | | | $p \neq 2$ |
| $I_4$ | $D_n < A_{2n-1}$, $n \geq 4$ | $\mu_1$ | $n-1 > k \geq 2$ | | $p \neq 2$ |
| $I_5$ | $D_n < A_{2n-1}$, $n \geq 4$ | $\mu_1$ | | | $p \neq 2$ |
| $I_6$ | $A_n < A_{n^2+n-2/2}$, $n \geq 3$ | $\mu_2$ | | | $p \neq 2$ |
| $I_7$ | $A_n < A_{n^2+3n/2}$, $n \geq 2$ | $2\mu_1$ | | | $p \neq 2$ |
| $I_8$ | $D_5 < A_{15}$ | $\mu_5$ | | | $p \neq 2$ |
| $I_9$ | $D_5 < A_{15}$ | $\mu_5$ | | | $p \neq 2,3$ |
| $I_{10}$ | $E_6 < A_{26}$ | $\mu_1$ | | | $p \neq 2$ |
| $I_{11}$ | $E_6 < A_{26}$ | $\mu_1$ | | | $p \neq 2,3$ |
| $I_{12}$ | $E_6 < A_{26}$ | $\mu_1$ | | | $p \neq 2,3$ |
| $II_1$ | $A_5 < C_{10}$ | $\mu_3$ | | | $p \neq 2$ |
| $II_2$ | $C_3 < C_7$ | $\mu_3$ | | | $p \neq 2,7$ |
| $II_3$ | $C_3 < C_7$ | $\mu_3$ | | | $p \neq 2,3$ |
| $II_4$ | $D_6 < C_{16}$ | $\mu_6$ | | | $p \neq 2$ |
| $II_5$ | $D_6 < C_{16}$ | $\mu_6$ | | | $p \neq 2,3$ |
| $II_6$ | $E_7 < C_{28}$ | $\mu_7$ | | | $p \neq 2$ |
| $II_7$ | $E_7 < C_{28}$ | $\mu_7$ | | | $p \neq 2,3$ |
| $II_8$ | $E_7 < C_{28}$ | $\mu_7$ | | | $p \neq 2,3$ |

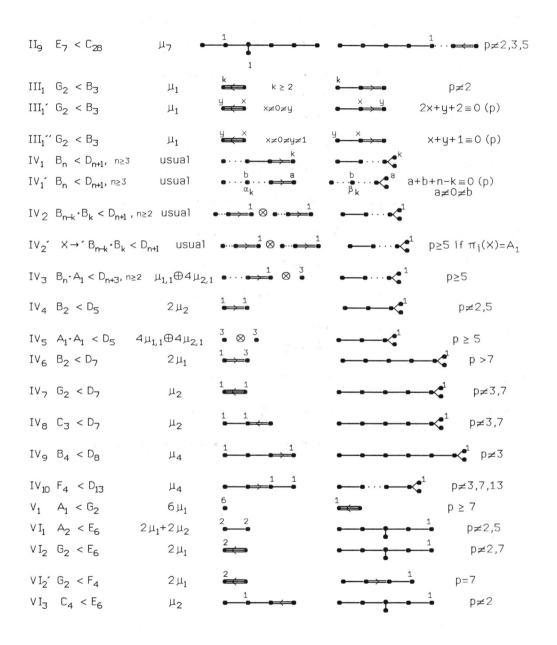

| | | | | |
|---|---|---|---|---|
| $II_9$ | $E_7 < C_{28}$ | $\mu_7$ |  | $p\neq 2,3,5$ |
| $III_1$ | $G_2 < B_3$ | $\mu_1$ |  $k \geq 2$ | $p\neq 2$ |
| $III_1'$ | $G_2 < B_3$ | $\mu_1$ |  $x\neq 0\neq y$ | $2x+y+2\equiv 0\ (p)$ |
| $III_1''$ | $G_2 < B_3$ | $\mu_1$ | 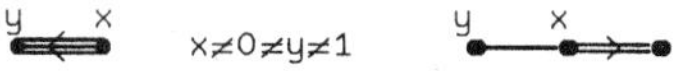 $x\neq 0\neq y\neq 1$ | $x+y+1\equiv 0\ (p)$ |
| $IV_1$ | $B_n < D_{n+1}$, $n\geq 3$ | usual |  | |
| $IV_1'$ | $B_n < D_{n+1}$, $n\geq 3$ | usual |  | $a+b+n-k\equiv 0\ (p)$<br>$a\neq 0\neq b$ |
| $IV_2$ | $B_{n-k}\cdot B_k < D_{n+1}$, $n\geq 2$ | usual | 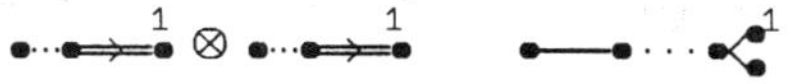 | |
| $IV_2'$ | $X \to' B_{n-k}\cdot B_k < D_{n+1}$ | usual |  | $p\geq 5$ if $\pi_i(X)=A_1$ |
| $IV_3$ | $B_n\cdot A_1 < D_{n+3}$, $n\geq 2$ | $\mu_{1,1}\oplus 4\mu_{2,1}$ |  | $p\geq 5$ |
| $IV_4$ | $B_2 < D_5$ | $2\mu_2$ |  | $p\neq 2,5$ |
| $IV_5$ | $A_1\cdot A_1 < D_5$ | $4\mu_{1,1}\oplus 4\mu_{2,1}$ |  | $p \geq 5$ |
| $IV_6$ | $B_2 < D_7$ | $2\mu_1$ |  | $p > 7$ |
| $IV_7$ | $G_2 < D_7$ | $\mu_2$ |  | $p\neq 3,7$ |
| $IV_8$ | $C_3 < D_7$ | $\mu_2$ |  | $p\neq 3,7$ |
| $IV_9$ | $B_4 < D_8$ | $\mu_4$ |  | $p\neq 3$ |
| $IV_{10}$ | $F_4 < D_{13}$ | $\mu_4$ |  | $p\neq 3,7,13$ |
| $V_1$ | $A_1 < G_2$ | $6\mu_1$ |  | $p \geq 7$ |
| $VI_1$ | $A_2 < E_6$ | $2\mu_1+2\mu_2$ |  | $p\neq 2,5$ |
| $VI_2$ | $G_2 < E_6$ | $2\mu_1$ |  | $p\neq 2,7$ |
| $VI_2'$ | $G_2 < F_4$ | $2\mu_1$ |  | $p=7$ |
| $VI_3$ | $C_4 < E_6$ | $\mu_2$ |  | $p\neq 2$ |

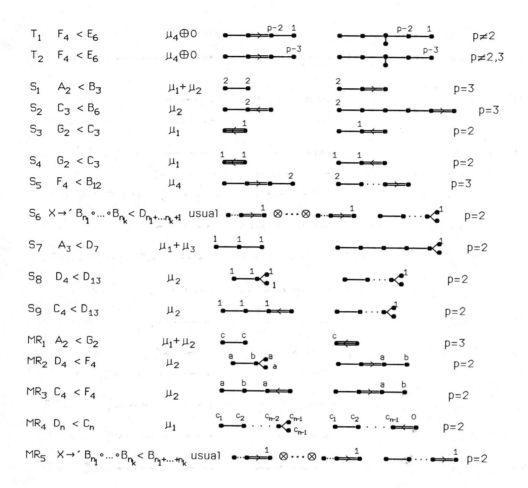

| | | | | | |
|---|---|---|---|---|---|
| $T_1$ | $F_4 < E_6$ | $\mu_4 \oplus 0$ | 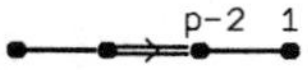 | 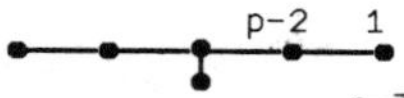 | $p \neq 2$ |
| $T_2$ | $F_4 < E_6$ | $\mu_4 \oplus 0$ |  | 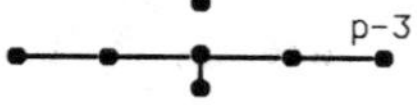 | $p \neq 2,3$ |
| $S_1$ | $A_2 < B_3$ | $\mu_1 + \mu_2$ |  |  | $p=3$ |
| $S_2$ | $C_3 < B_6$ | $\mu_2$ | 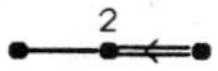 | 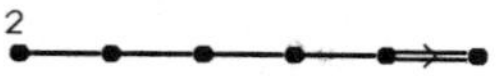 | $p=3$ |
| $S_3$ | $G_2 < C_3$ | $\mu_1$ |  | 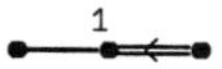 | $p=2$ |
| $S_4$ | $G_2 < C_3$ | $\mu_1$ |  | 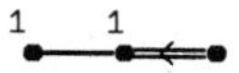 | $p=2$ |
| $S_5$ | $F_4 < B_{12}$ | $\mu_4$ | 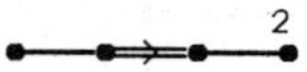 |  | $p=3$ |
| $S_6$ | $X \to' B_{n_1} \circ \dots \circ B_{n_k} < D_{n_1+\dots+n_k+1}$ | usual |  |  | $p=2$ |
| $S_7$ | $A_3 < D_7$ | $\mu_1 + \mu_3$ |  |  | $p=2$ |
| $S_8$ | $D_4 < D_{13}$ | $\mu_2$ | 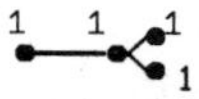 |  | $p=2$ |
| $S_9$ | $C_4 < D_{13}$ | $\mu_2$ | 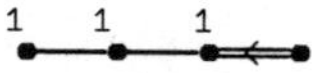 |  | $p=2$ |
| $MR_1$ | $A_2 < G_2$ | $\mu_1 + \mu_2$ |  |  | $p=3$ |
| $MR_2$ | $D_4 < F_4$ | $\mu_2$ | 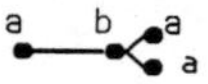 | 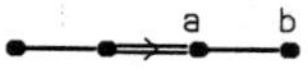 | $p=2$ |
| $MR_3$ | $C_4 < F_4$ | $\mu_2$ | 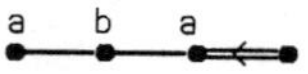 |  | $p=2$ |
| $MR_4$ | $D_n < C_n$ | $\mu_1$ | 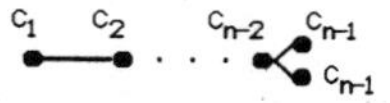 |  | $p=2$ |
| $MR_5$ | $X \to' B_{n_1} \circ \dots \circ B_{n_k} < B_{n_1+\dots+n_k}$ | usual |  |  | $p=2$ |

## REFERENCES


1. A. Borel, R. Carter, C. W. Curtis, N. Iwahori, T. A. Springer, and R. Steinberg, "Seminar on Algebraic Groups and Related Finite Groups," Lecture Notes in Mathematics, No. 131, Springer-Verlag, Berlin/New York, 1970.

2. A. Borel and J. Tits, Élements unipotents et sous-groupes paraboliques de groupes réductifs. I, Invent. Math. 12 (1971), 95-104.

3. B. Braden, Restricted Representations of Classical Lie Algebras of Type $A_2$ and $B_2$, Bull. Amer. Math. Soc. 73 (1967), 482-486.

4. N. Burgoyne, Modular Represenations of Some Finite Groups, Proc. Symp. Pure Math., Amer. Math. Soc. 21 (1971).

5. N. Burgoyne and C. Williamson, Some computations involving simple Lie algebras, Proc. 2nd Symp. Symbolic and Alg. Manipulation, ed. S. R. Petrick, New York: Assn. Computing Machinery, 1971.

6. R. Carter, "Simple Groups of Lie Type," John Wiley and Sons, London/New York/Sydney/Toronto, 1972.

7. E. B. Dynkin, Maximal Subgroups of the classical groups, Amer. Math. Soc. Transl. 6 (2) (1957), 245-378.

8. P. Gilkey and G. Seitz, Some Representations of Exceptional Lie Algebras, Geom. Ded., to appear.

9. J. Humphreys, "Introduction to Lie Algebras and Representation Theory," G.T.M. No. 9, Springer-Verlag, Berlin/Heidelberg/New York, 1972.

10. J. Humphreys, "Linear Algebraic Groups," G.T.M. No. 21, Springer-Verlag, Berlin/Heidelberg/New York, 1981.

11. V. Kac and B. Weisfeiler, Coadjoint Action of a Semisimple Algebraic Group and the Center of the Universal Enveloping Algebra in Characteristic p, Indag. Math. 38 (1976), 136-151

12. G. Seitz, The Maximal Subgroups of Classical Algebraic Groups, Memoirs Amer. Math. Soc., 365, 1987.

13. S. SMITH, Irreducible modules and parabolic subgroups, J. Algebra 75 (1982), 286-289.

14. R. STEINBERG, "Lectures on Chevalley Groups," notes by J. Faulkner and R. Wilson, Yale University, 1968.

15. R. STEINBERG, Representations of algebraic groups, Nagoya Math. J. 22 (1963), 33-56.

16. D. TESTERMAN, A Construction of Certain Maximal Subgroups of the Algebraic Groups $E_6$ and $F_4$, J. of Algebra, to appear.

The Ohio State University
Columbus, Ohio

## MEMOIRS of the American Mathematical Society

**SUBMISSION.** This journal is designed particularly for long research papers (and groups of cognate papers) in pure and applied mathematics. The papers, in general, are longer than those in the TRANSACTIONS of the American Mathematical Society, with which it shares an editorial committee. Mathematical papers intended for publication in the Memoirs should be addressed to one of the editors:


**Ordinary differential equations, partial differential equations, and applied mathematics** to ROGER D. NUSSBAUM, Department of Mathematics, Rutgers University, New Brunswick, NJ 08903

**Complex and harmonic analysis** to ROBERT J. ZIMMER, Department of Mathematics, University of Chicago, Chicago, IL 60637

**Abstract analysis** to MASAMICHI TAKESAKI, Department of Mathematics, UCLA, Los Angeles, CA 90024

**Classical analysis** to EUGENE FABES, Department of Mathematics, University of Minnesota, Minneapolis, MN 55455

**Algebra, algebraic geometry, and number theory** to DAVID J. SALTMAN, Department of Mathematics, University of Texas at Austin, Austin, TX 78713

**Geometric topology and general topology** to JAMES W. CANNON, Department of Mathematics, Brigham Young University, Provo, UT 84602

**Algebraic topology and differential topology** to RALPH COHEN, Department of Mathematics, Stanford University, Stanford, CA 94305

**Global analysis and differential geometry** to JERRY L. KAZDAN, Department of Mathematics, University of Pennsylvania, E1, Philadelphia, PA 19104-6395

**Probability and statistics** to RONALD K. GETOOR, Department of Mathematics, University of California at San Diego, LA Jolla, CA 92093

**Combinatorics and number theory** to CARL POMERANCE, Department of Mathematics, University of Georgia, Athens, GA 30602

**Logic, set theory, and general topology** to JAMES E. BAUMGARTNER, Department of Mathematics, Dartmouth College, Hanover, NH 03755

**Automorphic and modular functions and forms, geometry of numbers, multiplicative theory of numbers, zeta and *L* functions of number fields and algebras** to AUDREY TERRAS, Department of Mathematics, University of California at San Diego, La Jolla, CA 92093

**All other communications to the editors** should be addressed to the Managing Editor, RONALD GRAHAM, Mathematical Sciences Research Center, AT & T Bell Laboratories, 600 Mountain Avenue, Murray Hill, NJ 07974.

## General instructions to authors for
## PREPARING REPRODUCTION COPY FOR MEMOIRS

**For more detailed instructions send for AMS booklet, "A Guide for Authors of Memoirs." Write to Editorial Offices, American Mathematical Society, P. O. Box 6248, Providence, R. I. 02940.**

MEMOIRS are printed by photo-offset from camera copy fully prepared by the author. This means that, except for a reduction in size of 20 to 30%, the finished book will look exactly like the copy submitted. Thus the author will want to use a good quality typewriter with a new, medium-inked black ribbon, and submit clean copy on the appropriate model paper.

**Model Paper**, provided at no cost by the AMS, is paper marked with blue lines that confine the copy to the appropriate size. Author should specify, when ordering, whether typewriter to be used has PICA-size (10 characters to the inch) or ELITE-size type (12 characters to the inch).

**Line Spacing** – For best appearance, and economy, a typewriter equipped with a half-space ratchet – 12 notches to the inch – should be used. (This may be purchased and attached at small cost.) Three notches make the desired spacing, which is equivalent to 1-1/2 ordinary single spaces. Where copy has a great many subscripts and superscripts, however, double spacing should be used.

**Special Characters** may be filled in carefully freehand, using dense black ink, or INSTANT ("rub-on") LETTERING may be used. AMS has a sheet of several hundred most-used symbols and letters which may be purchased for $5.

**Diagrams** may be drawn in black ink either directly on the model sheet, or on a separate sheet and pasted with rubber cement into spaces left for them in the text. Ballpoint pen is *not* acceptable.

**Page Headings** (Running Heads) should be centered, in CAPITAL LETTERS (preferably), at the top of the page – just above the blue line and touching it.

LEFT-hand, EVEN-numbered pages should be headed with the AUTHOR'S NAME;

RIGHT-hand, ODD-numbered pages should be headed with the TITLE of the paper (in shortened form if necessary).

Exceptions: PAGE 1 and any other page that carries a display title require NO RUNNING HEADS.

**Page Numbers** should be at the top of the page, on the same line with the running heads.

LEFT-hand, EVEN numbers – flush with left margin;

RIGHT-hand, ODD numbers – flush with right margin.

Exceptions: PAGE 1 and any other page that carries a display title should have page number, centered below the text, on blue line provided.

FRONT MATTER PAGES should be numbered with Roman numerals (lower case), positioned below text in same manner as described above.

## MEMOIRS FORMAT

**It is suggested that the material be arranged in pages as indicated below.**
**Note: Starred items (*) are requirements of publication.**

**Front Matter** (first pages in book, preceding main body of text).

Page i – *Title, *Author's name.

Page iii – Table of contents.

Page iv – *Abstract (at least 1 sentence and at most 300 words).

*1980 Mathematics Subject Classification (1985 Revision).* This classification represents the primary and secondary subjects of the paper, and the scheme can be found in Annual Subject Indexes of MATHEMATICAL REVIEWS beginning in 1984.

Key words and phrases, if desired. (A list which covers the content of the paper adequately enough to be useful for an information retrieval system.)

Page v, etc. – Preface, introduction, or any other matter not belonging in body of text.

**Page 1** – Chapter Title (dropped 1 inch from top line, and centered).

Beginning of Text.

Footnotes: *Received by the editor date.

Support information – grants, credits, etc.

**Last Page** (at bottom) – Author's affiliation.



ABCDEFGHIJ–AMS–898